Tropical Babylons

The University of North Carolina Press

Chapel Hill & London

EDITED BY STUART B. SCHWARTZ

Tropical Babylons

Sugar and the Making of the
Atlantic World, 1450–1680

Designed by Heidi Perov
Set in Electra by Keystone Typesetting, Inc.
Sugarcane ornament from North Wind Picture Archives.

The paper in this book meets the guidelines for permanence
and durability of the Committee on Production Guidelines
for Book Longevity of the Council on Library Resources.

Frontispiece: Frans Post, painting of Brazilian sugar mill
and planter's residence. Courtesy of the Museum Boijmans
van Beuningen, Rotterdam.

This book was published with the assistance of the William R.
Kenan Jr. Fund of the University of North Carolina Press.

Library of Congress Cataloging-in-Publication Data
Tropical Babylons : sugar and the making of the Atlantic
World, 1450–1680 / edited by Stuart B. Schwartz.
     p. cm.
Includes bibliographical references and index.
ISBN 0-8078-2875-0 (cloth : alk. paper)
ISBN 0-8078-5538-3 (pbk. : alk. paper)
1. Sugar trade—Atlantic Ocean Region—History. 2. Plantations—
Atlantic Ocean Region—History. 3. Slavery—Atlantic Ocean
Region—History. 4. Capitalism—Atlantic Ocean Region—History.
I. Schwartz, Stuart B.
HD9100.5.T76   2004
338.4′76641′0918210903—dc22
2004001752

cloth   08 07 06 05 04   5 4 3 2 1
paper   08 07 06 05 04   5 4 3 2 1

THIS BOOK WAS DIGITALLY PRINTED.

FOR MANUEL MORENO FRAGINALS (1922–2001)
AND JOSÉ ANTÔNIO GONÇALVES DE MELLO (1916–2002)
historians of sugar whose enthusiasm for their
craft, generosity to fellow scholars, and love of their
homelands serve as examples to those who follow

Contents

Illustrations, Tables, Figures, and Maps

Figures

Maps

Acknowledgments

During the preparation and coordination of this volume a number of people provided invaluable guidance and assistance. Stanley Engerman and Herbert Klein read early versions of a number of chapters and provided helpful suggestions. George R. Trumble IV took on the task of translating Eddy Stols's essay into English. David Ryden helped in the preparation of some of the figures. Norman Fiering and the staff of the John Carter Brown Library aided in identifying suitable illustrations. Jeremy Mumford and Catharine Livingston helped in the editing process, and Anupama Mande provided her considerable expertise in the final copyediting stages. I am grateful to all of them as well as to the authors of the articles for their patience during the preparation of this volume, and to the many friends, colleagues, and critics who over the years have shaped my thinking about sugar's impact on the Atlantic world.

Note on Weights and Currencies

Considerable variations in the values of weights and currencies existed between Europe and its colonial possessions and within specific regions over time. The basic unit in the Luso-Hispanic world for the weight of sugar was the *arroba*. In Spain, the Canary Islands, and Spanish America, the *arroba* was normally about 25 pounds. Its value in Madeira was 28 pounds until 1504, when it was increased to 32 pounds. In Brazil, the 32-pound *arroba* was standard. Thus, in terms of comparison the Portuguese *arroba* was slightly over 20 percent heavier than the Spanish measure.

Spanish currency was based on the *maravedí*. There were 34 *maravedís* in a *real* and 8 *reales* in a *peso de ocho* or *castellano*, which was thus equivalent to 272 *maravedís*. The more valuable *peso ensayado* was equal to 450 *maravedís* or 1.65 *pesos de ocho*. Other coins also were in use; the *ducado* of 11 *reales* or 374 *maravedís* and the *escudo* valued at 350 *maravedís* (raised to 400 after 1566). In Portugal and its overseas settlements the standard currency was the *milréis*, or 1,000 *réis*, written 1$000. A coin of 400 *réis* called the cruzado also circulated.

The conversion rate between Spanish and Portuguese currencies varied over time. In 1637 a contract between Pedro Blanco de Ponte of Caracas and Mendes de Setúbal of Lisbon registered in Escribanias Juan Luis, Archivo Registro Principal, Caracas (16 June 1637), indicates that by the decade of the 1630s the standard rate of conversion was 40 Portuguese *réis* to each Spanish *real*. Thus there were 320 Portuguese *réis* to each Spanish *peso* and 1$000 *réis* equaled 3.125 Spanish *pesos*.

For a more detailed discussion of this subject readers are directed to John J. McCusker, "Les equivalents métriques des poids et mesures du commerce colonial aux xviie et xviiie siècles," *Revue Française d'Histoire d'Outre-Mer* 61:224 (1974): 349–65; and especially to his *Money and Exchange in Europe and America, 1600–1775: A Handbook* (Chapel Hill: University of North Carolina Press, 1978).

Tropical Babylons

CHAPTER ONE

Introduction

Stuart B. Schwartz

 It has in recent years become something of a commonplace to say that the origins of merchant capital and slavery in the Atlantic world were intimately and intrinsically tied to the production of sugar. The transference of sugarcane cultivation and sugar production from the Mediterranean to the Atlantic islands in the fifteenth century and then to the Americas in the sixteenth century is a story that has been often told, and its implications for the interwoven history of peoples on the continents of Europe, Africa, and the Americas have been the subject of great interest. Since the publication of Eric Williams's *Capitalism and Slavery* (1944), which argued that the slave-based economies of the Caribbean contributed directly, and even massively, to the British Industrial Revolution, scholars have become used to an association of sugar, slavery, and capitalism in which European capital and technology, American land, and African sweat were combined to produce profit in a commercial crop of great value.[1] The Williams thesis has become an issue of considerable debate and controversy and, right or wrong, his vision of the late eighteenth century, when about 90 percent of the West Indies' value to Europe was from sugar, has been read backward in time so that even from its origins, the production of sugar and the combination of the various factors that went into its making have been viewed as a foundational capitalistic enterprise.

Of course, that idea predated the Williams thesis. Karl Marx, by implication, had indicted sugar for "the turning of Africa into a warren of commercial hunting of black skins," as part of the "rosy dawn of the era of capitalist production," and a chief element in the process of primitive accumulation.[2] The fact that sugar production called for relatively large estates, a regimented labor force, which often consisted of enslaved workers, led to a view that the plantation regime, slavery and the Atlantic slave trade, and capitalism grew simultaneously, perhaps inevitably, as part of the same complex. The process of forming large estates using coerced labor in a semi-industrial productive activity geared toward export has sometimes been called the "sugar revolution," and while scholars have disagreed about the exact nature and timing of this "revolution," they have, nevertheless, tended to agree on the importance of the process on the areas where sugar became the principal staple.[3]

This historical vision of sugar as the quintessential capitalist crop and the satisfactions that the implied association of capitalism and slavery seemed to bring to a critical scholarship anxious to condemn both of those institutions as well as the nasty foodstuff they produced has been a powerful interpretative incentive. Being able to criticize capitalism, even at its origins, by its association with slavery and with a sweetener that caused hyperactivity in children, dental decay, and numerous other health problems and social ills was more than many critics could resist. But there were from the outset certain interpretative problems in that criticism. First, socialist Cuba's continued dependence on sugar agriculture and the Cuban state's mobilization of society at various moments to harvest the crop demonstrated that sugar agriculture could be adapted to a variety of social or political regimes, and that there was no necessary connection with slavery or with capitalism or any other particular mode of production. In the twentieth century sugar was produced in a socialist society in the Caribbean, and in the fifteenth century in a feudal society in the Mediterranean.[4] Second, with its origins in the Atlantic world, sugar production in the Americas was introduced by the Spaniards and Portuguese; in the fifteenth and sixteenth centuries Spain and Portugal could hardly be called capitalistic, and they lagged behind in the subsequent development of capitalism. Thus there would seem to be a certain contradiction, or at least irony, in the Iberian origins of the rise of mercantile capitalism and the plantation system in the Atlantic world. Finally, there were questions to be raised about both the nature of the "plantation" and the historicity of the "sugar revolution." What exactly did those terms mean? Had they changed over time? And, if so, what were the implications of those changes for our understanding of the history of slavery, capitalism, and sugar?

Let us begin with the term "plantation." In sixteenth-century Spanish and Portuguese, the term did not exist in its present meaning and was never used as such. Its use today in those languages is a neologism derived from English. Both the Portuguese and the Spanish tended to refer to the sugar-producing estates with their characteristic mill by the word for that machine: *engenho* in Portuguese, and *ingenio* in Spanish. Curiously, when the English began to establish sugar estates in Barbados they did not call them plantations, but used the Spanish term *ingenio* instead, even though they seem to have drawn primarily on the experience and expertise of Brazil where the term *engenho* was used. The etymology of the English term "plantation" changed over time from a synonym for colony in the seventeenth century to something more akin to its modern use in the eighteenth, although in English, Spanish, and Portuguese it also retained its meaning as any sort of farm. The word has come to mean a large agricultural enterprise, managed for profit, usually producing a crop for export, and often, because of its labor organization, hierarchically stratified.[5] Agricultural properties that grew sugarcane and produced sugar had peculiar characteristics because of the nature of the crop. Sugarcane once cut must be processed within forty-eight hours or else the juice in the cane dries. Thus sugar plantations were agro-industrial enterprises that combined the farming of the cane and the processing of its juice into sugar, giving the sugar mill its distinctive, quasi-industrial character.

According to Luso-Brazilian Jesuit Father António Vieira, it was "an incredible machine and factory."[6] Historically, sugar plantations implied large labor inputs under a regime of discipline that usually implied coercion, which sometimes took the form of slavery. In many ways, in its pure form, plantation organization seemed to foreshadow the modern factory. It was operated under a single authority, its labor was regimented and regularized, in part by the nature of the productive process—planting, growing, cutting, milling, cooking, cooling, crystallizing, sorting, packing, shipping—in part by the capacities and demands of the existing technology, and in some measure by social habits and expectations of command. With the exception of the skilled workers, who made up perhaps 10 or 12 percent of a sugar plantation's workforce, slaves were viewed as replaceable or interchangeable workers. Sugar plantations tended to combine skilled and unskilled workers and subordinate the labor of all of them to the goals of the mill. Slaves performed individual acts repeatedly and were separated from the means of production, the mill, and from the final product of their labors: sugar. Slaves did not make sugar; only the plantation did that. Their destiny was labor at the mill or in the field in continually repeated individual tasks so that the mill could make sugar.

In many of these characteristics the sugar plantation seemed to be a forerunner of the modern industrial units of the capitalist world, in operation long before capitalism had begun to emerge as a predominant economic system. A perceptive observer, Father António Vieira, who visited a Jesuit-owned plantation in Bahia, Brazil, during the 1630s, described his impressions: "People the color of the very night, working briskly and moaning at the same time without a moment of peace or rest, whoever sees all the confused and noisy machinery and apparatus of this Babylon, even if they have seen Mt. Etnas and Vesuvius will say that this indeed is the image of Hell." For Vieira and his contemporaries, Babylon symbolized sin, exile, and damnation, the inversion of Jerusalem, the symbol of peace and salvation. Vieira's metaphor sought to capture the essence of the sugar mill, but what he and his contemporaries were really seeing was, in some ways, simply a preview of the industrial future.

But such analogies may be overly simplistic. The traditional slave-based sugar estates, for example, did not prove to be very committed to technological innovation or to adapting mechanical improvements in order to lessen the burden and cost of labor. Their ratio of productive factors of capital, labor, land, and technology remained relatively stable over long periods of time. Moreover, unlike modern industry in which labor is essentially a variable cost, slavery turned labor into a fixed cost that needed to be financed, resupplied, and maintained. This had the effect of limiting owners' flexibility and ability to respond to changing market conditions. So long as planters could meet variable costs and some portion of their fixed costs, they sometimes remained in operation at considerable loss for long periods because to do otherwise would be to lose everything.[7] Then, too, plantership usually had a social as well as an economic basis. The desire to achieve status as a landowner and attitudes of paternalism toward slaves and dependents undercut the purely economic considerations of sugar planting. The relationship with labor, in fact, became an overriding concern of the sugar planters in many societies, and the conflicts between property and paternalism inherent in slavery penetrated deeply into their lives and psyches. While such attitudes and hesitations may have been also true to some degree of capitalism, the emphasis in these plantation-based societies was elsewhere.

Then there is the problem of measurement. Plantations were supposedly efficient operations in which inputs and outputs were considered in terms of profitability or "efficiency." But early modern accounting practices in agriculture rarely permitted such calculations, nor were planters interested in them. As Douglas Hall wrote of eighteenth-century West Indian planters, "They could indulge in little rational capital accounting, consequently they lacked the basic permissiveness of calculability of success or failure in their business. They seldom had any realistic idea of how the enterprise stood financially, or what its prospects were."[8] In the preceding centuries this was even more the case for planters, who, like the Portuguese in Brazil, usually calculated costs as what they spent each year without any differentiation between current expenses and capital investment, and thus rarely had an accurate idea of their annual return on investment. Most likely, given the many factors involved in sugar planting, some of which were beyond their control, planters viewed their operations in terms of a series of ratios and simply sought to optimize each of these as much as possible.

Geographer Ward Barrett's work, based on planter manuals and scattered plantation accounts, has demonstrated that the range of levels of productivity per worker and per unit of land were very inconsistent and did not illustrate a consistent pattern of improvement over time, nor were they geographically consistent. In Barbados, for example, productivity per slave ranged from an estimate of 1,170 kilograms (1649) to 720 kilograms (1690) to between 100 and 300 kilograms (1740–55), figures that do not support the idea that planters became more efficient over time, and which make it difficult to explain the success of Barbados.[9] Such inconsistent and contradictory figures complicate any attempt to view the earlier sugar estates as inefficient in comparison with those of the eighteenth century or Spanish and Portuguese ones as any less rational or productivity-oriented than later English and Dutch estates. A "sugar revolution" there may have been, but it is difficult to document it from existing accounts and manuals.

These contradictions and seeming anomalies have troubled many observers. How could there be capitalist plantations that predated capitalism? How could an archaic or even atavistic form of labor be mobilized by a progressive economic system which, by definition, turned labor into a commodity? As anthropologist Sidney Mintz points out, Eric Williams himself had noted the irony. Plantations, according to Williams, combined "the sins of feudalism and capitalism without the virtues of either."[10] It should also be noted that serious theoretical objections have been raised concerning the idea that capitalism grew from agrarian origins in general, and that even in the specific and peculiar case of sugar plantations, especially in this early period, traditional aspects of rural social organization proved stubbornly resilient. This aspect of the "agrarian question" has long troubled theorists of social change.[11] Mintz's own resolution of the problem was to claim that while not necessarily capitalist by themselves, plantations had provided the commodities that led to Europe's economic growth and accumulation of capital, and moreover, the plantation's precocious organizational forms provided an important step toward the eventual emergence of capitalist productive relations.[12] However, as Eric Hobsbawm has argued, it was not until the seventeenth century, with the development of full-blown plantation regimes in the Caribbean, that the overseas colonies had begun to create markets for European manufactures and thus closed the circuit necessary for capitalist development in Europe.[13]

One might also object to the organizational strategy inherent in this volume that emphasizes the commodity, sugar, as the point of analytical departure. We might have placed more emphasis on the imperial systems in which sugar plantations were embedded, or focused on the individual colonies or on specific communities as a way of understanding the history of the Atlantic economy.[14] Each of these approaches would have brought certain benefits, but emphasis on the commodity in this period of economic beginning is particularly useful.

Such an approach owes much to what has been called staple theory, which emphasized the way in which factors of production were allocated toward the commodity because of its relative market value, and then how this allocation determined the relationship between the staple-producing colony and the consuming metropolis.[15] After the economy is diversified, it is difficult to measure the effect of a single staple, such as sugar, or assess how it stimulated the economy, promoted linkages, or even contributed to real economic growth because of the presence of multiple crops as well as an artisanal-industrial or a large service sector. When population grows naturally, meaning not as a result of increased immigration or forced importation, its impact on the size of the market, the pressures it creates on resources, and the development it implies about a region's productive capacity all call for a model far more complicated than one contained in a staple and its market. But at this dawn of the Atlantic economy, a commodity approach may be particularly useful, especially in relation to plantation colonies that were clearly export driven. First, it enables us to cross imperial and colonial boundaries by examining each of the colonies not in relation to its particular national metropolis, but in relation to the Atlantic economy as a whole. Although the staple model traditionally did not incorporate relations with indigenous populations very well, that was not a problem in Madeira, which had no indigenous population, or in the Canaries and various places in the Caribbean where they were quickly decimated. The staple model helps to frame the movement of populations through immigration or the slave trade as well as the attraction of capital and technology toward the staple as rational responses to market opportunities across national or cultural boundaries. Each of the regional studies included in this collection provides materials that can be used to test a staple theory approach for individual areas, and together they may be taken as variants within a broader framework in which the two classic loci of traditional staple interpretations—a colony and its metropolis—are superseded by the combination of sugar-growing regions, on one hand, and the European consumer countries on the other.

Alongside the difficulties of conceptualizing plantation agriculture and the origins of the modern world economy, there has also arisen the question of the plantation complex and the "sugar revolution." Historically, sugar production has combined a series of elements: coerced labor, large estates, flows of capital and labor to the producing units and of its production back to European markets for the most part. Some scholars have referred to the peculiar mix of these elements in the production of sugar as the "sugar revolution," and it has been dated, defined, and interpreted in a variety of ways. While all of the classic elements of the plantation complex were in some ways already apparent and had been mounted in sugar production in the medieval Mediterranean and in the fifteenth-century Atlantic islands, they were not all present nor had they been articulated in a way in which economies of scale were made apparent. Forms of divided ownership and management such as sharecropping, a continual mix of free and slave labor, the use of local indigenous workers or even indentured Europeans, and a relatively unregimented work regime that used personalized quotas all had existed in various early plantation regimes. Thus authors have tended to exclude these precursor sugar economies within the framework of the sugar revolution. Instead, they tend to locate the origins of the "sugar revolutions" with the late seventeenth century in Caribbean islands such as Barbados, Guadaloupe, and Jamaica, where the English and French were able to combine their control of tropical island colonies with their access to relatively cheap African labor, their extensive merchant marines, and their ability to reach considerable European markets.[16] In these cases, as with the Dutch in Suriname during the seventeenth century or the Spanish in Cuba during the eighteenth century, an important social transformation took place with the turn toward sugar. In all these locales, the result of the process was a rapid transformation of the regions, often from a white or indigenous to a black population, from small farms to large plantations, from sparse to intensive settlement, and from small farmers and free workers to slaves. It has also been suggested that this process was accompanied by a specific and generally unhealthy demographic regime among the slaves characterized by an imbalance in the sex ratio strongly favoring males, high mortality and low fertility rates, and a negative rate of growth. These features certainly affected various social and cultural dimensions of slave life, interfering with the formation of family units, disrupting generational and gender relations, and creating long-term dependencies on the Atlantic slave trade from Africa.[17] Some historians, such as Philip Curtin, argue that the transference of the sugar plantation complex from the Mediterranean to Madeira or the Canary Islands or its development in São Tomé and Brazil constituted sugar revolutions as well; others, such as David Eltis, Richard Dunn, and Robin Blackburn, argue that it was really only with close attention to economies of scale, an expansion in the size of plantations, the institution of regimented "gang" labor for slaves, which imposed a simplified but repetitive and tightly controlled regime on every worker, and a system of close management with its positive effects upon profit levels that the "sugar revolution" really took place.[18] In other words, they claim that although the Canary Islands, Española, and Brazil exhibited forms of plantation arrangement earlier, it was only after the 1640s in Barbados, 1660s in Jamaica, and 1670s in Guadaloupe that we can speak of true sugar revolutions. The essay by McCusker and Menard in this volume argues that even in the case of Barbados, the term "sugar revolution" is a mis-representation of the formation of the sugar economy. In any case, it has also been suggested that the expansion of sugar was due, in great part, to the decline in shipping and handling costs caused by the expansion of European markets, but historians have found it difficult to document that decline limited.[19]

Whether revolutionary or not, both the earlier sugar regions as well as the later ones played a central role in the story of the Atlantic economy. It is for that reason that this volume concentrates on the early Atlantic sugar economies. Sugar's development in the Mediterranean and its spread westward, first to Iberia, then to the Atlantic islands of Madeira, the Canaries, and São Tomé, and from there to the Americas, has been well established in the literature.[20] But that story needs to be updated based on the most recent research based on the scattered and fragmentary records of the period. The historiography of the early sugar economies is simply not as full or as well-developed as that of the later ones. There are, for example, excellent studies of Jamaica in the eighteenth century and Cuba and Puerto Rico in the nineteenth that have examined the functioning of the sugar plantation system at its height and in the fullness of its development as part of the "sugar revolution."[21] The story of sugar has been carried forward in time and to other areas of the world as well, to Mauritius, the Philippines, Hawaii, and to Queensland. There is now hardly an island in the Caribbean or Indian Ocean that does not have a study dedicated to its history of sugar and coerced labor. Fine general books such as Noël Deerr's early and suggestive *The History of Sugar* (1949–50) and, more recently, J. H. Galloway's *The Sugar Cane Industry* (1989), or regional studies such as David Watts's *The West Indies* (1987), have drawn the monographic literature together and provided new syntheses. Anthropologist Sidney Mintz, after a distinguished academic career analyzing aspects of sugar's influence on the societies of the Caribbean where it was produced, then published *Sweetness and Power* (1985), a stimulating analysis of sugar's effects on the consuming societies. Studies continue. An international seminar on sugarcane meets regularly in Motril, (Granada), Spain, and publishes conference proceedings; the Center for Atlantic Studies in Funchal, Madeira, has also sponsored a number of meetings and publications in the area of sugar history.[22] But with all this activity and advancement of knowledge, understandings of the crucial first two centuries or so of sugar's development in the Atlantic world have been little changed in the general histories. The most cited study of the early Caribbean sugar economy, for example, is now about half a century old.[23] The lack of documentary sources, the difficulties of research on this early period, and preconceptions about the Iberians as managers have all influenced this state of affairs, and because of it, some fundamental questions about sugar and its relationship to the Atlantic economies and societies have remained unanswered—and sometimes unasked. Research, however, has continued. This collection is an attempt to bring some of those recent findings to the attention of a wider audience and to examine a number of the early centers of sugar production in a comparative perspective.

Each author has been given the freedom to develop his chapter independently and to concentrate on those themes that he felt most important; however, in the conceptualization of this volume, a set of general questions was developed to help make comparisons between areas easier. As editor and organizer, I asked the authors of the regional studies to pay particular attention to certain themes wherever possible, given the availability of sources, in the hopes of providing some data that would allow for future comparisons across regions. Such comparisons have been attempted for the eighteenth and nineteenth centuries for which a number of planter's manuals and local histories provide a good starting point, but for the early period such data is difficult to obtain.[24] Among the questions that the authors were asked to consider were the following:

1. What was the organization and size of the production units? Was the "plantation" inherent in the production of sugar, and were there other possible options for organization? How did scale transform the method of production, and did economies of scale improve the ability to compete in the international market?

2. What were the levels of productivity? What was the output per mill, for the region as a whole, and per slave, and did these ratios change over time?

3. What were the sources of capital and credit for the sugar economy in its early stages? What was the role of foreign investors and when did local capital begin to predominate in the industry?

4. What was the basic technology and how was it transferred from one region to another? Here the question of the introduction of the three-roller vertical mill, its timing and effects, is crucial. So too is the question of the circulation of specialists from one producing region to another.

5. What was the nature and composition of the labor force? Was slavery the predominant form of labor? When and how was a transition made from local laborers to Africans, and why?

6. Who were the mill owners and what was their social and political role in these societies?

7. What was the role of government in stimulating the sugar economy and what can be said about the commercialization and taxation of sugar?

Readers will notice that authors could often not respond to many of these questions given the nature of the available records. To a large extent, these chapters are based on primary archival sources, often unknown to previous researchers. The discovery of these sources has often been due to painstaking and creative research. For example, Genaro Rodríguez Morel's study of early Española is based on his extensive knowledge of the Archive of the Indies in Seville. There he discovered that important materials on the sugar estates of that island were located in collections of miscellany or in archival series relating to New Spain, because some of the sugar mill owners had migrated to participate in the conquest of Mexico and when they died their estates were probated there. Thus accounts of sugar estates in Española were to be found in the records of other colonies, and only by meticulous searching was he able to uncover these previously unused materials. His essay, like all the others contained here, is nested in an existing historiography and is engaged in conversation and debate with its predecessors.

This volume thus concentrates on the early history of sugar in the Atlantic up to the rise of Barbados as a major producer in the mid-seventeenth century, and with its emergence, the beginning of the sugar revolution. It dedicates chapters to most, but not all, of the principal producing regions and it seeks to view sugar agriculture as constitutive not only of an economic system but of a regime of social organization as well. Sugar followed the path of conquest and settlement in the Atlantic. The Madeira and Canary groups witnessed the first establishment of mills and the combination of factors that made a plantation regime possible. Unfortunately, little is known about the parallel situation on the island of São Tomé, close to the African coast, which also experienced a sugar boom in the sixteenth century but which eventually declined due to competition, a disease affecting the cane, and the destruction caused by slave resistance.[25] Early Caribbean sugar production was centered on the island of Española (today shared by the nations of Haiti and the Dominican Republic) that by the mid-sixteenth century produced about 80 percent of the sugar that reached Europe.[26] The system for producing sugar was also transferred to Cuba in 1511, to Puerto Rico in 1515, and to Jamaica, probably in 1519. Despite the relatively early transference of the plant and the technology of sugar to these other islands, their industries were slow to begin, due apparently as much to the protectionist policies of the producers on Española as to any local problems of climate, soil, or labor procurement.

About the early industry in Spanish Jamaica we know little, but about Puerto Rico, which exported about 15 percent of Caribbean sugar in this period, we know considerably more. Canes had been introduced there in 1515, but because of the pressures from Santo Domingo merchants and because of the capital investments that sugar demanded, it remained an elite activity and was far less popular than gold mining, which required only access to indigenous workers. Only in the 1530s did continuing attacks by hostile Caribs and a number of devastating hurricanes move the islanders to seek a new basis for the economy in sugar.[27] In 1535 the government provided a loan for the building of two mills. The industry that eventually developed was much like that on Española. The mills were relatively small units and the average number of slaves was about fifty. Puerto Rico, like the other island sugar economies, was faced with the rapid decline of the indigenous population and the difficulty of acquiring African slaves, especially after the New Laws of 1542, with their limitations on the employment of indigenous workers, increased the demand for African slaves in the mainland Spanish colonies. During the period 1561 to 1599, the island averaged an export of fifty tons a year, but by the time of a report in 1582 there were only eleven *ingenios* on the island and complaints about the effects of the shortage of laborers on the industry were already serious.[28] Sugar continued to be produced, but new crops such as ginger were already being sought as an alternative or supplement, and hides probably provided the most consistent island export in the period.[29] Still, sugar, by the effects it had on the social composition of the island through the importation of Africans and through the positions of power on the local municipal council held by sugar planters, exercised considerable influence on local affairs.[30]

Cuba was another matter. Its early sugar industry was frustrated by the sugar interests on Española and it was not until the crisis of production on that island that a Cuban sugar industry began to develop at the close of the sixteenth century, as Alejandro de la Fuente demonstrates in his essay contained in this volume.[31] But for a variety of reasons, Cuba remained a minor producer characterized by small units, a limited slave force, and low levels of production. In the period from 1560 to 1620, Cuban sugar made up only 1.4 percent of the sugar shipped to Seville, far exceeded by Puerto Rico (14 percent), New Spain (8 percent), and Española (74 percent).[32]

On the mainland after the conquest of Mexico, the Spanish also established sugar mills. Cortés apparently planted the first canes in 1523 and the industry developed primarily in tropical and semitropical zones. In New Spain, the areas of Vera Cruz, Morelos, and Campeche all produced sugar. By 1535 *ingenios* were operating in the area of Cuernavaca, and by 1599 there were three *ingenios* and six *trapiches* (smaller, animal-powered mills) in that region. The studies of Berthe (1966), Barrett (1970), Martin (1985), and von Wobeser (1988) demonstrate that the technology and organization of the early industry in New Spain was similar to that in the Spanish Caribbean.[33] Indigenous laborers were employed at first almost exclusively, but eventually by the 1540s, as restrictions on the use of Indian workers were enacted, Africans were introduced as slaves and a mixed labor force resulted. Sugar planters raised capital from religious institutions and acquired land and water rights by royal grants during a period of indigenous population decline in the sixteenth century.

New Spain's sugar industry presented some peculiar features. The need to irrigate fields in the area of Cuernavaca raised production costs, but in this, too, there had been Caribbean precedents in Española. Just as the decline of the indigenous population had facilitated the Spanish acquisition of land titles and water rights during the formation of the sugar economy, the recovery of that population and the growth of the general colonial population created a market for locally produced sugar. Almost invariably, especially after 1570, sugar produced in New Spain was consumed in that colony and generally did not enter the Atlantic trading system. The same could be said of the sugar produced in northern Peru in the region of Trujillo and the Lambayeque valley.[34] Here, too, the form of the manufacturing units and the system of production was similar to the contemporaneous island industries, but transportation and marketing costs for the bulky commodity made export unattractive. Thus New Spain and Peru did not figure in the international markets for sugar of the period, and for that reason they are not included here in the regional studies. Nevertheless, there is evidence that those industries were themselves not unaffected by the developing Atlantic sugar market. A comparative study of Mexico and Bahia, Brazil, by Barrett and Schwartz (1975), demonstrated that the price of Morelos sugar consumed in Mexico and of Bahian sugar sold in Europe essentially followed the same secular trends, illustrating the growing importance and influence of an Atlantic mercantile system (see figure 1.1).[35]

Prior to 1650, Brazil was the largest producer of sugar in the Atlantic world. Drawing directly on the expertise and experience of the Atlantic islands and employing foreign and local capital as well as large numbers of coerced indigenous laborers, the Brazilian industry had grown rapidly in the half century after 1570. It seems to have adopted the Madeiran model using small farmers and sharecroppers to supply cane, but to have expanded the scale of the producing units, many of which had eighty or one hundred slaves and which could produce over 140 tons a year. Brazil's productive capacity was over 20,000 tons a year and until the rise of Barbados after 1650, Brazil was the dominant producer in the world. Long after the post-1650 rise of new competitors, Brazil continued to be a major supplier.

A number of features emerge from the initial history of the sugar economies

FIGURE 1.1. Comparison of Sugar Prices in Brazil and Morelos, Mexico, 1500–1840

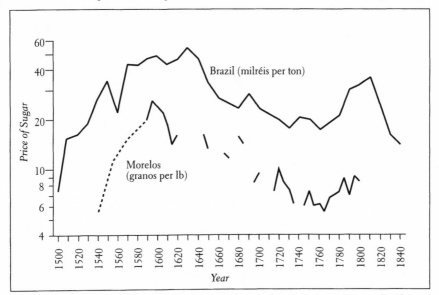

Source: Ward Barrett and Stuart B. Schwartz, "Comparación entre dos economias azucareras coloniales: Morelos México y Bahía, Brasil," in Enrique Florescano, ed., *Haciendas, latifundios y plantaciones en América Latina* (Mexico City: Siglo XXI, 1975), 532–72.

in the early Atlantic colonies. The sugar-producing zones seem to have developed serially, with each island or region supplanting a predecessor and then being replaced itself in turn. Why that should be and by what mechanisms did one region overtake another remains to be studied in depth, but the contributions in this volume suggest that economies of scale, access to labor supply and markets, and perfection of the plantation system itself all contributed to this process. Certainly such leapfrogging grew not only from comparative advantages in resources, market control, and agricultural skills and practices but also from the commercial and political restrictions that consumer nations imposed in order to stimulate their own plantation colonies.

Thus the story of each of the sugar regions is implicitly a history that is both comparative and interrelated at the same time. The production of sugar was a process that had been learned in the Mediterranean and then diffused from Spain and Portugal into the Atlantic. The techniques were, for the most part, common to all the sugar-producing regions. Let us take this short contemporary description by Cuthbert Pudsey, an English visitor to Brazil in the seventeenth century, on the making of white or "clayed" sugar as a guide to the basic techniques:

When the mill grinds, there is a wheel goes round either by water, or oxen which moves two great pillars which are made round, bound about with iron made to come close to each other, but not to touch. And betwixt these pillars they use to feed with reed by two or three slaves who pass and repass the reed. The juice distilling into certain trosses that convey it into the boiling kettles, that labor in their order, scumming and refining the sop until it come to the last which makes it sugar. They use then, after it is a little cool and thick, to put it in the forms and above these, after it have stood a space, they have a kind of science to make a mixture with wood ashes and oils which they cover above their forms to make their sugar white and to cause the dross to purge from it. Then after it hath stood about 4 months in the packhouse to purge, they take it out of the forms and break it and dry it, which makes it become as white as snow. That done they weigh it and put it in chest fit for sale.[36]

This seventeenth-century description of Brazil captures the basic elements in the process, but there were usually considerable regional variations and differences in the husbandry and production. Moreover, the early American sugar agricultures seem to have differed in some aspects from their European predecessors in their mobilization of productive factors, at least in degree if not always in kind. The scale of operations probably expanded considerably as the industry moved from southern Spain to the Atlantic islands and then to the Caribbean and Brazil. Caribbean producers tended to produce little of the white "clayed" sugars that the Brazilians specialized in and preferred to produce the darker, less-expensive, muscabado sugars. While there is some evidence of irrigation in Española and it was definitely used in Morelos, it was not characteristic of Brazil. Then, too, animal fertilizer was not used in Brazil, and plows were not a regular part of the agricultural process in either Mexico or Brazil, probably because of the nature of the coastal soils that planters preferred, making these elements unnecessary. Each region thus took the basic techniques and modified them to suit local conditions of climate, soil, power capacities, and labor supply. In each region there was a history of that process as local practice and knowledge were perfected and acquired. In this period, sugar making was an art, not a science, and it was learned by repetition and experience. Many old hands in the various sugar regions would have agreed with an eighteenth-century observer in Jamaica who wrote, "The Negro-boilers have no rule at all, and guess by the appearance of the liquor; and indeed it is wonderful to see what long-experience will do."[37] Those local histories within the basic story of the

The early modern sugar mill. This somewhat hypothetical view shows various stages of sugar-making, including the caldrons of the boiling house. Two mills, one water-powered, the other oxen-driven, are shown in simultaneous operation. This was an uncommon arrangement. From Simon de Vries, Curieuse aenmerckingen der bysonderste Oost en West-Indische verwonderens-waerdige dingen *(Utrecht, 1682). Courtesy of the John Carter Brown Library at Brown University.*

impact of sugar on the early history of European expansion and its effects on the peoples of Africa and America form the core of this volume.

In many of the Caribbean islands and at various points on the American mainlands, sugar became the dominant staple. In eighteenth-century Jamaica and St. Domingue (Haiti), in nineteenth-century Cuba, Puerto Rico, and Louisiana, a virtual "sugar revolution" took place as local economies and societies were transformed by the expansion of sugar cultivation, but that process cannot be generalized to describe sugar production's origins as well. This volume seeks to examine the early Atlantic sugar economies from Iberia to the Atlantic islands and then to the Caribbean and Brazil, up to the rise of sugar in Barbados circa 1640–60, as well as the entry of England, France, and other European nations as major consumers of the product and financiers of its production.[38]

The volume begins with William Phillips's survey of sugar in Iberia itself in a chapter that discusses the Islamic origins of the industry in Spain and its subsequent development in the fifteenth century. Phillips outlines the geographical distribution of the industry and aspects of its internal organization. Alberto Vieira then provides an extensive and detailed examination of the sugar industries in Madeira and the Canary Islands. By examining both island groups he is able to make a number of comparisons about the structure and dynamic of the two sugar industries that reveal the considerable influence of local conditions as well as the common impact of international market forces on both of them. Vieira's study also introduces the role of foreign capital in the formative stage of the industry, the existence of a class of non-mill-owning agricultors, the importance of water rights as well as land on Madeira, and the development of forms of coerced labor for sugar, all of which reemerge in the American sugar economies. What becomes clear from his account, however, and is in some ways a departure from older studies, is that on Madeira, at least, the full plantation complex was not yet apparent in the sixteenth century; slave forces were limited, properties were often small, and a class of small farmers who grew cane but did not own mills were also characteristic of the island.

There then follow three chapters that examine three important regional sugar economies—Española, Cuba, and Brazil. Genaro Rodríguez Morel's study of Española establishes the parameters of the Spanish American Caribbean sugar complex on the island that produced about 80 percent of the sugar sent from the Spanish empire in America to Europe. Española was also important because of its early use of indigenous workers whose high mortality quickly led to the importation of African slaves. Española served as a base for the spread of sugar cultivation to the other areas of Spanish settlement like Jamaica and Puerto Rico. Rodríguez Morel demonstrates how the planters and merchants in the sugar trade came to exercise considerable political and economic power on the island. He argues that an early form of the plantation system was established in the sixteenth century but that eventually it failed because of competition from Spanish merchants, the rise of competitive producing regions, and a number of local and international factors. In any case, he argues that the elements and economies of scale of the plantation system were achieved at times on Española. But it would seem that the subsequent development on other islands was not linear and that alternative levels of production were also possible. Alejandro de la Fuente examines the early Cuban sugar economy that started only at the close of the sixteenth century. In many ways it mirrored its predecessor on Española, but it was a relatively small-scale operation in comparison to it. Nevertheless, it did establish a base on which Cuba later developed a plantation economy in the eighteenth century.[39] In his study, de la Fuente presents new evidence about the transfer of technology between regions and about the formation of networks of local elites based on sugar. He is able to demonstrate how a political elite consolidated its position in conjunction with the sugar industry, even though that industry itself proved ephemeral. He suggests that the slave forces were smaller and the slaves were older on the average and lived longer than was the case during the great nineteenth-century Cuban sugar expansion, further evidence of the slow development of the plantation regime on the island.

In my own essay I present an overview of the Brazilian sugar economy based on my earlier published research, but I have added new materials to it.[40] Brazil came to dominate the Atlantic market for sugar until the mid-seventeenth century, and I examine both its peculiar productive arrangements that included the extensive use of sharecropping and other forms of contract, the increasing use of African slaves, and the predominant role of foreign shipping in Brazil's commerce. Additionally, I have included a brief overview of the Dutch experience with sugar making during their occupation of the Brazilian Northeast (1630–54). This episode is important in the history of sugar since it is sometimes argued that the Dutch transfer of technical knowledge and capital to the Caribbean after 1654 was crucial to the development of the industry in the islands after that date, but there is little evidence for that contention and the article by Menard and McCusker included here also argues against this claim.[41]

From these somewhat successive and overlapping regional studies, the following two chapters then move to broader, international themes that affected the Atlantic sugar economy as a whole. First, Herbert Klein examines the Atlantic slave trade in the period prior to 1650 and demonstrates that despite the synchrony, that trade was not closely tied to the rhythms of sugar's history. Data from this early period is admittedly difficult to obtain. It was a lacuna in Curtin's classic work and even though a careful combing of sources by Ivana Elbl has considerably improved our knowledge, the figures if not the patterns remain obscure.[42] Klein's innovative approach suggests that the relationship between the slave trade and the early sugar industry is considerably more problematic than has been previously thought. Then Belgian historian Eddy Stols looks at the European market for sugar prior to 1650 and at the manner in which this commodity was traded and consumed in Europe.[43] His approach is more cultural than economic, providing a welcome emphasis on the consumption and symbolic meaning of sugar in European society that balances the emphasis on production elsewhere in the volume. Stols demonstrates that the impact of sugar on European tastes was earlier and greater than most historians have realized, and his focus on Antwerp and Amsterdam as the epicenters of this impact demonstrates the international nature of the sugar economy and the importance of underlining demand as a crucial aspect of sugar's history. Finally, Russell

Menard and John McCusker return to a regional gaze in their examination of Barbados, the first non-Iberian colony to develop an important sugar economy. Their study demonstrates the continuity of practices and patterns with the Iberian colonies as well as the transfer of techniques from Brazil to Barbados, but it also emphasizes the specificities that allowed that small island to become a primary producer. For them, the introduction of sugar did not revolutionize Barbados since that island had already moved to the production of staples like cotton using slaves prior to sugar's introduction. Instead, by mobilizing and organizing sugar agriculture in relatively large, singly owned plantations, and employing slaves in closely organized work gangs, they argue that Barbados revolutionized sugar rather than vice versa. Barbados was thus a forerunner of a new epoch in sugar production in which the English, French, and Dutch colonies developed the classic forms of plantation agriculture.

Framing a set of comparative questions about these early sugar economies may help to form a clearer picture of the historical processes we are seeking to describe and analyze. Let me take just two aspects to demonstrate how these comparisons might lead to new questions and lines of inquiry. One might think that successive rise of sugar-producing regions implied a steady improvement in managerial skills and in efficiency as each region learned from the mistakes or successes of its predecessors. The chronology of the growth of sugar regions, however, was not necessarily paralleled by gains in efficiency or productivity. In Madeira at the close of the fifteenth century, mills averaged twenty-three tons a year. On São Tomé mills fell into a range that averaged approximately 15 to 25 tons. In Cuba about a century later the average was only 10.2 tons annually. Now the dimensions of the Brazilian mills come sharply into focus. In 1591, sixty-three mills in Pernambuco produced 5,500 tons or an average of 87 tons per mill; in Bahia in 1610 the average was 69 tons per unit. It would appear that the size of Brazilian operations was considerably larger than in the other industries. This differential is also implied by the larger number of slaves employed. Brazilian mills averaged about sixty slaves owned by each mill and perhaps forty to sixty held by dependent cane farmers for a total of 100–120 working at each mill. These figures suggest that economies of scale seem to have been important in Brazil's domination of the Atlantic sugar market in the early seventeenth century, but eventually external elements—war and politics—also played a role in determining the success of these industries. Brazilian sugar's difficulties caused by rising costs, heavy taxation, and political disruptions eventually undercut its ability to compete with new challengers, and as Menard and McCusker suggest they could also not compete with the new managerial strategies and political manipulations of trade and markets that gave the subsequent advantage to Barbados.

To take another issue for which we have some good information, we can turn to the question of slave labor and its productivity as a way of demonstrating how the data from different regions may be used in order to examine the peculiarities of each region. Various contemporaneous observers and modern authors have estimated productivity per slave in the early modern sugar economies. In his chapter, de la Fuente notes that Cuba between 1603 and 1610 had about twenty-five mills, each with about thirty slaves. Annual average production per mill was 900 *arrobas* or 10.2 tons. This would make productivity per slave about 341 kilograms. Brazilian planters often used a rough calculation of a crate of sugar with forty *arrobas* per slave (40 × 32 = 1,280 pounds = 582 kilograms), but such a level was not always reached. An actual accounting from the district of Serinhaem in Pernambuco made in 1788 revealed that productivity per slave in that year was actually about twenty-one *arrobas* or 305 kilograms, but that calculation included children, the aged, and the infirm. It probably does not represent the way in which planters usually made this calculation in their heads.[44] If only adult slaves are used in the Serinhaem case, the average per adult worker rises to 409 kilograms, and if only men are used as the denominator, the figure rises to over 545 kilograms.

While data for the early sugar industries are sparse and inconsistent, estimates and observations on eighteenth-century Jamaica are much firmer. Figure 1.2 provides a range of estimates. There, estimates ranged from 363 to 591 kilograms per slave, but the best data is provided by a census of twenty-five estates in St. Andrew parish in Jamaica made in 1753, which yields a figure of 422 kilograms per slave on average, or about midway in the range of other Jamaica estimates.[45] Historians would like to know if these averages changed over time and how technological innovations, economies of scale, improved management, and local practices might have influenced output and profitability. If we simply compare the figures from early Cuba and eighteenth-century Jamaica, it would appear that, despite all of the technical and practical improvements made over time and the creation of the "new plantation" in the English Caribbean islands, the gains in productivity per worker were only about 20 percent. To complicate the story, the productivity estimates offered here by Rodríguez Morel for early Española suggest much higher ratios than those obtained later in Cuba, Jamaica, or Brazil just as Ligon's early estimates for Barbados are far higher than later estimates for that island. As Barrett has argued earlier, sugar productivity per worker seems to fall into a range, but it is difficult to establish a trend over

FIGURE 1.2. Comparison of Slave Productivity in Sugar

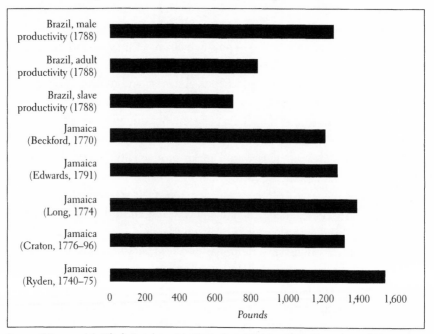

Sources: "Mapa geral da Vila de Serinhaem" (1788), Instituto Histórico, Geográfico, e Arqueológico Pernambucano, estante A, gaveta 5; W. Beckford, *A descriptive Account of the Island of Jamaica*, 2 vols. (London, 1790); Bryan Edwards, *The History, civil and commercial, of the British Colonies in the West Indies* (London, 1793); Edward Long, *The History of Jamaica*, 3 vols. (London, 1774); Michael Craton and James Walvin, *A Jamaican Plantation* (London, 1970); David Ryden, "Producing a Peculiar Commodity: Jamaican Sugar Production, Slave Life and Planter Profits on the Eve of Abolition, 1750–1807" (Ph.D. diss., University of Minnesota, 1999).

time or differentiated by region. Each of the regional chapters presented here provides some data to begin to address these questions.

Taken as a whole, these chapters are a beginning in a newer, more nuanced vision of the origins of the Atlantic economy and the role of sugar within it. What emerges here is an image of a constellation of sugar industries, developing in tag-team fashion, sharing similar technology and employing skilled workers and field hands, many of whom were enslaved. While all the elements of the later "plantation system" were present to a greater or lesser degree, the scale of operations, the regimentation of labor, and the effectiveness of production all seem to be of a proto-plantation nature, not yet fully developed or organized along "industrial" principles. Nevertheless, the early sugar industries were harbingers of the future, and contained the seeds of more "modern" forms of economic organization yet unborn. They demonstrated the adaptability of traditional agrarian practices and archaic forms of labor to a new kind of organization that provided the basis for a dynamic, if often destructive, political economy.

Atlantic history as a field has tended to emphasize the interaction of peoples, politics, and cultures of Africa, Europe, and the Americas and their islands. As that interchange began, sugar, as much as any commodity or activity, played a central role in influencing the nature of the interactions, and ultimately in making Africans and Afro-Americans crucial to Atlantic history as a whole. These essays demonstrate, however, that sugar itself determined very little and that the form its production shared throughout the Atlantic world owed much to decisions, actions, reactions, and interactions made on both sides of the Atlantic by men and women faced with choices.

The sugar mill, with its attendant labor force of slaves and its potential to generate great wealth for some by the oppression of many, came to symbolize the emerging Atlantic system. Like Jesuit Father Vieira, many observers who witnessed the back-breaking labor of the slaves, the continuous whirring of the mill day and night, the fires of the furnaces, and the steam rising from the kettles invoked the image of Hell to describe what they observed.[46] For Vieira, the sugar mills themselves were indeed tropical Babylons, but so too were the societies that grew around them, based on "the immense transmigration of the Ethiopian peoples and nations who from Africa continually pass to this America," who like the Children of Israel were carried into captivity to weep by the waters of Babylon, in an enslavement as much of the soul as of the body.[47] In the sermons of Father Vieira in the seventeenth century, and in the Jamaican reggae of Jimmy Cliff and Bob Marley in our own times, Babylon became the symbol of the social ills and inequities of Western society, and in the sugar colonies of the Atlantic world, the origins of those societies were inextricably intertwined with the legacy of slavery, plantations, and sugar. This volume seeks to trace that history and demonstrate its complexity.

NOTES

1. See the discussion in Barbara L. Solow and Stanley L. Engerman, eds., *British Capitalism and Caribbean Slavery: The Legacy of Eric Williams* (Cambridge: Cambridge University Press, 1987), Introduction, 1–23; see esp. Barbara L. Solow, "Capitalism and Slavery in the Exceedingly Long Run," 51–78.

2. Marx's often-repeated statement is cited in this regard in Jan de Vries, *The Economy of Europe in an Age of Crisis, 1600–1750* (Cambridge: Cambridge University Press, 1976), 139–41.

3. B. W. Higman, "The Sugar Revolution," *Economic History Review* 53, no. 2 (2000),

reviews the literature and discusses this concept and its application to various centuries and various places.

4. Philip D. Curtin, *The Rise and Fall of the Plantation Complex: Essays in Atlantic History* (Cambridge: Cambridge University Press, 1990), points out the feudal aspects of the plantation regime established in the eastern Mediterranean and in early Brazil. There is also an extensive literature that seeks to view the social and economic system established in Latin America as a peculiar form of feudalism articulated by a strong centralized state rather than in opposition to it or in place of it. See, for example, Marcelo Carmagnani, *L'America latina dal 500 a oggi: nascita e crisi di un sistema feudale* (Milan: Feltinelli, 1975).

5. A number of theoretical discussions of plantations as social systems are found in *Seminar on Plantation Systems of the New World*, Pan American Union monographs, no. 7 (Washington, D.C., 1959). See the classic formulation by Eric R. Wolf and Sidney Mintz, "Haciendas and Plantations in Middle America and the Antilles," *Social and Economic Studies* 6 (1957).

6. For similar examples, see Domingos de Loreto Couto, "Desagravos do Brasil e glorias de Pernambuco," in *Anais da Biblioteca Nacional do Rio de Janeiro* 24 (1902): 171–72; Richard Ligon, *A True and Exact History of the Island of Barbados* (London: H. Moseley, 1657); Jean Baptiste Labat, *Nouveau voyage aux isles de l'Amerique*, 2 vols. (Paris: P. Husson, 1724).

7. I have discussed this in more detail in the Brazilian case in Stuart B. Schwartz, *Sugar Plantations in the Formation of Brazilian Society* (Cambridge: Cambridge University Press, 1985), 223–25.

8. Douglas Hall, "Incalculability as a Feature of Sugar Production during the Eighteenth Century," *Social and Economic Studies* 10 (1961), cited in Ward J. Barrett, *The Efficient Plantation and the Inefficient Hacienda*, James Ford Bell Lecture 16 (Minneapolis: James Ford Bell Library, 1979), 7.

9. Barrett, *Efficient Plantation*, 21–22. See the table and the sources cited therein.

10. Cited in Sidney W. Mintz, *Sweetness and Power: The Place of Sugar in Modern History* (New York: Penguin Books, 1986), 60.

11. See Susan Archer Mann, *Agrarian Capitalism in Theory and Practice* (Chapel Hill: University of North Carolina Press, 1990). On the agrarian origins of capitalism, see Robert Brenner, "Agrarian Class Structure and Economic Development in Pre-Industrial Europe," *Past and Present* 70 (1976), and the debate it generated in T. H. Aston and C. H. E. Philpin, eds., *The Brenner Debate: Agrarian Class Structure and Economic Development in Pre-Industrial Europe* (Cambridge: Cambridge University Press, 1985). Brenner's original formulation did not include any discussion of agriculture in colonial settings.

12. Mintz, *Sweetness and Power*, 55. Immanuel Wallerstein, *The Modern World-System: Capitalist Agriculture and the Origins of the European World Economy in the Sixteenth Century* (New York: Academic Press, 1974), is less conflicted in ascribing capitalist production to this period and seeing sugar as a central part of the process, although as he says of this early period, "capitalists did not flaunt their colors before the world" (67). Wallerstein argues that slavery and "coerced cash-crop labor" were characteristic of the peripheries because they ensured the flow of surplus to the metropoles that ensured the development of capitalism.

13. Eric Hobsbawm, "The Crisis of the Seventeenth Century," in *Crisis in Europe, 1560–1660*, ed. Trevor Aston (New York: Basic Books, 1965). Hobsbawm does suggest that Brazilian sugar planters had precociously shown the way in the creation of a market in the sixteenth century (23).

14. See, for example, Peter Emmer, "The Dutch and the Making of the Second Atlantic System," in *Slavery and the Rise of the Atlantic System*, ed. Barbara L. Solow (Cambridge: Cambridge University Press, 1991).

15. My discussion is based on Jacob Price, "The Transatlantic Economy," in *Colonial British America*, ed. Jack P. Greene and J. R. Pole (Baltimore: Johns Hopkins University Press, 1984); and esp. John J. McCusker and Russell R. Menard, *The Economy of British America, 1607–1789* (Chapel Hill: University of North Carolina Press, 1985), which includes an analysis of the literature on this theme. See also Robert Baldwin, "Patterns of Development in Newly Settled Regions," *Manchester School of Economic and Social Studies* 24 (1956).

16. A classic formulation is Robert C. Beatie, "Why Sugar? Economic Cycles and the Changing of Staples in the English and French Antilles," *Journal of Caribbean History* 8–9 (1976). Hilary Beckles and Andrew Downes, "The Economics of the Transition to the Black Labor System in Barbados, 1630–80," *Journal of Interdisciplinary History* 18 (1987), takes on the problem of the transition in the labor force and provides a formal argument based on the supply and demand curves of indentured laborers and slaves. A useful and extensive overview, based on printed primary sources, of the introduction of sugarcane cultivation in the Americas in the sixteenth century is found in Justo L. del Río Moreno, *Los inicios de la agricultura europea en el Nuevo Mundo, 1492–1542* (Seville: Asociación Agraria-Jovenes Agricultores de Sevilla, 1991).

17. Russell Menard and Stuart B. Schwartz, "Was There a Plantation Demographic Regime in the Americas?," in *The Peopling of the Americas*, 3 vols. (Vera Cruz: International Union for the Scientific Study of Population, 1992), 1:51–66, suggested this as a generalized pattern. Similar arguments were made later in the specific case of Louisiana in Michael Tadman, "The Demographic Cost of Sugar: Debates on Slave Societies and Natural Increase in the Americas," *American Historical Review* 105, no. 5 (2000).

18. Robin Blackburn, *The Making of New World Slavery: From the Baroque to the Modern, 1492–1800* (London: Verso, 1997), makes a convincing argument about the "new plantations" of the mid-seventeenth century. See also David Eltis, *The Rise of African Slavery in the Americas* (Cambridge: Cambridge University Press, 2000); Richard Dunn, *Sugar and Slaves: The Rise of the Planter Class in the English West Indies, 1624–1713* (Chapel Hill: University of North Carolina Press, 1972).

19. Peter Emmer and Ernest van den Boogaart, "The Dutch Participation in the Atlantic Slave Trade, 1596–1650," in *The Uncommon Market: Essays in the Economic History of the Atlantic Slave Trade*, ed. Henry A. Gemery and Jan S. Hogendorn (New York: Academic Press, 1979).

20. See, for example, Charles Verlinden, *The Beginnings of Modern Colonization* (Ithaca: Cornell University Press, 1970); and J. H. Galloway, *The Sugar Cane Industry* (Cambridge: Cambridge University Press, 1989).

21. See, for example, Michael Craton, *Searching for the Invisible Man: Slaves and Plantation Life in Jamaica* (Cambridge, Mass.: Harvard University Press, 1978); Manuel Moreno Fraginals, *Sugar Mill: The Socioeconomic Complex of Sugar in Cuba, 1760–1860* (New York: Monthly Review Press, 1976); Laird Bergad, *Cuban Rural Society in the Nineteenth Century: The Social and Economic History of Monoculture in Matanzas* (Princeton: Princeton University Press, 1990); and Francisco Scarano, *Sugar and Slavery in Puerto Rico: The Plantation Economy of Ponce, 1800–1850* (Madison: University of Wisconsin Press, 1984).

22. From Motril, see *Producción y comércio del azúcar de caña en época preindustrial: Actas del tercer seminario internacional, 23–27 de septiembre 1991* (Granada: Diputación Provincial, 1993); *1492: Lo dulce a la conquista de Europa: Actas del cuarto seminario internacional sobre la caña de azúcar, 21–25 de septiembre 1992* (Granada: Diputación Provincial, 1994); *Paisajes del Azúcar: Actas del quinto seminario internacional, 20–24 de septiembre 1993* (Granada: Diputación Provincial de Granada, 1995). See also Alberto Vieira, ed., *Slaves with or without Sugar* (Funchal: Centro de Estudos Atlanticos, 1996).

23. Mervyn Ratekin, "The Early Sugar Industry in Española," *Hispanic American Historical Review* 34:1 (1953). See also Irene Wright, *The Early History of Cuba, 1492–1580* (New York: Macmillan, 1916).

24. Most important in this regard is Ward J. Barrett, "Caribbean Sugar Production Standards in the Seventeenth and Eighteenth Centuries," in *Merchants and Scholars: Essays in the History of Exploration and Trade*, ed. John Parker (Minneapolis: University of Minnesota Press, 1965); and his later *The Efficient Plantation and the Inefficient Hacienda* (cited in note 8).

25. A good brief summary is provided by Frédéric Mauro, *Le Portugal, le Bresil et l'Atlantique au XVII siècle* (Paris: SEVPEN, 1983), 213–14. See also Francisco Tenreiro, *A Ilha de São Tomé* (Lisbon: Junta de Investigaçoes do Ultramar, 1961); Robert Garfield, *A History of São Tomé Island, 1470–1655* (San Francisco: Mellen Research University Press, 1992); Fernando Castelo-Branco, "O comércio externo de São Tomé no seculo XVII," *Studia* 24 (1968). The widely differing estimates of its production are discussed in Schwartz, *Sugar Plantations*, 507. It would seem that average mill capacity on São Tomé fell into the range of fifteen to thirty tons, which suggests that its scale was considerably smaller than Brazil and more like that in Madeira or Cuba.

26. On sugar shipping from early Española, see Justo L. de Río Moreno and Lorenzo E. López y Sebastián, "El comercio azucarero de la Española en el siglo XVI: Presión monopolistica y alternativas locales," *Revista Complutense de Historia de América* 17 (1991).

27. My remarks are based on Genaro Rodríguez Morel, "Slavery and Sugar Plantations in Puerto Rico. XVI Century," in Vieira, *Slaves with or without Sugar*; and Elsa Gelpí Baíz, "Economía y sociedad: Estudio de la economía azucarera de Puerto Rico del siglo XVI" (Ph.D. diss., Universidad de Sevilla, 1993), subsequently published as *Siglo en blanco: Estudio de la economía azucarera en el Puerto Rico del siglo XVI (1540–1612)* (San Juan: Universidad de Puerto Rico, 2000), which provides export figures. A good brief overview is provided in Francisco Moscoso, *Agricultura y sociedad en Puerto Rico, siglos 16 al 18* (San Juan: Instituto de Cultura Puertoriqueña and Colegio de Agrónomos de Puerto Rico, 1999).

28. "Memoria y descripción de la isla de Puerto Rico, sometida por el gobernador, capitán Juan Melgarejo, 1582," cited in Moscoso, *Agricultura*, 61–62.

29. Justo L. del Río and Lorenzo E. López y Sebastián, "El jengibre: historia de un monocultivo caribeño del siglo XVI," *Revista Complutense de Historia de América* 18 (1992).

30. Gelpí Baíz, *Siglo en blanco*, 209–11.

31. See Irene Wright, "Why Cuba Developed Sugar Later Than Santo Domingo, Puerto Rico, Jamaica, and Mexico," *Louisiana Sugar Planter and Sugar Manufacturer* 54, no. 1 (1915). The classic account of Cuban sugar is Manuel Moreno Fraginals, *Sugar Mill*. Readers should consult the expanded Spanish edition, *El ingenio: Complejo económico social cubano del azúcar*, 3 vols. (Havana: Editorial de Ciencias Sociales, 1978).

32. Alejandro de la Fuente, "Población y crecimiento en Cuba (siglos XVI y XVII)," *European Review of Latin American and Caribbean Studies* 55 (1993).

33. Jean Pierre Berthe, "Xochimancas, Los trabajos ny los días en una hacienda azucarera de la Nueva España," *Jahrbuch fur Geschichte von Staat, Wirtschaft und Gesellschaft Lateinamerikas* 3 (1966); Ward Barrett, *The Sugar Hacienda of the Marques del Valle* (Minneapolis: University of Minnesota Press, 1970); Cheryl E. Martin, *Rural Society in Colonial Morelos* (Albuquerque: University of New Mexico Press, 1985); Gisela von Wobeser, *La hacienda azucarera en la época colonial* (Mexico City: Secretaría de Educación Publica and Universidad Nacional Autónoma de México, 1988). See also Jean Pierre Berthe, "Sur L'histoire sucrerie americane," *Annales: Economies, societies, civilisations* 14, no. 1 (1959); and Beatriz Scharrer Tamm, *Azúcar y trabajo: Technología de los siglos XVII y XVIII en el actual estado de Morelos* (Mexico City: Instituto Cultural de Morelos, 1997).

34. Susan Ramírez, *Provincial Patriarchs: Land Tenure and the Economics of Power in Colonial Peru* (Albuquerque: University of New Mexico Press, 1986).

35. Ward Barrett and Stuart B. Schwartz, "Comparación entre dos economías azucareras coloniales: Morelos, México y Bahía, Brasil," in *Haciendas, latifundios, y plantaciones en América latina*, ed. E. Florescano (Mexico City: Siglo Veintiuno, 1975),

36. Cuthbert Pudsey, *Journal of a Residence in Brazil*, in *Dutch Brazil*, ed. Nelson Papavero and Dante Martins Teixeira, 3 vols. (Petrópolis: Editora Index), 3:27.

37. John P. Baker, *An Essay on the Art of Making Moscovado Sugar* (Jamaica, 1775), cited in Barrett, "Caribbean Sugar Production," 161. I have demonstrated elsewhere in the case of Brazil how such knowledge was depreciated as scientific techniques and methods were introduced in the late eighteenth century. See Schwartz, *Sugar Plantations*, 335–36.

38. On the early history of the Lesser Antilles and their role in the development of the sugar economy, see Robert Paquette and Stanley Engerman, *The Lesser Antilles in the Age of European Expansion* (Gainesville: University Press of Florida, 1996).

39. An earlier study of a later period provides an excellent overview. See Alejandro de la Fuente Garcia, "Los ingenios de azúcar en La Habana del siglo XVII (1640–1700)," *Revista de Historia Económica* 9, no. 1 (1991).

40. See Schwartz, *Sugar Plantations*.

41. See Matthew Edel, "The Brazilian Sugar Cycle of the Seventeenth Century and the Rise of West Indian Competition," *Caribbean Studies* 9 (1969).

42. Philip Curtin, *The Atlantic Slave Trade: A Census* (Madison: University of Wisconsin Press, 1969); Ivana Elbl, "The Volume of the Early Atlantic Slave Trade, 1450–1521," *Journal of African History* 38, no. 1 (1997).

43. He has also written on Dutch and Flemish participation directly in the Brazilian sugar economy; see Eddy Stols, "Convivências e conivências luso-flamengas na rota do açúcar brasileiro," *Ler História* 32 (1997).

44. Mapa geral da vila de Seranhaem, 1788, Instituto Arqueológico, Histórico e Geográfico Pernambucano, estante A, gaveta 5.

45. David Ryden, "'One of the Fertilist Spotts': An Analysis of the Slave Economy in Jamaica's St. Andrew's Parish, 1753," *Slavery and Abolition* 21, no. 1 (2000).

46. *Sermões do Padre António Vieira*, 16 vols. (São Paulo, 1945 facsimile of the edition of 1748), 5:484–520. I have used this quotation from Vieira previously to make the same point in Stuart Schwartz, "A 'Babilonia' colonial: A economia açucareira," in *História da expansão portuguesa*, ed. Francisco Bethencourt and Kirti Chaudhuri, 5 vols. (Lisbon, 1997–99), 2:213–31. Vieira, in fact, favored the metaphor of Babylonia as a state of mind and a place of captivity from which slaves might emerge to reach the Jerusalem of salvation. See especially sermon XXVII in the series Maria Rosa Mystica, in *Sermões do Padre António Vieira*, 16 vols. (São Paulo: Editora Anchieta, 1944), 5:391–429.

47. António Vieira, "Sermam XXVII com o santissimo sacramento exposto," in *Sermões do Padre António Vieira*, 6:391–429. This sermon is among Vieira's strongest statements against the enslavement and mistreatment of African and Native American peoples.

CHAPTER TWO

Sugar in Iberia

William D. Phillips Jr.

For over a thousand years, favored valleys in the southern part of the Iberian Peninsula have supported fields of sugarcane, and Iberians have extracted juice from the canes and have refined sugar from it. Nevertheless, sugarcane production in Iberia, as in all the Mediterranean basin, was a marginal operation at best.

Sugarcane, originally a tropical crop, needs abundant water and warm growing conditions throughout the year. Neither of these conditions are available in Iberia, which was one of the most northerly regions where sugar has ever been grown commercially. The winters are cool and prevent the cane from reaching its optimal growing conditions. In some years, freezing temperatures can kill the cane. Water has proved to be a problem as well. The warm summer months, when the rapidly growing cane demands ample water, are just the time when rainfall is most sparse. The fields must be irrigated, and southern Iberia is a land of few major rivers. Fields of cane must nestle along the streams and in the deltas, where irrigation can bring the water to the canes. Sugar can be grown in Iberia, as a thousand years of records attest, but the cane produced has a low sugar content, which prevented it from competing on an equal plane with sugar from tropical or semitropical regions where more favorable growing conditions prevailed. An additional problem was a chronic lack of firewood for the sugar refineries. The question of the timing and extent of deforestation in the Mediterranean lands is still open, but it is clear that by the late fifteenth century a lack of firewood for boiling the cane juice hindered Iberian sugar production. By the late fifteenth century, sugar was entering the European markets from Madeira and the Canaries, both with climates better suited to sugar growing, and in the sixteenth century sugar from the Caribbean islands and the American mainland competed as well.

Even as a marginal product, Iberian sugarcane production has lasted over a millennium.[1] That alone would make its history worthy of study. Nonetheless,

Iberia's greater importance to the global story of sugar is as an intermediary stage in sugarcane's spread from the Mediterranean to the Atlantic and in its historical progression from an exotic medicinal product, greatly desired and highly priced, to a widely used commodity, with prices constantly tending to fall. Iberia owed that role, as so many others it has played throughout its history, to its position as a geographical crossroads, where the Mediterranean meets the Atlantic and where Africa and Europe approach at their closest point. Iberia provided a locale for the convergence of two paths of sugar's spread: the Muslim path that led from Mesopotamia to Egypt and then around the southern shore of the Mediterranean, and the Christian path that led from the Crusader states in Syria and Palestine to Cyprus and Sicily and from there to Valencia.

As we proceed, we should be aware that there are almost no sources that illuminate in detail the history of labor in the production of sugar in Iberia. Historians and other scholars have devoted great attention in the two decades to the histories of slavery in Spain and of sugar production. Nonetheless, historians of sugar have paid relatively little attention to the story of labor, and historians of slavery have seldom concentrated on sugar. From the few comments they have made, it is clear that sugar growing and refining in Iberia followed common practice in other areas, and free workers mainly provided the necessary labor. There may have been a few slaves involved here and there, but for the Mediterranean world, including Islamic and Christian Spain, in both the medieval and early modern periods, slavery was mainly on a small scale, with slaves employed primarily as additional workers in a system of free or semifree labor.[2] In short, Iberia was home to societies with slaves, not slave societies. Despite occasional earlier precedents, localized in space and limited in time, the close connection between slavery and sugar comes later in the colonial areas of the Atlantic.

The Muslim Path

The peoples of the Mediterranean in ancient and early medieval times did not know cane sugar, and their sweetening came from honey and fruit juices. Sugarcane originated in the Pacific islands or Southeast Asia and spread through India. The ancient Persians introduced it into lower Mesopotamia on the plain of Khuzistan, where the Muslims encountered it on their conquest of the region in the seventh and eighth centuries. The plain of Khuzistan was eminently suited for the cultivation of sugarcane. Located just to the north of the Persian Gulf, bounded on the north and the east by mountains and on the west by the lower Tigris, numerous streams watered the Khuzistan and allowed irrigation of the cane fields. Production flourished and the region paid taxes in kind to the caliph. In the eighth century, those taxes amounted to thirty thousand pounds of sugar annually. As the Muslims expanded sugarcane production, they also invented new processes for clarifying and crystallizing sugar, possibly in the cities of Ahwaz and Kjondisapour. Previously, the processing of sugar had been done by crushing the cane, extracting the juice, and boiling it down to a black paste. Sugar was used as a medicinal agent as well as a sweetener, and through the academic pharmaceutical research in Khuzistan, the method of adding potash (potassium carbonate) to clarify the sugar in the refining process was invented. From Khuzistan, sugar refining spread to Baghdad, which lasted as a refining center until the end of the Middle Ages. From Mesopotamia, sugarcane spread to Baghdad and then to Egypt in the eighth century. From Egypt, the Muslims introduced sugarcane to the southern lands of the Mediterranean basin.[3]

Sugarcane reached Spain with the Muslims. It was one of the many tropical and semitropical crops that the Muslims domesticated and spread throughout their lands. Some of the most lucrative crops, including rice and cotton as well as sugarcane, required irrigation in the Mediterranean basin, where summer rainfall was slight. Muslims, accordingly, expanded irrigation works and introduced new techniques from the Middle East.[4]

Even though the Muslims conquered most of the Iberian peninsula early in the eighth century and, from the time of the emir Abd ar-Rahman I in the mid-eighth century, were introducing and acclimating new crops in palace gardens in southern Spain, the first written evidence for sugarcane in Spain dates only from the tenth century. The *Calendar of Córdoba* first mentions it about the year 961, and even then the evidence is problematic. The *Calendar of Córdoba* followed an earlier Arabic model and therefore may more accurately reflect conditions in Egypt rather than Spain. Also, Córdoba has a continental climate, with winters too harsh for sugar, but the *Calendar* may refer to all territory under Cordoban control. Certainly in Muslim times, sugar was grown in a wide stretch of southern Iberian territory, from the wetlands of the lower Guadalquivir south of Seville to warm coastal valleys along the Mediterranean coast, from Málaga to Almería, and occasionally as far north as Castellón. Arab geographers mentioned that sugar was grown in the vega of Granada, but this is virtually the only mention of it inland.[5]

Given sugar's high demand for water, irrigation had to be provided if it were to be grown successfully in semiarid Iberia. Irrigation projects were present in Spain from Roman times, but they mainly provided supplementary irrigation for crops such as wheat, olive trees, and grape vines. All of them were cultivars that could survive without irrigation, which was indispensable for sugarcane and the other tropical crops that the Muslims introduced to Spain. The Spanish

Muslims employed a variety of techniques for irrigation, most of them of Middle Eastern origin, including canals along rivers, underground water conduits, spring-fed gardens, and *norias*, devices with belts of buckets bringing water to the surface.[6] Most of the cane fields of southern Spain were irrigated by diversions of river water through canals to nearby fields.[7]

Although the first record of sugarcane agriculture in Iberia dates from the tenth century, it may have been grown and consumed as cane for chewing even earlier. Only in the twelfth century was sugarcane manufacturing documented in the peninsula, and it seems likely that the machinery developed for milling and pressing olives was adapted to the needs of sugar manufacture.[8] The entire industrial complex for processing the canes and making sugar centered on a mill, which could be a water mill (*ingenio*) but most often was an animal-powered mill (*trapig* or *trapiche*). A twelfth-century description reads as follows: "Regarding the manner of making sugar from the canes, when the canes are mature in January, they are cut into small pieces, and, in order to obtain the juice, these, well pressed or chipped in the mill, are squeezed in wine presses or similar places in the mill. Their juice is placed in a clean cauldron to boil on the fire, and left until it clarifies, afterward it is brought back to a boil until part remains; and clay forms filled with it are placed to solidify in the shade and also the sugar is taken from there; and the residue of the cane after the pressing is kept for the horses as it is a very enjoyable food for them, with which they get fat."[9]

During Islamic times, sugar was a luxury product, used extensively in pharmacology and medicine and as a significant component of Andalusian cuisine. Muslim physicians, following Galen's approach, used it in an effort to balance the four humors. Honey and sugar, usually dissolved in water, were used to treat disorders of the respiratory, urinary, and digestive systems. An Egyptian allegorical tale of the fifteenth century shows the personification of sugar leaving the ranks of the army of medicine and joining the army of the foods, a story that reflects the increasing availability of cane sugar at that time.[10]

In fact, throughout the Islamic period sugar occupied an important position in cuisine.[11] Sweets, including candy and sweet baked goods and other confections, were popular throughout the Muslim world. Equally important was the use of sugar, along with fruits and other sweeteners, in meat dishes and vegetable recipes. This was a common feature of medieval cookery, in Christian lands as well as Islamic ones, although in modern times the cuisine of Europe has tended to shed sweet recipes for meats and vegetables and to confine sweetened foods to the dessert course, whereas in North Africa main courses of meats and vegetables sweetened with sugar and fruits have remained popular.

Before the late Middle Ages, not much is known of the trade in the sugar of al-Andalus. Presumably, home production supplied the local demand with what remained being sold in North African markets. By the thirteenth and fourteenth centuries, Italian merchants became interested in the western Mediterranean and, beyond Gibraltar, to the Atlantic, opening direct sea trade between Italy and northwest Europe. The history of the activities of the Genoese and Florentines is well known in both the Christian and Islamic portions of the Mediterranean basin. Italian merchants were especially active in the ports—Almería, Almuñécar, and Málaga—of the kingdom of Granada, and by the final quarter of the thirteenth century, they were dealing officially with the king of Granada. For the Genoese especially, Granada and its trade became an important focus.[12]

The authorities of Granada and the Genoese and Florentine merchants collaborated in the sale of Granadan sugar to markets in western Europe. The famous Florentine trading company of Francesco di Marco Datini left an invaluable archive of business accounts. Among them are the records of the sales of sugar in Montpellier, Avignon, and Paris for certain years in the late fourteenth and early fifteenth centuries. In all the accounts, sugar produced in the coastal portions of the Muslim kingdom of Granada and shipped from the port of Málaga occupied an important part of the sales.[13]

Granada came under Christian control in 1492, ending nearly eight hundred years of Muslim ruled lands in western Europe, and all Iberia was under Christian control. Sugar continues to be produced in the area of the former kingdom of Granada down to the present.[14] Before discussing the early modern and modern Spanish sugar industry, we must first discuss another area of Iberia that was producing sugar in the fifteenth century. That was Valencia on the Mediterranean coast, where sugarcane arrived by a separate, Christian route.

The Christian Path

It is not true, as it is sometimes said, that the first direct contact that Westerners had with sugar was in Syria and Palestine at the time of the First Crusade in the 1090s. Sicily and, as we have seen, parts of Spain already produced sugar, and Venice imported Egyptian sugar from the late tenth century. Some Europeans, and perhaps some Crusaders, had already tasted sugar. Nevertheless, sugarcane growing in the fields was new to many of the Crusaders and attracted attention of the chroniclers. Fulcher of Chartres reported the hardships that limited food supplies caused for the army and went on to say that "in those cultivated fields through which we passed during our march there were certain ripe plants which the common folk called 'honey-cane' and which were much like reeds. The name is compounded from 'cane' and 'honey,' whence the expression 'wood honey' I think, because the latter is skillfully made from these canes. In our hunger we chewed them all day because of the taste of honey. However, this helped but little."[15]

William of Tyre, in his description of the irrigation projects around Tyre, also mentioned sugar: "All the country round about derives immense benefits from these waters. Not only do they supply gardens and delightful orchards planted with fruit trees, but they irrigate the sugarcane also. From this latter crop sugar (*zachara*) is made, a most precious product, very necessary for the use and health of mankind, which is carried by merchants to the most remote counties of the world."[16]

Western sugar manufacturing began, and trade in sugar to the West increased as a result of the Crusades. The Crusaders maintained the cultivation and refining of sugarcane, both for their own use and for export. Throughout the period of the Crusader states, from 1099 to 1291, this afforded revenues for the Crusaders, and, more important in the long run, created additional demand in the West, as returning Crusaders and pilgrims took home samples of sugarcane and thus helped to spread the taste for it.[17]

When the Muslims drove the last of the Westerners from the mainland, the Christian refugees moved to the Mediterranean islands and took sugar production with them. The intensive cultivation of sugarcane and the processing of the juice spread to the islands and mainland areas held by the Christians. In most of those regions, Muslims had been growing sugar since the eighth or ninth centuries, but only in the late Middle Ages did it come to be exploited more fully.[18]

Among the first places for sugar to be developed was Cyprus. Sugarcane had probably been grown on the island since it was introduced by the Muslims in the seventh century, but it was only after the Crusader states had fallen that it became important in the island's economy.[19] Other islands in the Mediterranean were also sugar producers. The Venetians developed a sugarcane industry on Crete in the fifteenth century, but by then Sicily was in the forefront.[20]

Sicily was probably the most important Mediterranean producer of sugar in the late Middle Ages. Sugarcane had been produced shortly after the Muslims conquered the island in 878, and by the end of the ninth century Sicilian sugar was being sold in North Africa. Palermo was the center of sugarcane production in Sicily, and Ibn Hauqal wrote that "the banks of the streams around Palermo, from their sources to their mouths, are bordered by low-lying fields, upon which the Persian reed is grown."[21] In the eleventh century, the Normans took Sicily, and their conquest was motivated, in part, by sugar. In 1016 the Christians of Muslim-held Sicily had provided samples of sugar, among other products, to demonstrate the island's wealth and encourage the Normans to drive out the Muslims. The Sicilian sugar industry underwent a decline during the Norman period. It revived temporarily during the rule of Emperor Frederick II, who was also king of Sicily from 1197 to 1250. Frederick sponsored various reforms for the economy of Sicily and was particularly interested in sugarcane.[22]

Sicily was an important stage on sugar's Christian path to Spain, because King Jaume II of Aragón (1291–1327) imported plantings from there, together with a Muslim slave skilled in sugar techniques, to establish sugar production in Valencia. Jaume's second wife was Mary of the Cypriot ruling house of Lusignan, and a portion of her dowry was paid in sugar and sold by Barcelona merchants at the behest of the king. Jaume II's initiative came to fruition only slowly, with major production delayed until the 1380s. In 1433 the cathedral chapter of Valencia, endeavoring to establish their right to collect a tithe on sugar production, provided an important account of the growth of production in Valencia. As the canons reported the situation, Christian and Muslim farmers had planted sugarcane as a secondary crop since the 1380s and 1390s and had sold their cane in its raw state as a delicacy for children and adults. In 1407 the Valencian government gave financial aid to Nicolau Santafé, a sugar expert, to set himself up in Valencia. In the second decade of the fifteenth century, planting increased as grain gave way before cane in many places. The first evidence of a sugar mill in Valencia comes from 1417, when the master potter Thahir Aburrazach contracted to move to Burriana and make ceramic forms and vessels needed for the mill (*trapig de les canyes mels* in the Valencian dialect) owned by the merchant Francisco Siurrana. Nobles also invested in sugar mills. The knight don Galcerán de Vich built one at Jeresa and later another at Gandía, and the soldier-poet mosén Ausias March owned a sugar plantation and later built a mill on his property. By the 1430s Valencia had a number of mills in full operation, at the very time when eastern Mediterranean sugar production was faltering and commerce there was being threatened by the rise of the Ottoman Turks.[23]

Consequently, a large German merchant house, the *Ravensburger Handelsgesellschaft*, became interested in Valencian sugar. After its foundation in the 1380s, the company expanded rapidly and established branches in various parts of Europe, concentrating on commerce and avoiding banking, unlike other German houses such as the Fuggers and the Welsers. The Ravensburg company by 1420 had agents and warehouses in Valencia and exported Valencian sugar and other products. The Ravensburgers rented grain fields for the growing of sugarcane and contracted for labor in the harvest season. Some few slaves may have been among those contracted, for slave owners occasionally rented out their slaves' labor or permitted slaves to work on their own, but most workers were free people and many of them were Moriscos (that is, converted Muslims). Valencian sugar exports generally increased in the first half of the fifteenth century, and by 1460 increased profits encouraged the Germans to invest in production, on land purchased from Hugo de Cardona along the River Alcoy near Gandía. They built a mill and refinery managed by *maestre* Santafé, possibly the son of Nicolau de Santafé. The manufacturing complex prospered at first, and the quality of sugar produced there was extremely high. But in the 1470s difficulties intervened. A lawsuit involving Hugo de Cardona slowed production, and this coincided with the company's loss of some of its markets, difficulties in transportation, and competition from Madeiran sugar. The company's directors sold the facility in 1477 and a few years later rejected a proposal to reopen it.[24]

In Murcia, on Spain's southeastern coast between the sugar-producing regions of Valencia and Granada, there had been no sugar growing during Islamic times. During the fifteenth century, two efforts to establish sugar production in Murcia failed, clear indications of the difficulties encountered in trying to grow sugarcane in a climatologically marginal region. In the late 1430s, an effort led by a master Antonio, *maestre de fazer açúcar*, never reached production. Some twenty years later an association surfaced whose members intended to build a sugar mill in Murcia, but nothing came of the attempt.[25]

A German traveler, Hieronymus Münzer of Nuremberg, visited Iberia in 1494 and 1495. He commented on sugarcane in Valencia, which he "saw being produced in an establishment, and also the moulds in which they pour the molasses in order to make the sugar loaves, heavy labor that occupies a number of workers. We saw them clarify it, cook it, and elaborate candy sugar, an operation requiring a most thorough grading. We also saw the raw cane, [and] we enjoyed its juice."[26]

By the late fifteenth century, sugar from the new colonial areas, particularly from Madeira, began to enter the market, often at substantially lower prices than those of the Iberian producers. This foreshadowed the series of sugar booms that followed sugar's introduction into semitropical and tropical areas in the Caribbean and on the American mainland. Iberian production suffered in Valencia, the Granadan coast, and the lower Guadalquivir valley ceased to produce sugar on any but a minor scale.[27] Nonetheless, sugar production did continue in the peninsula long after the competition from Atlantic production began.

Sugar production continued in areas of the kingdom of Valencia in the early modern centuries. Several factors kept it alive in the sixteenth century. One was the general prosperity, buoyed by the beginning of the influx of American bullion into Europe. Another was the closing of the supply of Egyptian sugar after the Ottoman Turks took Egypt. Even though Valencian sugar faced competition from sugar from the Canaries, Madeira, and the West Indies, it still possessed the advantage of lower transportation costs. Sugar absorbs water from the air, and long sea voyages across the Atlantic lowered the quality of the sugar shipped.[28]

In the sixteenth century, the Duke of Gandía operated seven sugar mills in and near the town of Gandía. For labor, he followed the practice of his neighbors and used his own vassals for the work of growing the cane and refining the sugar. They provided the cane on a modified share-cropping arrangement, and provided seasonal labor for harvesting and transporting the canes. The cane fields were not extensive, and sugarcane was not a monocrop. The vassal farmers also had grain fields and garden plots, which occupied them in the long periods when they were not needed to tend the cane. Most of these workers were Moriscos, and sugarcane production prospered as long as they remained. Following the expulsion of the Moriscos in the early seventeenth century, Valencian sugar entered a long decline. The new settlers were old Christians who lacked the skills of the Moriscos in sugar production, and they were fewer in number as well. At the same time, deforestation was progressing, as sugar production's high requirements for firewood added to the demands of urban building projects and naval construction. The Valencian sugar industry limped along in the eighteenth century, and the death blow came with the harsh winter of 1754, when many of the canes were frozen and lost. Sporadic efforts to revive the industry in the eighteenth and nineteenth centuries came to nothing.[29]

Portugal was also crucially important in the spread of sugar production beyond Europe. By the fourteenth and fifteenth centuries, sugar was being produced profitably in the Algarve, and the desire to tap the wealth of Morocco's sugar plantations may have been one of the motivations for Portuguese expansion in Africa. Sugarcane's later career in the early modern period is less well served by scholarship. Portugal at that time certainly had a sizable population of slaves of African origin, and many of them worked as agricultural laborers, at times on farms producing sugarcane. Nonetheless, they seem to have been additional laborers in a regime in which free peasant labor predominated.[30]

Perhaps surprisingly, sugar production on Spain's southern coast, the littoral of the old kingdom of Granada, has continued up to the present day. Motril and Almuñécar were the most important towns with sugar mills and even expanded in importance in the eighteenth century.[31] This was a small-scale operation when compared with the extensive sugar plantations of the Americas. Labor in Spain, again in contrast to the Americas, was that of day laborers, who worked on seasonal contracts for the owners of the mills.[32]

TABLE 2.1. Documented Sugar Mills in Spain

Municipality	Sixteenth Century	Eighteenth Century
Adra	1	1
Motril	8	6
Petaura	1	1
Lobres	1	1
Salobreña	2	1
Almuñécar	4	2
Torrox	3	2
Manilva	1	1
Marbella	—	2
Maco	—	1
Nerja	—	1
Vélez Málaga	—	1
Jete	—	1

One attempt to establish a sugar plantation, complete with slave labor, took place along the Miel River near Algeciras early in the sixteenth century. Its promotors included a local nobleman, the marqués of Tarifa, and a Spanish merchant from the Canary Islands. The latter undoubtedly was familiar with the slave-run sugar establishments in the Canaries, but his efforts to establish something similar in metropolitan Spain failed due to local opposition. Citizens of Algeciras sued the would-be plantation proprietors. Their suit reveals that black slaves were working the fields along with white laborers, and it lays out their major concerns. The most important of these was that the flow of the river would be diverted to the sugar mills to the detriment of other millers and that the high demands for fuel for the boilers would quickly denude the available woodlands.[33]

If the Algeciras mill came to nothing, other mills dotted Spain's southern Mediterranean coast throughout the early modern period. They operated successfully through the sixteenth century, but they suffered a setback in the early seventeenth when the Moriscos were expelled from Spain. Moriscos owned and operated all the mills in Motril, for example, and the Old Christians who later acquired them lacked the necessary skills.[34] Nonetheless, there was a resurgence with time, and even something of an expansion in the eighteenth century. Table 2.1 shows those that have left documentation.[35]

These mills were small operations in comparison to the larger plantations in the Caribbean and on the American mainland. For a time, they successfully confronted the competition from the Atlantic sugar trade, in part because they were small and fit in easily with the local economy. Local farmers could be mobilized and paid to provide the necessary labor at harvest time. Cane tops provided food for local draft animals, and local pigs ate the residue left after the extraction of the cane. Nevertheless, not all the mills could resist indefinitely, and many closed over the course of the eighteenth century, victims of American sugar imports. By the early decades of the nineteenth century, only a few mills remained, but some of them lasted into the twentieth century.

In 1922, the poet Gaspar Esteva Ravassa published "Sugar Making," describing the area of Motril:

Under the vivid light of May
As the cutters' hatchets
Demolish the green thicket
At the sound of their songs, the trimmers
Remove the leaves and the tops.
They clean the canes which in beautiful mounds
Are the longed-for profit of the owner.
In extended lines the carts then
Carry them in bundles to the mill; and when
The superb scales tally their weight
They go from the ample cane carriers to the hole,
Forced by the magnificent wheel.
The three round spinning grinders
Swallow, consume, and leave bellowing
The poor cane converted into broth.
The broth boils, and being filtered and condensed
In a variety of beautiful devices
And when the swift turbine receives it
In a dense mass in which the fibrous matter shines,
Its metallic basket turns about
The axis with an almost fantastic spin,
The honey leaves for a new occupation
And the astonished eyes behold
The lustrous whiteness of the sugar,
Happy glory of the human palate.[36]

Sugar production in Spain could still inspire poets in the 1920s, and its persistence has inspired historians and archaeologists to study its past. In recent decades, scholars in the region have devoted great attention to reexamining sugar production through archaeology, with investigations of the sites of mills as well as the recovery of the vessels used to refine the sugar and the clay forms used to solidify it.[37] This effort has paid off for scholarship in a series of conferences detailing many facets of the history of sugar in southern Spain.[38]

Nonetheless, it is clear that Iberian and other Mediterranean sugar production paled in the shadow of cheaper sugar from the new colonial areas. Despite the long centuries of sugar production in Iberia and the Mediterranean world generally, sugarcane has always done better in tropical and semitropical climes similar to those in which it first developed. For the history of sugar, much more important than Iberia's role as a sugar producer was its role as a way station where two distinct paths in sugar's global spread converged. From that convergence, it spread into the Atlantic islands and from there to the Americas, where it reached heights of production undreamed of by the sugar producers of the Iberian peninsula.

NOTES

1. The recent interest in the history of Spanish sugar production has given rise to a number of publications, many of them first presented as papers at a series of five conferences organized by Antonio Malpica, who edited the conference proceedings: *La caña de azúcar en tiempos de los grandes descubrimientos (1450–1550): Actas del primer seminario internacional, Motril, 25–28 de septiembre 1989* (hereafter cited as *Actas Motril 1989*) (Motril: Junta de Andalucía and Ayuntamiento de Motril, 1990); *La caña de azúcar en el mediterráneo: Actas del segundo seminario internacional, Motril, 17–21 de septiembre 1990* (hereafter cited as *Actas Motril 1990*) (Motril: Junta de Andalucía and Ayuntamiento de Motril, 1991); *Producción y comercio del azúcar de caña en época preindustrial: Actas del tercer seminario internacional, Motril, 23–27 Septiembre 1991* (hereafter cited as *Actas Motril 1991*) (Granada: Diputación Provincial, 1993); *1492: Lo dulce a la conquista de Europa: Actas del cuarto seminario internacional sobre la caña de azúcar, Motril, 21–25 de septiembre de 1992* (hereafter cited as *Actas Motril 1992*) (Granada: Diputación Provincial, 1994); and *Paisajes del azúcar: Actas del quinto seminario internacional sobre la caña de azúcar, Motril, 20–24 de septiembre de 1993* (hereafter cited as *Actas Motril 1993*) (Granada: Diputación Provincial, 1995). For a brief but careful survey that appeared before the Motril conferences, see Alain Huetz de Lemps, "Une culture originale de l'Espagne: La canne à sucre," in Anne Collin-Delavaud and Alain Huetz de Lemps, *La canne à sucre: En Espagne, au Pérou, en Équateur* (Paris: Centre National de la reserche scientifique, 1983).

2. Scholars have long known that Spain and Portugal had slaves throughout the Middle Ages and the early modern centuries. See William D. Phillips Jr., *Historia de la esclavitud en España* (Madrid, 1990). Taking into account the steady growth of studies of slavery in Spain produced over the last decade, I am currently working on a completely revised version of that

book, to be published by the University of Pennsylvania Press. As the previous note indicates, sugar production in Spain has attracted numerous scholars, but almost none have had a great deal to say about the labor and laborers required. Antonio Malpica Cuello, "Arqueología y azúcar: Estudio de un conjunto preindustrial azucarereo en el reino de Granada: La Palma (Motril)," *Actas Motril 1990*, 124, indirectly comments on the lack of studies of labor in the Spanish sugar industry.

3. Andrew M. Watson, *Agricultural Innovation in the Early Islamic World: The Diffusion of Crops and Farming Techniques* (Cambridge: Cambridge University Press, 1983); Edmund O. von Lippmann, *Geschichte des Zuckers, seiner Darstellung und Verwendung* (Leipzig: Hesse, 1890); Noël Deerr, *The History of Sugar*, 2 vols. (London: Chapman and Hall, 1949–50); Wallace R. Aykroyd, *The Story of Sugar* (Chicago: Quadrangle Books, 1967). The latter book was published in London in 1967 as *Sweet Malefactor: Sugar, Slavery, and Human Society*. See also Sidney W. Mintz, *Sweetness and Power: The Place of Sugar in Modern History* (New York: Penguin, 1986).

4. Thomas F. Glick, "Regadío y técnicas hidráulicas en al-Andalus, su difusión según un eje Este-Oeste," *Actas Motril 1989*; Watson, *Agricultural Innovation*, summarized in Glick, "Innovaciones agrícolas en el mundo islámico," *Actas Motril 1990*.

5. Expiración García Sánchez, "El azúcar en la alimentación de los andalusíes," *Actas Motril 1989*; Sánchez, "Caña de azúcar y cultivos asociados en al-Andalus," *Actas Motril 1993*; Lucie Bolens, "La canne à sucre dans l'agriculture d'al-Andalus," *Actas Motril 1989*.

6. Glick, "Regadío." See also Thomas F. Glick, *Irrigation and Society in Medieval Valencia* (Cambridge, Mass.: Harvard University Press, 1970).

7. García, "Caña de azúcar y cultivos asociados"; Ignacio González Tascón and Joaquín Fernández Pérez, "El azúcar en el viejo mundo: El impacto en su elaboración," *Actas Motril 1989*, esp. 101–7.

8. González and Fernández, "Azúcar en el viejo mundo," 103–4.

9. Ibid., 104–5. My English translation is based on several Spanish variants that they provide.

10. Ana Labarta and Carmen Barceló, "Azúcar y medicina en el mundo islámico," *Actas Motril 1992*; Lucie Bolens, "L'affect et le doux: Essai d'interprétation en al-Andalus," *Actas Motril 1992*; David Waines, "Sugar in Andalusi 'Home Remedies,'" *Actas Motril 1992*; Amador Díaz García, "El azúcar en los textos árabes medievales," *Actas Motril 1990*; he mentions the Egyptian folk tale on p. 64.

11. In addition to the works cited in the previous note, see Manuela Marín, "Azúcar y miel: Los edulcorantes en el tratado de cocina de al-Warraq (s. IV/X)," *Actas Motril 1992*; and Rosa Khune, "El azúcar: Usos dietéticos y farmacéuticos según los médicos árabes medievales," *Actas Motril 1992*.

12. Adela Fábregas García, *Motril y el azúcar: Comerciantes italianos y judíos en el reino de Granada* (Motril: Ingenio, 1996), 43–45, 94–115, 119–32; Olivia Remie Constable, *Trade and Traders in Muslim Spain: The Commercial Realignment of the Iberian Peninsula, 900–1500* (Cambridge: Cambridge University Press, 1994).

13. Fábregas, *Motril y el azúcar*.

14. Antonio Malpica Cuello, "Medio físico y territorio: El ejemplo de la caña de azúcar a finales de la Edad Media," *Actas Motril 1993*.

15. Fulcher of Chartres, *A History of the Expedition to Jerusalem, 1095–125*, trans. Frances Rita Ryan, ed. Harold S. Fink (Knoxville: University of Tennessee Press, 1969), 130.

16. William of Tyre, *A History of Deeds Done Beyond the Sea*, trans. and ed. Emily Atwater Babcock and A. C. Krey, 2 vols. (New York: Columbia University Press, 1943), 2:6.

17. Jonathan Riley-Smith, *The Knights of St. John in Jerusalem and Cyprus, c. 1050–1310* (London and New York: Macmillan/St. Martin's, 1967), 425–27, 434; Wilhem von Heyd, *Historie du commerce du Levant au Moyen-Age*, trans. Furey Reynaud, 2 vols. (Leipzig, 1885–86; reprint, Amsterdam, 1967), 2:684; William D. Phillips Jr., "Sugar Production and Trade in the Mediterranean at the Times of the Crusades," in V. P. Goss and C. V. Bornstein, eds., *The Meeting of Two Worlds: Cultural Exchange between East and West during the Period of the Crusades* (Kalamazoo: Medieval Institute Publications, 1986).

18. Jean Richard, *The Latin Kingdom of Jerusalem*, trans. Janet Shirley, 2 vols. (Amsterdam: North Holland Publishing, 1979), 1:73, 125, 176, 351; Joshua Prawer, *The Crusaders' Kingdom: European Colonialism in the Middle Ages* (New York: Praeger, 1972), 363–64. On p. 254 Prawer mentions archaeological remains of sugar factories. Deerr, *History of Sugar*, 1:76–77.

19. Heyd, *Commerce du Levant*, 2:686; David Jacoby, "Citoyens, sujets et protégés en Chypre," *Byzantinische Forschugen* 5 (1977): 174–77; Freddy Thiriet, *La Romanie vénitienne au Moyen Age: Le développment et l'exploitation du domaine colonial veneziana* (Paris: E. de Boccard, 1959), 333; Gino Luzzatto, *Studi economica di Venezia dall'XI al XVI sècolo* (Venice, 1961), 54, 64, 196; Frederick Lane, *Venice: A Maritime Republic* (Baltimore: Johns Hopkins University Press, 1973), 144–45; Sidney M. Greenfield, "Cyprus and the Beginnings of Modern Sugar Cane Plantations and Slavery," *Actas Motril 1990*.

20. Deerr, *History of Sugar*, 1:79, 83; Charles Verlinden, *The Beginnings of Modern Colonization*, trans. Yvonne Freccero (Ithaca: Cornell University Press, 1970), 96–97; Thiriet, *Romanie vénitienne*, 417–18.

21. Quoted in Deerr, *History of Sugar*, 1:76.

22. Carmelo Trasselli, "Produzione e comercio delle zucchero in Sicilia dal XII al XIX sècolo," *Economia e Storia* 2 (1955); Henri Bresc, "La canne à sucre dans la Sicilie médiéval," *Actas Motril 1990*; see also Bresc's *Un monde méditerranéen: Économie et société en Sicile, 1300–1450*, 2 vols. (Palermo and Paris: Ecole française de Rome, 1986), 1:227–52, 474–75. Bresc shows that sugar production in Sicily was an occupation for free laborers, not slaves.

23. José Pérez Vidal, *La cultura de la caña de azúcar en el Levante español* (Madrid, 1973), 14–15, 37–38, 41–45; Carmen Barceló and A. Labarta, "La industria azucarera en el litoral valenciano y su léxico (Siglos XV–XVI)," *Actas Motril 1990*.

24. Pérez, *Caña de azúcar*, 41–45; Bolens, *Motril 1989*, 50.

25. María Martínez Martínez, "Producción de azúcar en Murcia: Un proyecto fracasado del siglo XV," *Actas Motril 1992*.

26. Jerónimo Münzer, "Relación del viaje," in J. García Mercadal, *Viajes por España y Portugal*, 3 vols. (Madrid: Aguilar, 1952–62), 1:340.

27. García, "Caña de azúcar y cultivos asociados," *Actas Motril 1993*, 68.

28. Pérez, *Caña de azúcar*, 18–21, 44–45.

29. Ibid., 50, 54–56, 83–87, 133–38; C. Barceló and A. Labarta, "La industria azucarera en el litoral valenciano y su léxico (siglos XV–XVI)," *Actas Motril 1990*.

30. Cláudio Torres, "A industria do açúcar nos alvores da expansão atlântica portuguesa," *Actas Motril 1990*; Henrique Gomes de Amorim Parreira, "História do açúcar em Portugal," *Anais: Estudos da história da geografia a expansão portuguesa* 7 (1952). See the comments of A. C. de C. M. Saunders, *A Social History of Black Slaves and Freedmen in Portugal, 1441–1555* (Cambridge: Cambridge University Press, 1982), 69–71. More generally, see José Ramos Tinhorao, *Os negros em Portugal: Uma presença silenciosa* (Lisbon: Caminho, 1988); and Didier Lahon, *O Negro no coração de Imperio: Uma memoria a resgatar, Séculos XV–XIX* (Lisbon, 1999).

31. Margarita María Birriel Salcedo, "Azúcar e estado: El intento del monopolio del azúcar granadino," *Actas Motril 1990*; Salcedo, "La producción azucarera de la Andalucía mediterránea, 1500–1750," *Actas Motril 1991*; Francisco Andújar Castillo, "Una estructura de poder: El monopolio de la producción y comercialización del azúcar en Adra (siglos XVI–XVII)," *Actas Motril 1992*; Antonio Malpica Cuello, "La cultura del azúcar en la costa granadina," *Actas Motril 1989*; Juan Luis Castellanos, "El azúcar de Motril en la conyuntura del siglo XVII," *Actas Motril 1991*; Justo del Río Moreno, "Refinerías de azúcar en Sevilla (S. XVI–XVII)," *Actas Motril 1990*.

32. Francisco Andújar Castillo, "Una estructura de poder: El monopolio de la producción y comercialización del azúcar en Adra (siglos XVI–XVII)," *Actas Motril 1992*, 185–86.

33. Malpica, "Medio físico y territorio: El ejemplo de la caña de azúcar a finales del la Edad Media," *Actas Motril 1993*.

34. Malpica, "Arqueología y azúcar," 138.

35. From Margarita M. Birriel Salcedo, "La producción azucarera de la Andalucía mediterránea, 1500–1750," *Actas Motril 1991*.

36. F. J. García Marcos, M. V. Mateo García, and D. Fuentes González, "Indices de mortandad del léxico cañero en la costa granadina," *Actas Motril 1992*; the poem appears on pp. 233–34.

37. Malpica, "Archeología y azúcar"; Fernando Amores Carredano and Nieves Chisuert Jiménez, "Sevilla y América: Interpretación del hallazgo de un grupo de formas de azúcar del siglo XVI en la cartuja de Santa María de las Cuevas (Sevilla)," *Actas Motril 1990*; Cláudio Torres, "A industria do açúcar nos alvores da expansão atlântica portuguesa," *Actas Motril 1990*; Josep A. Gisbert Santonja, "En torno a la producción y elaboración de azúcar en las comarcas de la Safor-Valencia y la marina Alta-Alicante, s. XIV–XIX," *Actas Motril 1990*.

38. See note 1 for a list of the conferences.

CHAPTER THREE

Sugar Islands

The Sugar Economy of Madeira and the Canaries, 1450–1650

Alberto Vieira

Europe was always quick to name its islands according to the products that they supplied to its markets. Thus some were called the islands of pastel (dyestuff), and others the islands of wine. Madeira and some of the Canary Islands, given the role that sugar played in their economies and in the life of their people, became known as sugar islands. These island groups played an essential role in the transfer of sugar from the Mediterranean to the Caribbean along what could be called the "sugar route."

This chapter traces the parallel evolution of sugar agriculture on the islands of Madeira, Gran Canaria, Tenerife, La Palma, and Gomera from the fifteenth to the seventeenth century. The focus is on the productive and commercial cycles of this product as well as on the essential questions of land, water, and slavery that determined much of the history of sugar in its Atlantic island stage. Madeira is the point of departure for this study for a couple of reasons: sugar agriculture was first introduced in Madeira, from where the industry spread to other areas, including the Canaries; and the surviving documentation from Madeira enables us to better understand the impact of sugar on society and economy in ways that could eventually fill in gaps in the documentary record of the Canary Islands as well.

The System of Landed Property and Water Rights

The process of the occupation and settlement of Madeira and the Canaries was not identical. Between 1439 and 1497 the two islands of the Madeiran archipelago were a dominion (*senhorio*) of the Order of Christ, which established as its representatives three captains, namely João Gonçalves Zarco at Funchal (1450), Tristão Vaz at Machico (1440), and Bartolomeu Perestrelo at Porto

Santo (1446). In the Canaries, there were both royal islands (Gran Canaria, La Palma, and Tenerife) and those under lordly or seigniorial control (Fuerteventura, Lanzarote, La Gomera, and El Hierro). Moreover, in the Canarian archipelago, an indigenous population existed, not only slowing the process of occupation but also confronting the colonists with rival claimants to the distribution of lands among those autochthonous people who accepted Castilian sovereignty.[1]

An understanding of the system of property requires an in-depth study, based on documentary sources, of relations based on the ownership and production of the limited arable land. For Madeira, some tax registers for sugar growers exist, but for the Canaries such information can only be found in land distribution (*repartimiento*) and notarial records.[2] The system of property in both archipelagos was defined by the distribution of land to the settlers and later by sale, exchange, or redistribution. Although there were many similarities, the process of settlement on each island varied due to their unique features. The Crown granted the captains and governors the power to distribute lands to settlers and conquerors according to their participation in the process and to their social rank.[3] All these donations or grants were made according to norms established by the Crown, based on the model established during the resettlement of the Iberian peninsula. These grants also included information, which was not always accurate, concerning the social status of the recipient, area of cultivation, improvements to be made, and a time table for cultivation.

On the Portuguese islands, the Crown and later the lord of the island, Prince Henry (Infante Dom Henrique), regulated the distribution of lands from the very beginning. At first, the monarch, Dom João I, instructed the captains that the lands should be "conveyed unencumbered and without any rent to those of high quality and others who possess the means to use them well and stripping timber and in breeding livestock."[4] Later, João Gonçalves Zarco, using the prerogatives bestowed upon him and his descendants, held a significant portion of the land in Funchal and Ribeira Brava. Other grants were made under the Alfonsine regulations to those who had the capacity to develop them; failure to do so resulted in losing their right of possession. In the Canaries too the social distinction between the grantees was apparent. Following the *cédula real* of 1480, Pedro de Vera made grants to the conquistadors "according to their merits."[5] It is important to note that not all the Canary Islands had an ecosystem that was ideal for sugar cultivation, unlike Madeira where the chroniclers noted the abundance of water and wood.

In Madeira, from the second half of the fifteenth century, leases of *aforamento and meias* became common and they evolved in the sixteenth century into sharecropping contracts. This was a specific situation in Madeira, which had the characteristic of consuetudinary law. We should note that the various contracts of lease (*arrendamento*) that have survived are not uniform in the arrangements between the contracting parties. In some, the lord contributed to improvements; in others this was left to the *colono* or renter, reserving possession at the end with no penalties. The norm was a contract of limited duration obliging the renter to pay an annual fee or one half of the product. In the Canaries, there were several different contract arrangements (leases, sharecropping, mortgages) for the use of the land similar to those of Madeira;[6] it is important to mention the contract of *complantación*, according to which the proprietor of the land, in order to begin cultivation, ceded the land for a fixed period and only after that period was rent paid.

Given the importance of water for the sugar crop, its possession and distribution were essential elements of the organization of the economy. In Madeira this was not a problem at first because of the abundance of water, but in the Canaries, scarcity immediately generated concern. Thus there were land grants with and without water. Water ran in the streams (*ribeiras*) abundantly in the north. In the south during the summer, the streams were almost all diverted to the *levadas* (water-course) or irrigation and aqueduct systems.[7] It was, in fact, in the stream beds and their margins that the history of the island was played out. The principal parishes contained the headwaters of one or more streams. Funchal, the principal settlement of the island, is traversed by three streams. Streams and their sources were considered public domain in the earliest documents about the island. In the areas of greatest population concentration and of intensive land use, such as Funchal, the water of the stream beds was not sufficient to meet the requirements of the residents. Thus in 1485, Duke Dom Manuel recommended that the waters of the Ribeira of Santa Luzia be used only for sugar mills, flour mills, and their associated activities and for no other reason. It was with Dom João II that water rights were definitely defined in a way that lasted until the nineteenth century. In the letters of 7 and 8 May 1493, he established once and for all that waters were common patrimony to be distributed by the captain and officers of the municipal council to all proprietors, since "without the waters the lands cannot be exploited." From this point water was public property to be used by those who held lands and needed it. Still, from the end of the fifteenth century, water was negotiated in the same way as land. It was with the regulations (*regimento*) of Dom Sebastião (1562) that the early system was changed. Water could be sold or rented, which then caused a distinction between property and land with water.[8] The tradition of building *levadas* made the Madeirans their most famous builders, and they took this skill wher-

ever they went, first to the Canaries and then to America.[9] The skill and ingenuity of the Madeirans in this occupation was reflected in the request of Afonso
de Albuquerque, who asked that the king send Madeirans to cut the wood to
make the *levadas* with which sugarcane was irrigated, "in order to change the
course of the River Nile."[10]

In the Canaries, except for the islands of Gomera and La Palma, water was
less accessible. It was the patrimony of the king or lord who then distributed it to
the settlers. The "dulas" were established "according to the measurement of the
said lands and the division made," and above all, according to the agriculture for
which they were destined, sugar having a preferential status. In this way, the
grants (*datas*) of land shed light on the cultures to be initiated and the system
that controlled the distribution of woodland and water. Thus we have "grants of
irrigation" (*regadio*) and of "dryland" (*secano*). Those who sought to invest in
infrastructure by building an *engenho* were guaranteed thirty *fanegas* of irrigated
land. In Tenerife, for the first decade of the sixteenth century we have twenty-
four cases in which the building of a water or animal powered mill was ordered
to be done within two or three years. In the Canaries, the most important
element was the rights to water, since they defined the ability to exploit the land,
and thus its utility. The lands granted for cane fields were made with the obligation to construct a water-powered mill. In this context, the lands near the
stream beds or *barrancos* were greatly sought and were reserved for the principal
settlers.[11]

According to Virginia Rau and Jorge de Macedo, "the production of sugar
benefited broad sectors of the population, including among the producers not
only small and medium farmers, but also shoemakers, carpenters, barbers, merchants, surgeons, and millers as well as noble functionaries, municipal officers,
and others who lived on the margins of this rich production. All these small
producers took advantage of the system on the island to make their tiny production profitable."[12] Historian Vitorino Magalhães Godinho reinforced this char-
acterization of Madeiran social reality by noting the concentration of cane fields
in the hands of a small number of islanders.[13] The situation in the first half of the
sixteenth century was different in that the limited number of owners indicates
that the cane fields were concentrated in the hands of privileged island social
groups: the aristocracy, merchants, and artisans, local and royal functionaries. At
both times this group of proprietors represented only about 1 percent of the
island population.[14] This tendency toward concentration accelerated from the
fifteenth to the sixteenth century as the number of proprietors decreased in the
regions near the "partes do fundo" (embracing the districts of Ribeira Brava,
Ponta do Sol, and Calheta). Moreover, the continuity of ownership was marked,

MAP 3.1. The Sugar Industry on the Canary Islands

since changes by sale, dowry, or lease were reduced. The stability of property depended primarily on its entail (*vinculação*). Thus, between 1509 and 1537, 18 percent of the cane fields of the zones of the "partes do fundo" were entailed while in Funchal about 17 percent were so encumbered, an amount representing about 38 percent of the production of that captaincy.

For the Canary Islands we lack the documentation to conduct a similar analysis of the ties between the proprietors of the cane fields and the mills. We do know that the mill owners were favored from the outset even though they were guaranteed thirty fanegas of land. We know of eleven grants in Tenerife. Among these were the "haciendas" of the Adelentado in Daute, Icod, and El Realejo; Tomás, Justiniano, Bartolomé Benítez, and the Duke of Medina Sidonia in La Orotava; Cristóbal Ponte and Mateo Vina in Daute; Blasyno Inglesco de Florentino and Juan Felipe in Güimar, Tenoya; and Lope Fernández in Taganana. Along with the haciendas of Argual and Tazacorte, Juan Fernández de Lugo Señorino developed one of the most important properties. In 1508 its ownership was taken over by Jácome Dinarte, who in the following year sold it to the Welsers, who in turn sold it in 1513 to Jácome de Monteverde. The size of his property is based on an observation of Gaspar Frutuoso, who stated that the mills could operate from January to July with enough cane to produce 7,000 to 8,000 *arrobas* of sugar. The information on production is scattered and does not permit a definite conclusion. Thus in La Orotava the mill that belonged to Pedro de Lugo and had been owned by Tomás Justiniano produced 556 *arrobas* in 1535 and 1,112 in 1536. In Daute, the two sugar mills of Mateo Viña produced 5,000 to 6,000 *arrobas*. Finally, the hacienda El Realejo of the Adelentado produced in 1537–38 some 9,000 *arrobas* of sugar. In Gran Canaria, a sugar mill at Telde produced 1,190 *arrobas* in 1504.[15]

The Production of Sugar

Sugarcane's first experience outside of Europe demonstrated the possibilities of its rapid development beyond the Mediterranean. Gaspar Frutuoso testified to this: "This plant multiplied in the land in such a way that its sugar is the best that is known in the world and it has enriched many foreign merchants and a good part of the settlers of the land."[16] This reality attracted both foreign and national capital, which explains its rapid increase. Although sugar had been a secondary activity at the beginning of the occupation of the islands, it became for a short time the predominant agricultural product there.

With the support and protection of the lord and the Crown, sugar occupied Madeira, taking over the arable in two areas: a warm southern strip from Machico to Calheta, sheltered from prevailing winds (*alisios*), where the cane fields rose up the slopes to 400 meters of altitude; and the captaincy of Funchal, which contained most of the best sugar lands within its borders. Machico had only a small area appropriate for cane. With external investments, state and local protection, and markets in the Mediterranean and in northern Europe, sugar expanded rapidly on the island. By the mid-fifteenth century chroniclers such as Cadamosto and Zurara took note of the situation.[17] There was a period of growth from 1450 to 1506 despite a depression from 1497 to 1499. It was especially rapid from 1454 to 1472, during which production grew at a rate of 13 percent per year, and then from 1472 to 1493, when that rate was 68 percent per year or an increase of 1,430 percent in that period. Recovery after the depression of 1497–99 was rapid. The high point was reached in 1506, after which rapid decline began. In the captaincy of Funchal production fell by 60 percent between 1516 and 1537. In Machico, the fall was slower and resulted from the impoverishment of the soil, but after 1521 the decline was the result of several factors, and by 1525 levels were more or less what they had been in 1470. By the 1530s the sugar economy on the island was in full crisis and the inhabitants were abandoning their cane fields and turning toward the planting of vineyards.

Many explanations for the sugar crisis have been offered, most of them based on external factors. Nevertheless, Fernando Jasmins Pereira in his *Açúcar madeirense* has offered a different view, arguing that the crisis resulted from ecological and socioeconomic conditions on the island itself: "The decline of Madeiran production is principally due to the impoverishment of soils, which given the limited area available for agriculture, inevitably reduced the productive capacity."[18] According to this view, the Madeiran crisis was not the result of the competition from the Canaries, Brazil, the Antilles, and São Tomé alone, but was caused by internal factors such as the lack of fertilizers, soil exhaustion, and climatic changes. Competition from other areas, plague in 1526, and labor shortage aggravated the situation. In addition to these factors, there is evidence that a species of insect damaged the cane in 1593 and 1602. Thus the last quarter of the century witnessed a turn to more profitable agricultures such as wine. In 1571 Jorge Vaz from Câmara de Lobos spoke of a property "that had always been in cane and I now order that it be planted in grapes so that it can yield more."[19]

The Canaries have been seen as an area of competition with Madeira, but it was the Madeirans themselves that promoted sugar there. It was during the crisis on Madeira that technicians linked to the sugar industry went to the Canaries and cane plantings arrived in Gran Canaria, Tenerife, La Palma, and Gomera, but not to the other islands due to their sterility, as Gaspar Frutuoso tells us. The surviving documentation provides scattered information about levels of

MAP 3.2. Sugar Production on the Island of Madeira, ca. 1590

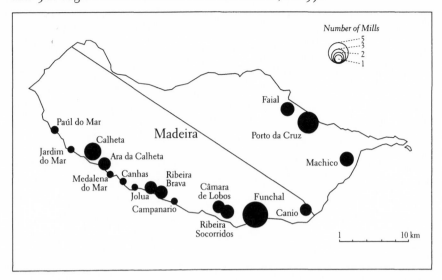

production. In 1507 Tenerife produced 34,545 *arrobas* and La Palma 2,727. We know that in 1506 Gomera yielded 1,100 *arrobas* to its lord, and a reference to Gran Canaria for 1534 mentions 80,000 *arrobas.*[20]

Traditionally, historians have argued that after the middle of the sixteenth century competition from others producers and the uncontrolled expansion of viticulture caused a crisis in sugar. Manuel Lobo Cabrera does not agree, and has held that there was a certain flourishing in the reign of Philip II.[21] He believes that the crisis resulted mostly from Caribbean competition and, above all, from the closing of the northern market, particularly of Antwerp, due to Philip's military policies in Europe.

During the seventeenth century the cane fields on the islands gradually declined in importance. Only on Madeira does there appear to have been a slight recovery when Brazilian production slowed, but this seems limited to the area around Funchal. That is substantiated by a tax record of 1600, which listed 108 owners of cane fields, most of them from this area. This is almost the only evidence of sugar production on the island until other tax records of 1689.[22] By the year 1600 on Madeira, the retreat of the industry is obvious. Medium-size properties had been replaced by very small ones. The great majority (89 percent) produced only from 5 to 50 *arrobas*, indicating an activity aimed at household use for the making of conserves, jams, and sweets. Up to 1640 this decline was made ever more apparent by the increasing presence of Brazilian sugar in the port of Funchal, to the extent that measures were taken in 1616 to ensure that there would be an equitable sale of sugars from both places. Dutch occupation of sugar producing areas in Brazil caused a rebirth of some sugar production on the island to meet the market demand for jams and preserves. In 1643 there were not enough functioning *engenhos* to handle production of the cane fields. In accord with a royal provision of 1 July 1642, the Crown sought to promote cane cultivation by exempting mills from paying the *quinto* tax for five years or half of it for ten years. Various owners took advantage of this benefit, but when Brazilian production recovered in the following decade and Brazilian sugar reappeared in the port of Funchal, the former situation returned.[23] Madeiran sugar once again lost out to the competition. As late as 1658 there was an attempt to stimulate the industry by reducing the tax on production to one-eighth, but the crisis was inevitable. Added to this was the fact that from 1643 to 1675 the *quinto do açúcar* tax was not properly collected as was noted in the latter year. In an *alvará* of 15 October 1688 the Crown ordered that taxes on sugar should be limited to an eighth of production as the most effective way of stimulating the industry.[24]

The existing historiography of land ownership and distribution in Madeira has focused almost entirely on the judicial conditions of land distribution and ownership and has not been concerned with whom, and under what conditions, the grants of land (*sesmarias*) were awarded, the nature and changes of the land-owning system, and the ways in which differing levels of fertility may have influenced this system.[25] Madeira, because it was unoccupied when discovered, provided a kind of experiment for European colonization beyond the continent, and the techniques and processes of its settlement provided a model for the other Atlantic islands and for Brazil.[26]

The system of property on both the Canarian and Madeira archipelagos was defined by the distribution of land to colonists and then by sale, exchange, or later grant. In both cases, with variations depending on local conditions, the process was similar. The Crown gave to captains and governors the power to distribute lands to colonists and conquerors according to their actions in the conquest or settlement and with regard to their social status. All grants were made according to norms established by the Crown and following the models previously defined in the resettlement of the peninsula. In both archipelagos the grants required the improvement or development of the land within a set period of time, which decreased as settlement grew. After 1433, the time period decreased from ten years to five years in Madeira. In the Canaries, the first colonists in Gran Canaria were given a period of six years to develop their lands, while the grants made at Tenerife at the end of the sixteenth century provided only two to three years. That these grants were intended to stimulate colonization is demonstrated by the requirements to construct a house on the property, to reside there within five years, and, in the case of single men, to marry.

The process in the Canaries differed from that on the Madeira archipelago, which from 1439 to 1497 was controlled by the Order of Christ. The Canaries were another matter. In that island chain there were royally controlled (*realengo*) islands (Gran Canaria, La Palma, Tenerife) as well as those under seigniorial control (Fuerteventura, Lanzarote, la Gomera, El Hierro).[27] Moreover, the Canaries had an indigenous population that slowed down the process of occupation and placed the settlers in conflict with islanders who accepted Castilian sovereignty.[28]

On the Portuguese islands, the distribution of lands was from the beginning regulated by the Crown and later by the lord of the island, *Infante* Dom Henrique. The king, Dom João I, ordered the captains to grant the lands "free and without any pension."[29] Later, João Gonçalves Zarco, making use of his prerogatives as captain, reserved for himself and his descendants an important tract of land in Funchal and Ribeira Brava. Other grants were made according to the regulations of Dom Afonso to those who were required to improve them; those who lacked the ability or resources to do so lost their right of possession. This created the basis for social differentiation among the first colonists and opened the door to the growth of large-scale properties. In the Canaries also there was social differentiation among those receiving land grants. In compliance with a royal *cédula* of 1480, Pedro de Vera was required to make these grants to conquistadors "according to their merits."[30]

After 1433, with the donation of the lordship of the islands to Dom Henrique, he had the power to distribute lands but was required to respect the previous concessions, demonstrating that the regulation of land distribution was done by the king. Dom Henrique ceded this power to the captains.[31] The grants of Dom Henrique confirmed the royal regulations and stipulated that the lands could be granted for a period of five years, after which the right of possession ended and the lands could be given out in a new concession, a significant departure from the former royal concessions. With this the social differentiation of the grantees disappeared and the period to initiate cultivation was shortened. Both demographic pressure and the scarcity of lands to distribute caused this change.

In the following decades the granting of lands in *sesmarias* and the legitimating of occupation generated a number of conflicts that called for the legislative intervention of the lord or the judicial arbitration of his *ouvidor* (senior judge). For example, conflicts arose over the use of fire to clear forests because of the prejudicial effects on the neighboring cane fields. Finally, between 1501 and 1508, the concession of lands in *sesmaria* was ended except for the lands that could be developed as cane fields or vineyards.[32]

On both archipelagos the power of the captains and the governors to distribute lands created innumerable problems. On Madeira, the lord sent Dinis de Goa in 1466 as his representative with full powers to resolve all disputes, including those involving land and water. In the Canaries as well, similar disputes over land grants moved the Crown to send representatives to regulate and legitimate concessions in 1506 and 1509.

The Sugar Mills

The processing of sugarcane was done with the technology common in the Mediterranean world. The availability of water power led to a generalized use of water mills. On Madeira, the first mill for which we have evidence is that of Diogo de Teive, registered in 1452. In those areas without access to appropriate water power, animal or human force was used; those mills were called *trapiches* or *almanjaras*. We know little about the technical aspects of those mills. We do know, according to Giulio Landi, that in the third decade of the sixteenth century one of them operated more or less by the same system used for pressing oil from olives: "The places where with great activity and skill sugar is made are in great properties and the process is the following: first, after the cut cane is carried to these places, they are placed underneath a millstone moved by water which presses and squeezes the cane, extracting all the juice."[33]

A question that has provoked the greatest debate has to do with the evolution of the technology of sugar making, particularly the development of the cylinder mill. The primitive *trapettum* was used in ancient Rome to press olives and sumach and was, according to Pliny, invented by Aristreu, God of Shepherds. But this became an inefficient method on the large plantations and was succeeded by the mills arranged with an axle and cylinders. It is here that opinions differ. One version holds that this was a Mediterranean discovery. Noël Deerr and F. O. von Lippmann attribute the discovery to Pietro Speciale, a prefect in Sicily; Spanish historiography favors Gonzalo de Velosa, a *vecino* of the island of La Palma who presented his invention in 1515 on the island of Santo Domingo. David Ferreira Gouveia ascribes this innovation to Diogo de Teive on Madeira in 1452. Others look to the origins of the invention in China. The three-cylinder sugar mill developed later in Brazil, where it was considered a Portuguese invention, always linked to the Madeirans who resided there.[34] On Madeira, the first reference to axles for the mill date from the last quarter of the fifteenth century. In 1477 Alvaro Lopes received authorization from the captain of Funchal to "make a sugar mill of mill stone and presses or in another form. . . . This *engenho* should be water-powered with its building and a boiling house." In 1505 Valentim Fernandes referred to the white wood used in the making of "axles and screws for sugar mills." To this was also associated the inventory of the mill of António Teixeira at Porto da Cruz, which mentioned "wheels, axles, presses, furnaces, and *speeches.*"[35]

For the Canaries, Guillermo Camacho y Pérez Galdós describes this *engenho* as being constructed of three cylinders. The author bases this statement on a contract of 1511 between Andrés Baéz and the Portuguese Fernando Alonso and Juan González to cut three axles, one big and the others smaller for a water wheel and its equipment. Twenty years later, we have the inventory of the mill of Cristóbal de Garcia in Telde, where wheels and axles are mentioned. Nevertheless, J. Pérez Vidal remains of the opinion that the first system used in the Canaries was like an olive press, a Renaissance invention with "little rollers."[36]

The word *trapiche* later entered the vocabulary of sugar to designate all types of mills composed of cylinders used to grind sugarcanes. Around Funchal, near Arucas, there is a place with this name, proving the existence of this type of mill. In the Canaries, land grants (*data de terras*) distinguished between water- and animal-powered mills. On Madeira, the hydrologic conditions were favorable to the general use of water mills, of which the Madeirans became expert builders. Moreover, the conditions were created for the development of this agriculture with the innumerable water courses and the large forests that could provide fuel for the furnaces and lumber for the construction of the axles for the mills. All the social and economic interactions created by sugar were dominated by the mill, but this did not mean that the development of cane fields only took place in their shadow. Here, even more than in Brazil, there were many proprietors without the financial resources to set up the basic industrial operation of a mill and thus remained dependent on the services of others.[37] In an estimate of the production of the captaincy of Funchal in 1494, there were only fourteen *engenhos* listed for a total of 209 agricultors holding 431 cane fields.

It is not easy to establish the exact number of mills in the islands. The information is in many cases contradictory. Thus for Madeira in 1494 there are references to only fourteen sugar mills, whereas in another document of 1493 eighty sugar masters are mentioned, indicating a higher number of mills. German historian Edmund von Lippmann referred to one hundred fifty sugar mills in Funchal at the beginning of the sixteenth century, a number that does not seem to conform to a reasonable estimate of production given the size of the arable or the number of cane fields. Later, at the close of the sixteenth century,

Gaspar Frutuoso referred to thirty-four sugar mills, nine of which were in the captaincy of Machico and the rest in Funchal.[38] In the seventeenth century the numbers of mills was smaller. Thus Pyrard de Laval referred in 1602 to seven to eight working sugar mills. In the decade after 1649 there is notice of only four sugar mills, two constructed in 1650. This decline called for new incentives such as loans and tax exemptions from the *quinto* for five years. These were aimed mostly at Funchal and Câmara de Lobos, which implied that there were even harder times for sugar growers in Calheta, Ponta do Sol, and Ribeira Brava who did not receive such favors.

Trying to establish the number of mills in the Canaries presents a similar problem, since information is imprecise and scattered. Perhaps the most exact is that of Thomas Nichols in 1526 and of Gaspar Frutuoso in the last decade of the sixteenth century. Still, while the data provided by the former seems trustworthy, Frutuoso does not seem to merit much confidence.[39] He noted twenty-four mills on Gran Canaria while Tenerife had only three. Also of note is that on Gomera and La Palma, islands under lordly control (*señorio*), the mills were for the most part property of the lord, who then leased them to Genoese and Catalan merchants.[40]

The price of setting up an industrial operation of this type was beyond the capacity of many agricultors. The evaluation made of a mill for the inventory of António Teixeira of Porto La Cruz in 1535 placed its value at 200 *milréis*. Another document of 1547 set a value of 461 *milréis* on the cane fields, mill, and the water needed by them. In 1600, in Funchal, João Berte de Almeida sold to Pedro Gonçalves da Câmara an *engenho* valued at 700 *milréis*. In 1644 the mill of Gaspar Bethencourt in Ribeira de Socorridos was valued at 500 *milréis* and in the previous year that of Baltesar Varela de Lira was sold for 422 *milréis*.[41] For the Canaries, we have similar dispersed estimates for the cost of building a mill. In 1519 the mill of Miguel Fonte in Daute was evaluated at 4,641,320 *maravedís*. There was considerable variation here as well. In 1556 the mill of Valle de Gran Rey was priced at 1,237,417 *maravedís*, while in 1567 one in La Orotava was sold for 6,000,000. For Gran Canaria we have the mills of Francisco Riberol in Agaete y Galdar valued at 300,000, that of Francisco Palomar in Agaete at 750,000, and that of Constantino Carrasco in Las Palmas at 450,000. In La Orotava we have more precise construction costs of various aspects of a mill's infrastructure taken from the inventory of Alonso Hernández de Lugo's mill made in 1584. Its total value was 1,125,252 *maravedís*.[42]

Production levels for the Atlantic island mills were different from the sugar mills of the Americas. For Madeira at the end of the fifteenth century we have a listing of only seventeen sugar mills for a total of 233 cane field owners (see

TABLE 3.1. Madeiran *Engenhos*

Area	No. of Mills	Arrobas	Average per Mill
Funchal	2	16,545	8,273
Partes do Fundo	15	66,906	4,460
Total	17	80,451	5,563

table 3.1). This does not include those who operated in the area of Caniço and Câmara de Lobos.

Taking into account only the "partes do fundo" region, we note that each mill would have a production of almost 5,000 *arrobas* or about sixty-three tons, a rather high figure given the state of the available technology.[43] Moreover, these mill owners were not among the most important owners of cane fields. Only Fernão Lopes had some 1,600 *arrobas*. There were cane farmers with a higher production but who did not own mills themselves. In the first half of the sixteenth century these levels fell by two-thirds, to an annual average of 1,479 *arrobas* per mill (see table 3.2).

Sugar mill owners constituted a minority of the total number of sugar producers, and in this period of profound changes in the structure of production, the disparity between them was growing. In the early sixteenth century, there were 269 owners of cane fields and 46 owners of sugar mills.

The difference between cane farmers and mill owners is very clear. A great proprietor of cane fields was not synonymous with a mill owner. In the sixteenth century, some mill owners were among the principal producers, but most grew much less, as for example was the case of João de Ornelas, who in 1530 declared a production of only seventy *arrobas* on his sugar mill in Funchal. The existence of the two groups, cane farmers and mill owners, created the peculiar dynamic of sugar production on Madeira.

TABLE 3.2. Madeira Sugar Production in the First Half of the Sixteenth Century

Area	No. of Mills	Arrobas	Average per Mill
Funchal	17	17,863	1,051
Ribeira Brava	6	13,524	2,254
Ponta do Sol	5	8,012	1,602
Calheta	10	19,204	1,920
Machico	8	9,409	1,176
Total	46	68,012	1,479

TABLE 3.3. Estimated Canary Islands' Sugar Production in 1520

Island	Sugar Mills	Sugar (*Arrobas*)
Gran Canaria	38	152,000–190,000
Tenerife	16	64,000–80,000
La Palma	4	16,000–20,000
La Gomera	6	24,000–30,000
Total	64	256,000–320,000

In the Canaries, particularly on Gran Canaria and Tenerife, the situation appears to have been different. Here, great property was synonymous with the existence of a sugar mill, a result of the process of how land had been distributed, and the average production per mill seems to have been higher than those of Madeira. Gaspar Frutuoso referred to two mills of the Ponte family in Adeje (Tenerife) that produced 8,000–9,000 *arrobas*, while that of Juan de Ponteverde in La Palma was at around 7,000–8,000. For Gran Canaria, he indicated that the twenty-four mills produced on an average of 6,000–7,000 *arrobas*. From rental contracts of mills we know that Don Pedro Lugo in El Realejo produced in 1537–38 an average of 4,500 *arrobas* and another mill in La Orotava produced 1,122 *arrobas*. In the seventeenth century, the tithes paid by the seven mills operating on Gran Canaria, Tenerife, and La Palma provide an idea of annual production for the period after 1634. Macías presents new information about the sugar in the Canary Islands, with the estimated production in 1520 (see table 3.3).[44]

Slaves and Sugar

In the encounter between the force of will of the first European settlers and the rugged terrain of the islands, the colonists constructed a Europe in the Atlantic. Madeira, thanks to its geography, became defined by a specific agrarian appearance, quite distinct from the great open spaces of the continent. The excessive division of agricultural lands, the only possible way of making use of the arable, and the distribution of population in both the south and north of the island influenced the system of cultivation and the ownership of the land. The large initial grants of land were divided as the population grew and as agriculture developed. The early extensive use of the land gave way to intensive cultivation based on innumerable terraces constructed by owners, renters, or sharecroppers. Given this situation, it is difficult if not impossible to imagine great sugarcane properties comparable to those of the Americas. There, the cane fields advanced outward from the mills and were always indissolubly linked to them. This was not the pattern in Madeira, where many people owned cane fields but few owned mills. Still another peculiarity of Madeira was the concentration of sugar mills in areas with the easiest access to the external world that is principally around Funchal, even though it was not always the area of greatest importance in cane cultivation. This peculiar arrangement in the production of sugar influenced the use of slaves. In Madeiran agriculture it is necessary to distinguish two groups of proprietors: those who had leased their lands to renters or dependents, and those who were full proprietors. This double form of ownership promoted the development and use of contacts of sharecropping (*contrato de colonia*) beginning in the sixteenth century. On the other hand, the reduced size of the cane fields meant that a sugar mill was not always nearby nor were numerous slaves always necessary. The use of slaves must be seen in relationship to the structure of landholding on the island. In direct ownership and in leased arrangements the role of slaves was clear enough, but the same cannot be said for the *colonia* contracts.[45]

In the Canaries as well, on the islands of Gran Canaria, Tenerife, La Palma, and Gomera, one must take the geographical and agricultural environment into account in establishing a link between the slave and the sugar economy and the extent of the cane fields. The conquest itself produced the first slaves, taken as prizes of war from among the original inhabitants or Guanches. Later, the proximity to Africa favored access to the market for black slaves, who eventually assumed a role of importance in the society. Moreover, unlike Madeira, the evolution of landholding depended on the initial process of conquest. Large estates developed, although they were later broken up as a result of death, dowries, and sale. The available information drawn from notarial records reveals this process and the perpetuation of some important large estates (*fazendas*) associated with sugar mills. This process can also be noted on Tenerife and on La Palma.[46]

The presence of slaves in the formation of the island societies from the fifteenth century onward was not a phenomenon isolated from the social and cultural context of the Atlantic. The lack of laborers for new cultivation, the need for workers in sugarcane agriculture, the active role of the Madeirans in the opening of the Atlantic world, and the proximity of Africa all played a role in shaping slavery. Madeira, because of its location near the African continent and because it was much involved in the exploration, occupation, and defense of Portuguese areas there, was wide open to this advantageous trade in slaves. The Madeirans marked the first centuries by their efforts to acquire and trade in this powerful and promising commodity. The first slaves who arrived in Madeira and contributed to its economic takeoff were Guanches, Moroccans, and Africans. On the one hand, the sugar harvest called for access to laborers, which implied slaves in the absence of free workers. On the other hand, the proximity of the market for slaves in West Africa and the involvement of islanders in this commerce made the islands one of the first destinations for these slaves, and they remained so until the growth of other regions. Note, for example, the relationship between the curve of sugar production and the manumission of slaves in which the numbers of freed slaves evolved according to the state of the sugar economy. As sugar production declined in the last quarter of the sixteenth century, the number of manumissions rose. An opposite movement took place in the first quarter of the seventeenth century, probably associated with a rise in sugar production stimulated by the Dutch occupation of Pernambuco. But this island recuperation was brief and the number of manumissions increased again in the second half of the century. The number of manumissions was not the highest in the principal cane-growing areas, but rather in Funchal, Câmara de Lobos, and Caniço. In the Canaries this relationship was also apparent. Lobo Cabrera notes that on Gran Canaria after the mid-sixteenth century there was a decline in the number of slaves, perhaps the result of the competition from American sugar. Proprietors determined the role and concentration of slaves. On Madeira, Funchal had 86 percent of the owners and 87 percent of the slaves, reaching its highest levels in the sixteenth century. Within the captaincy of Funchal, the district of the city had 74 percent of the owners, of which the two urban parishes—Sé and São Pedro—held 64 percent, the rest being distributed among the captaincy of Funchal (23 percent), Machico (11 percent) and Porto Santo (2 percent).

When we compare the distribution of the slaves in the sugar mills, we can see some distinct differences with the patterns in the Americas. In the Antilles and South America the numbers of slaves per mill was frequently over 100 and there were cases of mills with far more. On Madeira they usually did not exceed 30 per mill over all, the largest mean distribution being 77 per mill in Funchal and 24 in Ribeira Brava.[47] In a total of 502 sugar producers, only 78 (16 percent) owned slaves. For the seventeenth century, the number of owners with slaves was higher (39 percent), but there seems to be no direct relationship between the levels of production and the number of slaves. Thus, for example, Maria Gonçalves, the widow of António de Almeida, had the largest number of slaves reported but she produced very little sugar.

On the Canaries a parallel situation existed. On Gran Canaria, documents reveal properties with 30 to 35 slaves. The average size on Tenerife and La Palma was about half that on Gran Canaria, but on Tenerife there may have been a few rare properties with up to 100.[48] Note that on Madeira the highest number reported by João Esmeraldo was 14 slaves on the *fazenda* of Lombarda at Ponta do Sol. The majority of producers (63 percent) had about 5 slaves. Taking into consideration the labor necessary for sugar making, we must assume that the majority of workers on the *engenhos* were free, not slaves. The largest number we have been able to establish were the 20 slaves on the property of Ayres de Ornelas e Vasconcelos, but this was for both father and son.

On Madeira, the tendency was for a low average number of slaves per owner. Over half (58 percent) of the owners held only 1 or 2 slaves and no more than 11 percent of the owners held more than 5 captives. Those with more than 10 slaves were less than 2 percent of the owners, and once again these were found in the area of Funchal. In general we can say that this was small-scale slaveholding and 89 percent of the owners held from 1 to 5 slaves. Moreover, the link between slavery and sugar was weak. Of 104 persons who owned both slaves and land, only nine had cane fields. The majority of the rest owned wheat fields and vineyards.

For the Canaries, analysis of the existing data reveals a different arrangement. On Gran Canaria in the city of Telde, the majority of the slaves was held by cane farmers and mill owners and was thus directly employed in sugar. Here the family of Cristóbal Garcia de Moguer stands out. Owner of a mill, he had 60 slaves in his service, 37 of them at the mill, including a kettleman (*calderero*) and a cane-field specialist (*canavieiro*). This situation was also found in Galdár, Guia, Arucas, Agümes, and Agaete, all regions of cane cultivation. Around Tenerife we know only that Alonso Fernández Lugo had 28 slaves in 1525. In Daute there were two important slave owners—Cristóbal de Ponte and Gonzalo Yanes. In 1506 the sugar mill of Icod had 25 slaves. In the seventeenth century the situation changed, at least in Las Palmas, where the slave owners were found mostly in the service sector, a fact that suggests that slavery was a more patriarchal or household-related institution there.[49] On the island of La Palma, strongly associated with sugar is where the highest concentration of slaves was found, reaching 29.9 percent of the population.[50] There were also slaves on La Gomera, but at present it is impossible to determine the exact number.

Slaves were always linked to sugar cultivation on these islands, but never in the same proportion as was found on São Tomé and Brazil. The scattered evidence drawn from the documentation of Madeira and the Canaries attests to this. In 1496 the Crown noted this relationship on Madeira by prohibiting the sale for debts of real estate, slave men or women, animals or mill equipment, allowing only the charges to be made on production (*novidades arecadadas*). In another document of 1502 concerning irrigation, the king noted that it was the custom of proprietors to send "the slaves and the salaried men in their service to irrigate their fields."[51] The link between slaves and the work of cultivating and preparing the cane fields can also be seen in the existing documentation. That slaves did other jobs at the mills is also evident. The regulations of the *alealdadores* (those who checked the quality of sugar) of 1501 that mentions masters and *alealdadores* who made "broken sugar" (*açúcar quebrado*) would be subject to strict penalties; for slaves who were caught, their masters paid the fines.[52] Slaves sometimes served as assistants to the skilled workers or sometimes were themselves the skilled specialists. In 1482, in a suit over "tempered sugar," two sugar masters, Masters Vaz and André Afonso, testified. The first stated that while he had been away in the Canary Islands, his slave had tempered the sugar; the second said that in his absence this job had been done by a youth who worked on salary. In other words, slaves not only made sugar but also served as "officials" at the mills, that is, as skilled technicians. First, the Canarian slaves who served there as sugar masters are notable because there were limitations placed on their leaving the island in 1490 and 1505. From this period we have only two references to two "master" slaves on Madeira, and we cannot tell if they were Guanches. In 1486, Rodrigo Anes, "O Coxo," from Ponta do Sol freed his slave Fernando, a *mestre de engenho*, that is, probably a builder of mills. In the testament of João Vaz, he refers to his slave, Gomes Jesus, as a "sugar master." Later in 1605 a certain Jorge Rodrigues, a freedman, sought compensation of three *milréis* for the service he had performed at the *engenho* of Pedro Agrela de Ornelas.[53] The French traveler Jean Moquet reported in 1601 that the slaves had an important role on the *engenhos* and that he had seen "a great number of black slaves who worked in sugar near the town." The only peculiarity of slave service on the Madeiran mills was the fact that they worked alongside free men and freedmen, especially salaried employees. In 1578 António Rodrigues, a worker, declared in his will that he had worked, presumably for wages, under the direction of Manuel Rodrigues, the overseer of the *engenho* of Dona Maria.[54]

For the Canaries, recent studies, especially those of Manuel Lobo Cabrera, have revealed similar evidence for the sixteenth and seventeenth centuries. In the sixteenth century, the links between slavery and work in the field and at the mill is clear. There is reference to the house of blacks (*casa de negros*) as part of the infrastructure of the mills, implying their presence. Slaves did the most varied tasks at the mill: *molederos* (cane millers), *prenseros* (pressmen), *bagaceros* (bagasse removers), and *caldereros* (kettlemen). They might be owned by the mill owner or rented from other owners. Such rental contracts for mill service are common in the Canaries. There was also a strong presence of freedmen as skilled specialists and as workers.[55] We should also note that in the Canaries field cultivation was often done by cane farmers (*esburgadores de cana*) and by renters, so it was possible for a proprietor to hold extensive cane fields without directly needing to own the slave laborers. This system was common on the island of Tenerife and it must have had some effect on the weight of a slave presence on the society.[56] Still, many owners had slaves to perform these tasks. A free man who leased property during the sugar harvest almost always had a few slaves who acted as his assistants. Thus, slaves might be lacking as integral part of the property of those who owned fields and mills, but that does not mean that they were absent from the process. On the other hand, slaves were sometimes attached to the land. In 1522, in La Orotava (Tenerife), a city councilman rented out a cane field for five years and along with it three slaves who had to be clothed and fed by the renter.[57] This took place frequently on La Palma and Gran Canaria.

In sum, on Madeira, as happened in the Canaries, the labor force used at the mills was mixed, made up of slaves, freed, and free persons who did a variety of tasks and, when compensated, were paid in money or sugar. Some slaves belonged to the proprietor of the mill, but others worked for wages under rental contracts. In Brazil there was also a mixed labor force, but slaves predominated. They were considered to be property of mill owners, cane growers, or those who rented them out. The difference in the proportion between slave and free workers is the primary difference in the industry from one side of the Atlantic to the other.

The Price of Sugar

It is difficult to establish the evolution of sugar prices in the island markets because the existing documents needed to reconstruct a price series are few and scattered.[58] For Madeira it is possible to bring together sufficient data for the third decade of the sixteenth century, and the same can be done in the Canaries for the island of Tenerife. Moreover, there are other factors that influenced the price of sugar, such as the chronic lack of specie on the islands, and the use of sugar therefore as a means of exchange. In the fifteenth and sixteenth centuries this led to its constant devaluation. Sugar was used as a means of exchange in both island groups, but more commonly in the Canaries.[59]

We must also keep in mind that the law of supply and demand conditioned sugar prices over the course of the year. There were monthly fluctuations depending on the stage of the sugar cycle and the presence of ships in the port.[60] Thus we find the highest prices in the months of June and July, when the year's first sugar became available and when merchants had the most funds at hand.

White sugar had two prices, depending on whether it had been "cooked" once or twice. On Madeira, in 1496, one price was almost double the other. Of 15,000 *arrobas* from the first processing, only 10,000 would remain after the second, which had a strong effect on the final price.[61] Moreover, it reduced the volume of product and thus tended to maintain the value of sugar when there was an excess on the market.

In the decade of the 1470s the price of sugar declined. This is confirmed by the actions of the lord (*senhor*), who, after 1469, sought to impose a monopoly on commerce. The Madeirans' opposition to a similar solution led the duke, Dom Manuel, to try something new. Thus in 1496 he fixed the price at 350 *réis* for "once-cooked" sugar and 600 for sugar that had been processed a second time. Two years later, he established a maximum quota for export at 120,000 *arrobas*. This was at a time of sugar's decline. The first sugars sold at Machico were priced at 2000 *réis* per *arroba*. By 1469 the price was at 500 *réis* for "once-cooked" sugar and 750 for twice-processed sugar. In 1472, the price rose again to 1000 *réis* per *arroba*, but this increase was short-lived and the result of currency devaluation. In 1478 matters returned to normal. Prices continued to fall until the beginning of the sixteenth century, and only with the price revolution did the situation change on both archipelagos. On both Madeira and the Canaries it is clear that, after the 1530s, the competition of American sugars began to have an effect. The situation in the Canaries, however, reversed itself once again in the 1540s, probably due to inflation.[62]

Various subproducts and lower grades, as well as preserves and sugared fruits, were also produced. These were important on both archipelagos. At Tenerife, for example, lower grades (*escumas* and *rescumas*) were sold for half the price of white sugar, while on Madeira and Gran Canaria that was only true of *rescumas*, since the *escumas* were more highly valued. On Gran Canaria in the sixteenth century, 20 percent of 2,500 *arrobas* of sugar was refined, 60 percent white, 12 percent escumas, and 8 percent rescumas. A similar distribution existed on Madeira, from 1520 to 1537.[63]

Sugar and Atlantic Commerce

The social and economic developments in the Atlantic islands were directly related to the demands of the Euro-Atlantic world. This was true for the islands: first, as a peripheral region of European business, adjusting their economic growth to the needs of the European market and the European shortages of foodstuffs; later as consumers of continental production, trading at a disadvantage with Europe; and finally as an intermediary between the Old World and the

New. By the beginning of the sixteenth century, the "Mediterranean Atlantic" was defining itself as the point of contact and aid for commerce with Africa, the Indian Ocean, and America. All this created a network of interests between the bourgeoisie and the aristocracy in power in the peninsula during the process of occupation and the economic development of the new societies. This peninsular component was reinforced by the participation of a Mediterranean bourgeoisie attracted by new markets and by the rapid and easy expansion of their operations. A group of Italians, with links to great Mediterranean commercial groups, actively participated in the exploration, conquest, and occupation of the new Atlantic space. Thus they were interested in the conquest of the Canaries archipelago, the Portuguese expeditions of geographic exploration, and commerce along the West African coast. Their penetration in the island world gained them a position in the society and economy established there. The investment of merchant capital, whether national or foreign, was essential to the new economy and generated new wealth for these commercial endeavors. Commerce was thus the common denominator for the products introduced, and that most valuable product in the new economy was sugar.

Madeira was the most important entrepot. Exploration became linked to commerce, and from the mid-fifteenth century an active trade with Portugal was maintained, at first in woods, *urzela* (cudbear, a dyestuff), and wheat, and later in sugar and wine. This trade eventually spread to North European and Mediterranean cities with the appearance of foreigners interested in the sugar trade. Spaniards and Italians in the Canaries established an active trade with the Iberian peninsula after the mid-sixteenth century. After the conquest, Italians, Portuguese, and Castilians controlled the island trade. English and Flemish merchants layed out the routes of the Nordic trade in a second stage of this commercial development. By the end of the sixteenth century, Tenerife and Gran Canaria emerged as the primary producers.

The insular sugar trade, concentrated on Gran Canaria, Tenerife, Gomera, La Palma, and Madeira, was the principal link to the European market. On Madeira, this trade became dominant between 1450 and 1550, but on the other islands it grew at the beginning of the sixteenth century and became dominant only by the 1530s.

According to Vitorino Magalhães Godinho, the Madeiran sugar trade "oscillated between liberty strongly restricted either by the Crown or by powerful capitalist interest groups on one hand and overall monopoly." Thus commerce remained free only until 1469 when a fall in prices led to the intervention of the *senhor* and the exclusive control by the Lisbon merchants. Madeirans used to trading with foreigners did not appreciate this change. Nevertheless, in 1471

Infante Dom Fernando decided to establish a monopoly company, a move that resulted in a bitter conflict on the island between the contractors and the local government, which represented the sugar producers. Twenty-one years later, the island still faced a difficult situation in the sugar market and led the Crown in 1488 and 1495 to reestablish its monopoly control, establishing rules for the planting, harvesting, and marketing of sugar in 1490 and 1496. But this policy, designed to protect the income generated by sugar, ended in a disaster, and in 1498 a new policy was instituted by which a production limit (*escapulas*) of 120,000 *arrobas* was set among various European purchasers.[64] With some changes this system remained in place until 1508, when the system of free trade was restored. The charter of the captaincy of Funchal stipulated in 1515 that sugar "can be carried to the east and the west or to any other place that merchants and shippers desire without any impediment."[65] The situation in the Canaries was quite different. There the sugar trade had been opened to all agents and markets, the only restrictions being imposed by political and religious considerations, especially in regards to Flanders and England at the end of the sixteenth century.[66] The intervention of local municipal councils and the Crown was felt only in quality control, not in the marketing and production as was the case in Madeira.

The Sugar Merchants

The early development of the sugar economy attracted the first wave of foreign merchants to Madeira, a process that was only limited by ordinances against their residence on the island. Still, by the mid-fifteenth century the Crown was extending special privileges to Italians, Flemish, French, and Breton merchants, allowing them to remain on the island in order to gain access to European markets. This was considered destructive to the interests of Portuguese merchants and the Crown and led the lord to prohibit the permanent residence of foreigners. The question was raised at the Cortes of Coimbra in 1472–73 and that of Evora in 1481, when the Portuguese bourgeoisie complained against the effective monopoly of the sugar trade held by Genoese and Jewish merchants. The king, compromised by the advantageous position held by the foreigners, reacted ambiguously and tried to safeguard the existing concessions, but responded favorably to the petitions of his subjects to limit the residence of foreign merchants by making them secure licenses. On Madeira, residence was impossible without these, and resale in the local market was prohibited to foreign merchants. The *câmara* of Funchal sought to expel the foreigners in 1480 but were prevented by the lord. In 1489 Dom João II recognized the function of foreign merchants and ordered that foreigners be considered "natives and residents (*vezinhos*) of our kingdoms."[67]

By the 1490s, difficulties in the sugar market once again stimulated a xenophobic policy. Foreigners were given three or four months between April and September to do their business and were not allowed to have shops or agents in the city, but by 1493 Dom Manuel recognized the negative effects of such restrictions on the Madeiran economy and removed them all, allowing the foreigners eventually to become involved not only in commerce, but in administration and landholding on the island.[68]

The "white gold" of sugar attracted Italians, Flemish, and French merchants to Funchal. The Italians, chief among them Florentines and Genoese, were on the island from the mid-fifteenth century as the principal sugar merchants; their activities also extended into landholding, a situation made possible by purchase and marriage. In the decade of the 1470s through a contract established with the island's lordship, they had already established a predominant majority position. They were represented by Baptista Lomellini, Francisco Calvo, and Micer Leao. In the last quarter of the century, Christopher Columbus, João Antonio Cesare, Bartholomew Marchioni, Jerónimo Sernigi, and Luís Doria joined together. This group was followed by a more numerous one in the beginning of the sixteenth century and linked the resident Italian community together in the sugar trade. Foreigners came to depend on a group of agents or representatives to maintain the scope of their commercial operations in the islands; men like Gabriel Affaitadi, Luca Antonio, Cristóvão Bocollo, Matia Minardi, João Dias, João Gonçalves, and Mafei Rogell. While the first group was primarily made up of Italians, the second included representatives of some of the island's principal families.

The merchant-bankers of Florence were particularly important in making the commercial and financial arrangements for Madeiran sugar in European markets. From Lisbon, where they enjoyed royal confidence, they created an extensive network of ties that linked Madeira to the principal European ports. They obtained almost exclusive control from the Royal Treasury through their contract to collect royal duties. Figures such as Bartolomeu Marchioni, Lucas Giraldi, and Benedito Morelli had a direct effect on the sugar trade in the beginning of the sixteenth century. These merchants and their agents kept the network functioning. For example, Benedito Morelli, in 1509–10, maintained on the island agents such as Simão Acciaiuolli, João de Augusta, Benoco Amador, Cristóvão Bocollo, and António Leonardo. Marchioni, in 1507–9, was represented by Feducho Lamoroto. João Francisco Affaitadi, from Cremona, the Lisbon agent of one of the most important commercial families, actively participated in this trade between 1502 and 1526, by means of contracts of purchase and sale of the sugar collected by the Crown as duties (1516–18, 1520–21, and 1529) and in payments in sugar in exchange for pepper. He also did this in partnership with other merchants through agents on the island. This group of merchants penetrated insular society where their royal privileges favored their linkages to the land and office-holding elites. Their appearance among the municipal councilors and treasury officials indicates their position in the sugar economy. Men like Rafael Cattano, Luís Doria, João and Jorge Lomelino, and João Rodrigues Castelhano, among others, acquired some of the best and most productive lands and were counted among the most important owners of cane fields.

The French and the Flemish, following the Italian example, were attracted to the island as well by the sugar trade, but their interest remained only in the commerce of sugar and not in its production; thus they did not set down roots in local society as the Italians did. João Esmeraldo was the exception. The French played an active role in the sugar trade while the Flemish played a secondary role. The French acquired large amounts of sugar in Funchal, Ponta do Sol, Ribeira brava, and Calheta, shipping it in French ships to a number of French ports. Some of these merchants incorporated Madeira into a network that linked the Canaries to Nordic and Andalusian ports.

The *escapulas* or sugar quotas up to 1504 and the sugar collected as royal duties were funneled to European markets either by direct delivery, by free trade, or in exchange for pepper. This sugar was handled by merchants or by the commercial consortia in Lisbon in which Italians, such as João Francisco Affaitadi e Lucas Salvago, played a central role. The Italian-controlled network based in Lisbon dominated the sugar trade in the first three decades of the sixteenth century, but by the 1530s it was somewhat in decline as foreign merchants, faced with the instability of the Madeiran sugar market, began to seek other trades. After the Italians, the Portuguese and Spanish traders were the most important, while the northern merchants did not play much of a role. This is additional evidence that the Flemish sugar route remained under the control of the Portuguese factory in Antwerp. During the period between 1490 and 1550, exclusive Italian control in the first decade and predominance in the next two was replaced by Portuguese, Castilian, and French traders. Among the foreign merchants the trade was concentrated in a few hands. The five leading merchants in the period handled over 70 percent of the sugar shipped, or over 10,000 *arrobas* each, while among Portuguese merchants only one shipped over 1,000 *arrobas*. The Cremonese noble João Francisco Affaitadi, who headed the Lisbon operations of his family business, became the principal merchant in the Madeiran sugar trade from 1502 to 1529, handling more than seven times the amount of all the Portuguese merchants together.

The network of the sugar trade at Funchal was created and motivated by foreigners, Germans or Italians, who arrived after an advantageous stop in Lisbon. They controlled the major consortia in the sugar trade even though their fixed residence was often Lisbon, Flanders, or Genoa. Their operations depended on representatives and agents on the island whom they chose first from among their relatives, next from their compatriots with roots on the island, and last from locals or Portuguese. The number of local agents was a gauge to the importance of the firm. The Welsers and Claaes operated in the Funchal market through agents in Lisbon like Lucas Rem and Erasmo Esquet, who then had representatives in Funchal to deal with day-to-day operations. These men in turn had little to do with local society and often dealt with more than one foreign merchant firm, just as the firms often used multiple agents.

By the second half of the seventeenth century, Madeiran sugar was replaced by the Brazilian product. Madeirans and Azorians played a part in this commerce, supplying wine and vinegar in return for sugar, tobacco, and brazil wood and eventually even entering into the slave trade. For this the Madeirans created their own network of trade through Madeirans stationed in Angola and Brazil. Diogo Fernandes Branco was a perfect example of this new situation. He specialized in the export of wine to Angola in exchange for slaves that he then sold in Brazil for tobacco and sugar. A household industry, employing many women in the city and surrounding areas, developed on the island in which these products were transformed into conserves and other sugar by-products, all of which were organized by merchants, such as Fernandes Branco, according to requests they received. The principal ports for these goods were the north of Europe: London, St. Malo, Hamburg, La Rochelle, and Bordeaux. Fernandes Branco served as the direct representative for merchants in a number of these ports, sending wines and sugar products in return for manufactured goods since money and bills of exchange were rarely sent to Madeira. His correspondence reveals his own network of contacts in Lisbon and in Brazilian ports. He seems to have specialized in supplying wine to Angola and Brazil and sugar to the dinning tables of Europe. His activities reveal the structural position of Madeira in the second half of the seventeenth century as an entrepot between the interests of the commercial bourgeoisie of the Old and New Worlds. Funchal was a key piece in this puzzle, a place where small merchants awaited an opportunity to enter into these trades. Angola and Brazil were two other locales for this activity, as was Barbados from time to time, until it eventually assumed a dominant position with the rise of English commercial hegemony in the Atlantic world.[69]

The Canaries also witnessed the active participation of foreign merchants through the fifteenth and sixteenth centuries. Portuguese, Genoese, and later Flemish and French merchants were involved in the conquest and occupation of the islands, in the creation of their social and economic base, and in the development of commercial networks. The Genoese, well-established in Andalusia, participated actively in the trade of *urzela* and slaves in the archipelago. Blocked in their Mediterranean trade by the Muslims and by Italian rivals, they sought in the "Atlantic Mediterranean" a new site for their activities. Madeira, Gran Canaria, and Tenerife in the fifteenth and sixteenth centuries thus became their Atlantic homeland where they settled as residents (*vezinhos*), becoming in the process powerful landowners, merchants, and moneylenders. We can identify three types of foreigners: (1) conquerors who took part in the winning of the Canaries as warriors or financiers of expeditions; (2) settlers who developed after the conquest benefiting from the process of occupation; and (3) merchants who handled local exchanges and then the commerce in sugar and manufactured items, aided to some extent by their resident compatriots.

Conquerors and settlers became important in the new societies of Tenerife and Gran Canaria as hacendados. Such was the case of Cristóbal Ponte and Tomás Justiniano, who, next to the Lugos, were the richest men on the island. F. Clavijo Hernández considers Tenerife the center of Genoese mercantile operations. They financed the conquest, the planting, and the harvesting of the sugarcane. A similar role was played on Gran Canaria by Francisco Riberol, Antonio Manuel Mayuello, Bautista Riberol, and Jácome Sopranis, whose importance was symbolized by their patronage of the principal chapel of the Franciscan convent and by the designation of one of the streets as the "street of the Genoese." As in Madeira, their influence spread into local administrative life as functionaries or as the holders of government tax contracts, as in the case of Juan Leandro and Luís de Couto, who in 1524 collected the royal third.[70] To this group of legal residents (*vezinhos*) we must add the more numerous merchants who were simply passing a period on the island. According to the count by Guilherme Camacho y Pérez Galdós, they considerably outnumbered the resident merchants.[71] On Tenerife, the situation was inverted. There the *vecinos* made up 57 percent of the resident merchants. The majority of *vecinos* dedicated their activity to sending sugar to Europe and importing manufactures to the islands. Most had shops on the Andalusian coast and operated through a network of agents and representatives. Francisco Riberol, one of the principal Genoese, for example, sometimes resided in Seville and sometimes on Gran Canaria, where he had considerable interests in the sugar industry. While the Genoese were the principal representatives of the Italian merchant community on the islands, there were also Lombard's like Jácome de Carminatís and Flor-entines like Juanoto Berudo, one of the conquerors of La Palma.

The Flemish community had equal importance in Canarian society and economy. Despite their occasional presence in the fifteenth century as mer-chants or conquerors, it was really in the early sixteenth century that they began to arrive in the archipelago in force. Attracted by the commerce in sugar and dyestuff, they established an important export trade, and their activities ex-tended into all aspects from sales to loans of capital and goods to export trade. In this way, they created a net of relations throughout the islands from their bases on Gran Canaria, Tenerife, and La Palma.[72] Tenerife attracted the largest num-ber of merchants from the Low Countries, most of whom were visitors rather than residents on the island. Like the Genoese, the Flemish also penetrated island society and achieved the status of residents (vecinos), becoming tied to the principal local families and directing trade circuits with Bruges or Antwerp, their cities of origin.[73] Only on La Palma did a small community develop, which played a major role in local matters.

In the Canaries, companies (partnerships) developed not only in the com-mercial sector, but in transport and production as well. For example, in 1513 the Welsers acquired cane field in Tazacorte (La Palma), which were later passed on to their agents Juan Bissan and Jácome de Monteverde. On Gran Canaria partnership contracts were common between cane farmers and merchants or between cane farmers and canavieiros (those who weeded the cane fields). In Las Palmas, Santa Cruz, and Garachico partnerships were formed by local and foreign merchants to do business with three primary markets; the northern and Mediterranean ports, the African coast, and the Americas. This was generally done through Seville or Cádiz using the offices of resident agents. Three Bar-celona merchants formed a company in 1536 to trade in Canarian sugar and slaves using Cádiz as redistribution point. Another Barcelona-based group was established in 1574.[74] In these relations between the Canaries and Andalusia, family ties predominated, with relatives often serving as agents in the islands. By the first quarter of the seventeenth century, the picture was changing because of political considerations, the English were gone, and there were fewer Flemish and Genoese.[75]

Commerce in White Gold

Sugar provided the major element in the trade between Madeira and Europe in the fifteenth and sixteenth centuries, and it played a similar role in the Canaries beginning in the sixteenth century. On Madeira and some of the Canary Islands it was the basis of wealth and the commodity that could be used to acquire food and manufactured goods. But during this period the sale and value of sugar oscillated because of conditions in the markets where it was consumed and because of competition from other producing regions. The producers' expenses were varied. Direct sales, sometimes pledged before the harvest, were often used to pay existing debts or were made in exchange for goods and services. On Madeira, registers of taxes, the "books of the fourth and the fifth," reveal how producers disposed of their sugar. [76] In the Canaries, different types of contracts are registered in notarial records. These reveal the principal buyers as well as the use of sugar to pay for services. For Madeira in the first half of the seventeenth century we can see how the sugar was distributed by mill owners and cane farmers. There, 81,280 *arrobas* was sold to 2,492 buyers; an indication of a distribution to small buyers and a situation quite different from the monopoly control that had characterized the high point of sugar's growth in the previous century.

Engenho owners and cane farmers usually used the product of their harvests to pay for the salaried laborers they employed. From 1509 to 1537 there are references to the payment in sugar for a variety of services and purchases. The accumulation of profits by the sugar producers and their redistribution into the local economy had an effect on the life of the island and on the development of its artistic and architectural context.[77] In the Canaries, there was also an advance of goods and services against the expected harvest, a system that tended to subordinate the producers. Here too, despite regulations to the contrary, the payment of workers in the harvest was made in sugar, which led to its circulation as a means of exchange.[78]

For over a century, sugar was Madeira's principal item of trade with the outside world. The difficulties of penetrating the European market led the Crown to control this trade, which after 1469 was done under the permanent supervision of the lord proprietor and the Crown. This situation remained in place until 1508, when the contract system was abolished. The northern ports, especially Flanders, dominated the sugar trade, receiving half of the established quotas (*escápulas*). Similarly, the Italian ports dominated the Mediterranean trade. If we compare the quotas of 1498 with the sugar shipped from 1490–1550 (see table 3.4), we can see the major difference lies in the share taken by the Italian cities, perhaps because of their role in redistributing this sugar to France and the Levant.

Madeiran sugar was being carried primarily to the Flemish and Italian markets; Portugal itself, the ports of Lisbon and Viana do Castelo, was only in third place, receiving about 10 percent of the total. From about 1511, Viana do Castelo

TABLE 3.4. Export of Madeira Sugar to Europe, 1490–1550

Destination	Quota of 1498		Market, 1490–1550		Merchants	
	Arrobas	%	Arrobas	%	Arrobas	%
Flanders	40,000	33	105,896	39	11,375	2
France	9,000	13	500	—	8,469	2
England	7,000	6	1,438	1	1,072	—
Italy	21,000	30	140,626	52	407,530	80
Portugal	7,000	6	20,657	10	23,798	5
Turkey	15,000	13	2,372	1	—	—
Others	—	—	32	—	68,185	13

became important, redistributing sugar to Spain and northern Europe. From 1535 to 1550, of the fifty-six ships entering Antwerp with Madeiran sugar, sixteen had sailed from Viana. From 1581 to 1587, Viana was the only Portuguese port receiving Madeiran sugar. For the Mediterranean, Cádiz and Barcelona played a similar role as the major ports for the trade with Genoa, Constantinople, Chios, and Agues Mortes.[79]

Export statistics for the period 1490–1550 demonstrate that about 39 percent of the trade went to Flanders and 52 percent to Italy, but Italian merchants actually shipped about 78 percent of all Madeiran sugar. The early difficulties for foreign traders were surmounted by the 1480s as some became residents involved with both production and commerce of sugar. Data for the late sixteenth century is more difficult to locate, but from 1581 to 1587 the island exported just under 200,000 *arrobas*.

In the early sixteenth century the sugar market was expanding. Madeira in the previous century had been almost alone as a producer, but now the Canaries, the Barbary coast, São Tomé, and later Brazil and the Antilles were also making sugar. This competition affected the sugar market. Madeira, however, maintained its preferential status and in the markets of Florence, Antwerp, and Rouen its sugar still commanded the highest prices. Perhaps this situation explains the frequent references to stops in Madeira of ships trading with São Tomé, the Canaries, and North Africa. It may also explain why there is a reference to the sale of Madeiran sugar in Tenerife in 1505.[80] Normand shipping also favored Madeira, although after 1539 São Tomé began to overtake it as a supplier to northern markets.

As competitors arose, the routes of trade shifted away from Madeira. Cane fields were abandoned, the industry of sweets and conserves was endangered, and activity in the port of Funchal atrophied. As this happened, the commerce of the Canaries picked up, providing an active competition in northern and Mediterranean markets. Both archipelagos sent their products to the markets of London, Antwerp, Rouen, and Genoa. Madeira's only advantage was that being first as an exporter of sugar and wine, it had won the preference of many sellers and consumers.

Canarian sugar began to arrive in quantity in European markets. Between 1549 and 1555 fifty-eight ships traveled between Antwerp and the Canaries. According to A. Cioranescu, the commerce of Tenerife was most intense with the low countries, limited only by warfare and religious conflicts. Santa Cruz was more oriented toward the sending of wine and dyestuff to England, a result of the opening of Bristol to trade with the Canaries, as had been proposed in 1538 by Charles V. On Gran Canaria, the northern trade, particularly with Flanders, was based on sugar, although the Flemish did not become important in it until the decade of the 1550s.[81]

Italian merchants based in Cádiz and Seville played a leading role in developing the Canarian sugar trade. They established themselves on Tenerife, Gran Canaria, and La Palma and used Cádiz as the central distribution point in the Mediterranean. The conquest of northern markets came later. In fact, the first shipment of Canarian molasses to Antwerp in 1512 did not please the buyers.[82] Only by the 1530s were Flemish buyers anxious to get Canarian sugars, partly because of the collapse of the Madeiran market, and partly because of the Flemish community established on the islands by that time. The trade with the northern ports was facilitated by Portuguese from Lisbon, Vila do Conde, and Algarve who had learned the routes and skills in the Madeira trade. On Gran Canaria and Tenerife as earlier in Madeira, the Italian-Flemish merchant community was the axis of trade with the European markets for sugar. On all these islands, the communities overlooked religious differences to unite for the common cause of selling sugar, and together they dominated the sugar trade.

Good information on Canarian sugar exports is difficult to find, but it seems clear that the relatively low number of sailings to Italy from the islands can be explained by the fact that Andalusian ports, especially Cádiz, served as intermediary destinations, playing a role similar to Viana do Castelo in the Madeiran trade with northern Europe. Canarian trade with northern Europe was often direct. Gran Canaria, for example, sent various grades of sugar and conserves to Rouen and Antwerp.

By the mid-sixteenth century competition from Brazilian sugar began to have an effect on the Atlantic islands. Madeira turned to the Brazilian product to stimulate its own trade. José Gonçalves Salvador has stated that the islands served as "a trampoline for Brazil and the Rio de la Plata" in the period 1609–21.[83] He also made clear that this relation might be direct or indirect through Angola, São Tomé, Cape Verde, or the Guinea coast. From the close of the sixteenth century, the trade in Brazilian sugar used the ports of Funchal and Angra dos Reis for legal and contraband exports to Europe. Pressures on the Crown and appeals from Madeirans led to its limitation. Thus in 1591 unloading Brazilian sugar in Funchal was banned, an action that seemed to have little effect since the minutes of the town council of Funchal for 17 October 1596 asked for the full application of this law. After 1596 there is evidence of an active role in defense of local sugar production by local authorities. Violations of these restrictions were punished by a fine of 200 *cruzados* and a year of penal exile.[84]

Constant pressure from businessmen in Funchal involved in this commerce led to a consensual solution. In 1612 a contract was established between the merchants and the town in which the merchants were allowed to sell a third of this Brazilian sugar, which after 1603 had been completely restricted from sale; violators were punished by loss of the cargo and a 200 *cruzado* fine. After 1611 this changed and sale of Brazilian sugar was allowed after local sugar had all been sold. Thus slaves and boatmen were threatened that any movement of sugar without expressed authorization by the municipal council would be punished by a fine of fifty *cruzados* and two years of penal exile.

After the Portuguese restoration of independence in 1640, commerce with Brazil faced further regulations. First, there was the creation of monopoly through the Brazil Company in 1649 and its creation of a convoy system. Madeira and the Azores after 1650 were allowed to send two ships a year with a capacity of 300 *pipas* to trade for tobacco, sugar, and wood. Later a limit was set at 500 crates of sugar. Two ships were sent every year with licenses from the Conselho da Fazenda and were supposed to benefit all the island's merchants. Some ships claiming to be victims of shipwreck or corsair attacks landed crates of sugar, perhaps attempting to avoid the prohibitions. Infractions were punished with prison terms.[85] For the seventeenth-century Canaries we only have export figures for Gran Canaria in the first quarter of the century.[86] By that time, the relative importance of Seville and the French ports had become inverted.

Place of Madeira in the World of Sugar

Madeira, archipelago and island, played a singular role in European expansion. Various factors in the fifteenth century made it a kind of Atlantic "lighthouse" to orient and guide further maritime activity. This role as a base of communications and the development of its agriculture of sugar and wine allowed Madeira to overcome the isolation of its location. It also served as a point of reference for the Atlantic in terms of its social organization and in the role of slavery within it. As Sidney Greenfield has observed, Madeira served as a trampoline between "Mediterranean sugar production" and American "plantation slavery." In this, Greenfield was simply following the arguments developed by Charles Verlinden in the 1960s, arguments that now must be modified due to recent work on slavery on the island.[87] In truth, Madeira was the social, political, and economic starting point for the Portuguese Atlantic and for "the world the Portuguese created" in the tropics.

It was Columbus who opened the New World and traced the route for sugar's expansion to it. He was no stranger to this product, having been involved in its commerce on Madeira. Prior to his personal relationship on the island, he had been, like many of the Genoese merchants, dealing in Madeiran sugar. Tradition has it that the first cane plantings he brought to America came from La Gomera in the Canaries, which at that moment was involved with sugar's expansion while the industry was already well established on Madeira.

Madeira's soils made sugarcane cultivation through intensive agriculture profitable. Madeira made production on a large scale possible as prices began to reflect by the late fifteenth century. In 1483 Governor Don Pedro de Vera, wishing to make the conquered areas of the Canaries productive, sought to bring sugar plantings from Madeira. Portuguese took an active part in that conquest and brought this new area into the world economy by acquiring lands as settlers, by working for wages as specialists in sugar making, or by constructing sugar mills and setting them in motion. On La Palma, for example, we can refer to Lionel Rodrigues, *mestre de engenho*, who earned that title after twelve years of work on Madeira.[88] The Canaries would later play a similar role for the Spanish Indies. Thus, in 1519 Charles V recommended to the governor Lope de Sosa that he facilitate the departure of sugar masters and specialists for the Indies.[89]

Sugar had moved southward to Cape Verde and São Tomé, but it was only São Tomé's water, forest, and land that were suitable for its expansion. In 1485 the Crown recommended that João de Paiva proceed with the planting of sugarcane. For the making of sugar there are references to "many masters from the island of Madeira." It was on São Tomé that the sugar structure, which eventually passed to the other side of the Atlantic, developed. From the sixteenth century, the competition from the Canaries and especially São Tomé naturally led to a reaction from Madeiran producers who complained to the Crown in 1527.[90] The Crown promised to respond in the following year, but no decision seems to have been made.[91]

Meanwhile across the Atlantic, the first steps in the distribution of land in

Brazil were being made. Once again, the presence of Madeiran cane and Madeiran sugar specialists can be noted. The Crown drew on them to create the industry's infrastructure. In 1515 the Crown had asked for the good offices of anyone who might build a mill, and in 1555 João Velosa, called by many a Madeiran, built one at royal expense. To develop the industry in Brazil, specialized laborers would be needed and Madeira was the principal source. Thus in 1537 *engenho* carpenters on the island were prohibited from traveling to the lands of the Moors.[92]

With such restrictions and facing the slow decrease in island sugar production, many Madeirans headed for the Brazilian cane fields, where they served as specialists and proprietors in Pernambuco and Bahia. Some Madeirans such as Mem de Sá and João Fernandes Vieira, the liberator of Pernambuco in the mid-seventeenth century, became important mill owners. The ties between Brazil and the island and sometimes through it to European markets continued. In 1599, for example, Cristóvão Roiz of Câmara de Lobos on Madeira declared having close to 100 *milréis* invested in three sugar masters in Pernambuco in partnership with two other investors.[93]

As the Atlantic sugar market revealed the existence of areas of better conditions and larger capacity, the island sugar industry was irretrievably lost. Cane fields slowly disappeared and were replaced by vineyards. Only the economic conjuncture in the second half of the nineteenth century would permit their return. But this situation proved ephemeral and even then was only possible with a protectionist policy. The cane fields lost their ability to produce sugar, the "white gold" of the islands, but in its place they made cane brandy and liquor. The rum and aguardente produced today are the heirs of the sugarcane culture of Madeira and the Canaries.

NOTES

Abbreviations

AEA   *Anuario de Estudios Americanos*
AHM   Arquivo Histórico da Madeira
ANTT   Arquivo Nacional da Torre do Tombo
ARM   Arquivo Regional da Madeira
CHCA   *Colóquio de Historia Canario Americana*
CMF   Câmara Municipal do Funchal
DAHM   *Das Artes e da História da Madeira*
PJRFF   Provedoria e Junta da Real Fazenda do Funchal
RGCMF   Registro Geral da Câmara Municipal do Funchal

1. See Antonio Rumeu de Armas, *La conquista de Tenerife, 1494–96* (Santa Cruz de Tenerife: Aula de Cultura de Tenerife, 1975); Elías Serra Ráfols, *Alonso Fernández Lugo: Primer colonizador español* (Santa Cruz de Tenerife: Aula de Cultura de Tenerife, 1972); and Alfonso García Gallo, "Los sistemas de colonización de Canarias y América en los siglos XV y XVI," *I Colóquio de historia canario americana* (Las Palmas: Cabildo Insular de Gran Canaria, 1977).

2. This documentation results from the accounting organized by each mill as can be inferred from a document of 1550 (*Provisão e regimento*) for the taxing of sugar, 12 June 1550, *Arquivo histórico da Madeira* 19, no. 98 (1990): 119–24. See two works by José Pereira da Costa and Fernando Jasmins Pereira, *Livros de contas da ilha da Madeira, 1504–1537* (Coimbra: Biblioteca Geral da Universidade, 1985), and *Livros de contas da ilha de Madeira. Registro da produção de açúcar* (Funchal: Centro de Estudos de História do Atlântico, 1989). For the Canaries, the documentation is limited to questions of land distribution. See Pedro Cullén del Castillo, ed., *Libro rojo de Gran Canaria, gran libro de provisiones y reales cédulas* (Las Palmas: Cabildo Insular de Gran Canaria, 1995); Elias Serra Ráfols and Leopoldo de la Rosa Oliveira, eds., *Reformación del repartimiento de Tenerife en 1506 y colección de documentos sobre el adelantado y su gobierno* (Santa Cruz de Tenerife, 1963); Francisca Moreno Fuentes, *Las datas de Tenerife, libro V de datas originales* (La Laguna: Universidad de La Laguna, 1978); Francisca Moreno Fuentes, *Las datas de Tenerife (libro primero de datas por testimonio)* (La Laguna: Universidad de La Laguna, 1992); and Eduardo Aznar Vallejo, *Documentos canarios en el registro del sello (1476–1517)* (La Laguna: Instituto de Estudios Canarios, 1981). In recent years, some books of the provincial notary's records of Las Palmas and Santa Cruz de Tenerife have been published.

3. Compare the provisions of the letter of Dom João I with those referred to by José de Vieira y Clavijo, *Noticias de la historia de las islas Canarias*, 3 vols. (Santa Cruz de Tenerife: Goya Ediciones, 1950–52), 681.

4. "On the concession of lands to the first settlers of the island of Madeira (1426)," in João Martins da Silva Marques, *Descobrimentos portugueses: Documentos para a sua história* (Lisbon: Instituto para Alta Cultura, 1944), supplement to vol. 1, bk. 19, 109.

5. *Cédula regia* (4 February 1480); del Castillo, ed., *Libro Rojo de Gran Canaria*, 1–2.

6. See J. Peraza de Ayala, "El contrato agrario y los censos en Canarias," *Anuario de historia del derecho español* 25 (1955); and Aznar Vallejo, *Documentos canarios en el registro*, 239–42.

7. In the Canaries, water, which was always scarce, played a central role in the occupation of the islands. Although the situation was quite different from that found in Madeira, the policy toward water developed in the same direction, moving from common rights to private control over time. See J. Hernández Ramos, *Las heredades de aguas en Gran Canaria* (Madrid: n.p., 1954); Antonio M. Macías Hernández, "Aproximación al proceso de privatización del agua en Canarias, c. 1500–1879," in *Agua y modo de producción* (Barcelona: Crítica, 1990), 121–49; and José Miguel Rodríguez Yanes, *El agua en la comarca de Daute durante el siglo XVI* (Santa Cruz de Tenerife: Aula de Cultura del Cabildo Insular de Tenerife, 1988).

8. See also Arquivo Histórico Ultramarino, Madeira e Porto Santo, no. 3281 (5 November 1813), published by E. C. Almeida, *Archivo da Marinha e Ultramar: Madeira e Porto Santo* (Lisbon, 1907), 1:223–25, 238. Register books of the distribution of water exist only from the

eighteenth century: Arquivo Regional da Madeira, *Câmara de Santa Cruz*, no. 135; *Câmara da Ponta do Sol*, no. 181; *Câmara do Porto Santo*, no. 46, 124; Biblioteca Nacional de Lisboa, cod. 8391. In an *alvará* of D. Henrique of 18 August 1563, cited in E. C. Almeida, *Archivo da Marinha e Ultramar: Madeira e Porto Santo* (Lisbon, 1907), 238, the position of evaluator was created to determine the price of water; see J. José de Sousa, "As levadas," *Atlântico* 17 (1989).

9. Felipe Fernández-Armesto, *The Canary Islands after the Conquest: The Making of a Colonial Society in the Early Sixteenth Century* (Oxford: Clarendon Press, 1982); Leoncio Alfonso Pérez, *Miscelanea de temas canarios* (Santa Cruz de Tenerife: Cabildo Insular de Tenerife, 1984).

10. Afonso de Albuquerque, *Comentários de Afonso de Albuquerque*, 2 vols. (Lisbon: Casa da Moeda, 1973), pt. 4, chap. 7, 39.

11. *Reformación del repartimiento de Tenerife en 1506 y colección de documentos sobre el adelantado y su gobierno* (La Laguna: Instituto de Estudios Canarios, 1953), 144. See also Fernández-Armesto, *Canary Islands*, 48–68; Eduardo Aznar Vallejo, *La integración de las islas Canarias en la corona de Castilla (1478–1526): Aspectos administrativos y económicos*, 2d ed. (Las Palmas: Cabildo Insular de Gran Canaria, 1992), 229–45; Jiménez Sánchez, *Primeros repartimientos de tierras y aguas en Gran Canaria* (Las Palmas, 1940); A. Guimera Ravina, "El repartimiento de Daute (Tenerife), 1498–1529," *III CHCA* (1980); and Benedicta Rivero Suárez, *El azúcar en Tenerife 1496–1550* (La Laguna: Instituto de Estudios Canarios, 1990), 19–33.

12. Virginia Rau and Jorge Macedo, *O Açúcar na Madeira no século XV* (Funchal: Junta Geral do Funchal, 1992), 22.

13. Vitorino Magalhães Godinho, *Os descobrimentos e a economia mundial*, 2d ed., 4 vols. (Lisbon: Ed. Arcádia, 1983), 4:81.

14. To calculate this percentage, we must take into account the number of owners in 1494 and between 1509 and 1537. For an estimate of the population, we take into account the 15,000 inhabitants for 1500 and 19,172 for 1572. See Fernando Augusto da Silva, *Elucidário madeirense*, 3 vols. (Funchal: Junta Geral do Funchal, 1960), 3:103.

15. Tithe records and account books are not available for the sugar mills. Existing data has been derived from notary records in the archives of La Palma and Santa Cruz de Tenerife. Data on production is only available for the period from 1634 to 1813. See J. R. Santana Godoy, "Acerca de un recuento decimal de los azucares de las islas confeccionado por Millares Torres, 1634–1813," in *Historia general de las islas Canarias*, ed. Augustín Millares Torres, 15 vols. (Las Palmas: Edirca, 1979), 4:151–55; and J. R. Santana Godoy, "La hacienda de Daute 1555–1606," *Revista de historia de Canarias* 38, no. 174 (1984–86). Reference to the four sugar mills of the Adelantado is found in Guimera Ravina, "El repartimiento de Daute (Tenerife)." See also Oswaldo Brito, "Argenta de Franquis una mujer de negocios," *Santa Cruz de Tenerife* (1979); C. Negrin, "Jácome Monteverde y las ermitas de su hacienda de Tazacorte en La Palma," *Anuario de estudios atlânticos* 34 (1988); Ana Viña Brito, "Apoximación al reparto de tierras en La Palma a raíz de la conquista," *VII CHCA* (1990), 473; Viña Brito, "Los ingenios de Argual y Tazacorte (La Palma)," *Producción y comercio del azúcar de caña en época preindustrial: Actas del tercer seminario internacional, Motril, 23–27 de septiembre 1991* (Granada: Diputación Provincial, 1993), 75–93; Gaspar Frutuoso, *Livro primeiro*

das saudades da terra (Ponta Delgada: Instituto Cultural de Ponta Delgada, 1984), 53, 58, 71; Eduardo Aznar and Ana Viña Brito, "El Azúcar en Canarias," *La caña de azúcar en tiempos de los grandes descubrimientos (1450–1550): Actas del primer seminario internacional, Motril, 25–28 de septiembre 1989* (Motril: Junta de Andalucía and Ayuntamiento de Motril, 1990), 173–88; and A. Macías, "Canarias, 1480–1550: Azúcares y crecimiento económico," in *História do Açúcar. Rotas e mercados* (Funchal: Centro de Estudos de História do Atlântico, 2002).

16. Gaspar Frutuoso, *Saudades da terra*, 6 vols. (Ponta Delgada: Instituto de Ponta Delgada, 1963), 1:113.

17. António Aragão, *A Madeira vista por estrangeiros, 1455–1700* (Funchal: Direcçao Regional dos Assuntos Culturais, 1982), 37; Gomes Eanes de Zurara, *The Chronicle of the Discovery and Conquest of Guinea*, trans. Charles Raymond Beazley and Edgar Prestage, 2 vols. (New York: B. Franklin, 1963), 1:17.

18. Fernando Jasmins Pereira, "Açúcar madeirense," *Estudos políticos e sociais* 7, no. 13 (1969): 158.

19. Isabel Drummond Braga, "A Acção de D. Luís de Figuereido de Lemos: Bispo do Funchal, 1585–1608," *III Coloquio internacional de historia da Madeira* (Funchal: Centro de Estudos de História do Atlântico, 1992): 572; ARM, Julgado de residuos e capelas, fs. 499v–500v (30 May 1571), 52v–88 (20 August 1583).

20. José Sánchez Herrero, "Aspectos de la organización eclesiástica y administración económica de la diocesis de Canarias a finales del siglo XVI," *AEA* 17 (1971).

21. Manuel Lobo Cabrera, *El Comercio canario-europeo bajo Felipe II* (Funchal: Centro de Estudos de História do Atlântico, 1988), 7, 115–16.

22. This comes from the *recollection do oitavo*. See Arquivo Nacional da Torre do Tombo, Provedoria e Junta da Real Fazenda do Funchal, no. 980, 525–39.

23. Among those who benefited were Diogo Guerreiro, Inácio de Vasconcelos, António Correa Bethencourt, and Pedro Betancour Henriques. See ANTT, PJRFF, 965a, f. 7, 181–82, 222, no. 966, fol. 8v; ANTT, PJRFF, 396, f. 63v, 969, fol. 48–48v. See also Fréderic Mauro, *Le Portugal, le Brésil et l'Atlantique au XVIIe siècle (1570–1670): Étude économique*, 2d ed. (Paris: Foundation Calouste Gulbenkian; Centre Cultural Portugais, 1983), 248–50; and ANTT, *Convento de Santa Clara*, bk. 19, letters of 10 February 1649 and 18 October 1649.

24. Taxes on sugar production are a key to evaluating the state of the industry. On Madeira there was first the *quarto* (one-fourth) and then the *quinto* (one-fifth), which was collected from each producer. In the Canaries, the most important tax was the tithe (*diezmo*), which was collected by the church. The register books for the tithe have disappeared and all that remains is the information gathered by A. Millares Torres for the period 1634–1813. During this period there are seven sugar mills listed on the islands of Tenerife, Gran Canaria, and La Palma. See Paulino Castañeda Delgado, "Pleitos sobre Diezmos del Azucar en Santo Domingo y Canarias," *II CHCA* (1979), 2: 247–72; and Rivero Suárez, *El Azúcar en Tenerife*, 179–86. The tithe was not collected as one-tenth of the cane produced, but rather as one out of each twenty *arrobas* of white sugar. This led to conflicts that were resolved in 1543 by a brief of Pope Paul III, who established the tithe as one-tenth of all sugar produced before the division made between mill owners and dependent cane farmers.

25. For Madeira documentation of productivity is available for each sugar mill. See regiment to the recollection of the sugar (12 June 1550), AHM 19, no. 98 (1990): 119–24; and Costa and Pereira, *Livros de contas da ilha da Madeira*. For the Canaries, the documentation is limited to questions of land distribution. See, for example, del Castillo, ed., *Libro rojo de Gran Canaria*; Serra Ráfols and Rosa Oliveira, eds., *Reformación del repartimiento de Tenerife*; Moreno Fuentes, *Las datas de Tenerife, libro V de datas originales*; Moreno Fuentes, *Las datas de Tenerife (Libro primero de datas por testimonio)*; and Aznar Vallejo, *Documentos canarios en el registro*.

26. On the evolution of landed property, there are few studies for Madeira and the ones that exist are limited on the question of the land grants. See Fernando Jasmins Pereira, *Elementos para a história económica de Madeira* (Funchal: Centro de Estudos de História do Atlântico, 1991), 22–35, 88–95; Maria de Lourdes Freitas Ferraz, *A ilha da Madeira sob o domínio da casa senhorial do infante D. Henrique e as suas descobertas* (Funchal: Secretaria Regional do Turismo e Cultura, 1986); Manuel Pita Ferreira, *O arquipélago da Madeira. Terra do senhor infante* (Funchal: Junta Geral do Funchal, 1959); and Joel Serrão, "Na Alvorada do mundo atlântico," *Das artes e da história da Madeira* 64, no. 31 (1961). On the Canaries the question of land distribution is better documented. See, for example, Vicente Suárez Grimón, *La propriedad pública, vinculada y eclesiástica en Gran Canaria en el crisis del antiguo regimen*, 2 vols. (Las Palmas: Cabildo Insular de Gran Canaria, 1987), vol. 1; and Fernández-Armesto, *Canary Islands*.

27. See Alberto Vieira, "O senhorio no Atlântico insular oriental: Analise comparada da dinâmica institucional da Madeira e Canárias nos séculos XV e XVI," *III Jornadas de estudos sobre Fuerteventura y Lanzarote* (Puerto del Rosario: Cabildo Insular de Fuerteventura, 1989), 1:33–48.

28. Armas, *La conquista de Tenerife*; Serra Ráfols, *Alonso Fernandez Lugo primer colonizador*; Alfonso Garcia-Gallo, "Los sistemas de colonización de Canárias y América en los siglos XV y XVI," *I CHCA* (1977).

29. Silva Marques, *Descobrimentos portugueses*, vol. 1 supplement, bk. 19, 109.

30. Del Castillo, ed., *Libro rojo de Gran Canaria*, 1–2; ANTT, *Santa Clara*, maço 1, no. 47 (1454).

31. ANTT, *Livro das ilhas*, f. 550v.

32. ARM, Registro geral da câmara municipal do Funchal, vol. 1, fs. 204–9, 249–52, 287–88, 289v–91.

33. António Aragão, *A Madeira vista por estrangeiros*, 87.

34. Noël Deerr, *The History of Sugar*, 2 vols. (London, 1940–50); Edmund von Lippmann, *História do Açúcar, desde a época mais remota até o começo da fabricação de açúcar de Beterraba*, 2 vols. (Rio de Janeiro: Le Uzinger, 1941–47); Fernando Ortiz, *Los primitivos técnicos azucareros de América* (Havana, 1955), 13–18. See, for comparison, Moacir Soares Pereira, *A origem dos cilindros na moagem de cana* (Rio de Janeiro, 1955); David Ferreira Gouveia, "O açúcar da Madeira: A manufactura açucareira madeirense (1420–1550)," *Atlântico* 4 (1985): 268–69; and Alberto Vieira, *A Madeira, a expansão e história da tecnología do açúcar*, in *História e tecnologia do açúcar* (Funchal: Centro de Estudos de História do Atlântico, 2000), 7–20.

35. ANTT, *Convento de Santa Clara*, maço 13, no. 1 (4 July 1477); António Baião, *O manuscrito de Valentim Fernandes* (Lisbon, 1940); A. Artur, "Apontamentos históricos de Machico," *DAHM* 1 (1949): 8–9. Historians are not certain about the date of Teixeira's inventory: should it be based on the date of his testament on 7 September 1535, or should it be calculated from the date of his wife's testament on 13 September 1495?

36. Guillermo Camacho Pérez-Galdós, "La hacienda," 29; Archivo Historico y Provincial de Las Palmas, *Protocolos*, no. 733, f. 81; A. Millares Torres, *Historia general de las islas Canarias* (Las Palmas: Edirca, 1977), 120–21; Luís Pérez Aguado, *La caña de azúcar en el desarrollo de la ciudad de Telde* (Telde: Ayuntamiento de Telde,1982), 5–27; Pérez Vidal, "El Azúcar," *II Jornadas de estudios Canarios-América* (Santa Cruz de Tenerife: Caja General de Ahorros, 1981), 177; Manuel Lobo, "El ingenio en Canarias," in *História e tecnología do açúcar* (Funchal: Centro de Estudos de História do Atlântico, 2000), 105–15.

37. On 20 March 1499, AHM 17, no. 227 (1973): 386–87, this situation was noted and the possible negative implications for the collection of the *quinto* tax.

38. *Livro segundo das saudades da terra* (Ponta Delgada: Instituto Cultural de Ponta Delgada, 1979), 99–135; von Lippmann, *História do açúcar*, 13.

39. See A. Cioranescu, *Thomas Nichols, mercader de azúcar, hispanista y herege* (La Laguna: Instituto de Estudios Canarios, 1963); and Gaspar Frutuoso, *Livro primeiro*.

40. Gloria Díaz Padilla and José Miguel Rodríguez Yanes, *El señorio en las Canarias occidentales, la Gomera y el Hierro hasta 1700* (Santa Cruz de Tenerife: Cabildo Insular de El Hierro, 1990), 319–20.

41. A. Artur, "Apontamentos históricos de Machico," *DAHM* 1 (1949): 1, 8–9; ARM, *Capelas*, caixa 8 (19 January 1547); ARM, *Misericórdia do Funchal*, no. 40, fs. 49–58 (11 September 1600); ANTT, *Convento de Santa Clara*, caixa 4, no. 11 (20 December 1644); ARM, *Misericórdia de Funchal*, no. 42, fs. 249–51 (25 March 1645).

42. Díaz Padilla and Rodríguez Yanes, *El señorio*, 320; Aznar and Viña Brito, "El azúcar en Canarias," 185; Archivo Historico y Provincial de Tenerife, *Protocolos: Juan de Anchieta*, no. 455, fs. 82ff., in Fernando Gabriel Martín Rodríguez, *Arquitectura domestica canaria* (Santa Cruz de Tenerife: Aula de Cultura, 1978), 298–304.

43. Editor's note: The Madeira *arroba* was equivalent to 28 *arratéis* (lbs.) until 1504, when it was changed to 32 *arratéis*.

44. A. Macías, "Canarias, 1480–1550."

45. These contracts have merited a number of studies. See, for example, Fernando Augusto da Silva and Carlos Azevedo Menezes, "Colonia, contrato de," *Elucidário madeirense*, 1:290–92; Jorge de Freitas Branco, *Camponeses da Madeira* (Funchal: Publicaçoes D. Quixote, 1987), 153–87; and João José Abreu de Sousa, "O convento de Santa Clara do Funchal: Contratos agrícolas (século XV a XIX)," *Atlântico* 16 (1988).

46. Manuel Lobo Cabrera, *La esclavitud en las Canarias Orientales en el siglo XVI: Negros, moros y moriscos* (Santa Cruz de Tenerife: Cabildo Insular de Gran Canaria, 1982), 165; Manuel Lobo Cabrera and Ramón Díaz Hernández, "La población esclava de Las Palmas durante el siglo XVII," *AEA* 30 (1984): 4. See also Rivero Suárez, *El azúcar*, 43–81; and Oswaldo Brito, *Augusta de Franquis una mujer de negocios* (Santa Cruz de Tenerife: Cabildo Insular de Tenerife, 1979).

47. Lobo Cabrera, *Esclavitud en las Canarias Orientales*, 211–12.

48. Manuel Lobo Cabrera, "Esclavitud y azúcar en Canarias," in *Escravos com e sem açúcar*, ed. Alberto Vieira (Funchal: Centro de Estudos de História do Atlântico, 1990), 106–9.

49. Manuel Lobo Cabrera, "La población esclava de Telde en el siglo XVI," *Hacienda* 150 (1982): 6, 70–71; Lobo Cabrera, *La esclavitud en las Canarias*, 200; A. Cioranescu, *Historia del Puerto de Santa Cruz de Tenerife* (Santa Cruz: Viceconsejería de Cultura y Deportes, 1993), 110; Pedro Martínez Galindo, *Protocolo de Rodrigo Fernández (1520–1526)* (La Laguna: Instituto de Estudios Canarios, 1988), 107; Manuel Marrero, "La esclavitud en Tenerife," *Revista de historia* (La Laguna: Universidad de La Laguna, 1966), 77; Manuel Lobo Cabrera, "La poblacion esclava de Las Palmas," *AEA* 30 (1984).

50. M. Garrido Abolafia, *Los esclavos bautizados en Santa Cruz de La Palma (1564–1600)* (Santa Cruz de la Palma: S.E., 1994); Manuel Lobo Cabrera e Pedro Quintana Amdrés, *Población marginal en Santa Cruz de La Palma, 1564–1700* (Madrid: Ediciones La Palma, 1997).

51. ARM, RGCMF, vol. 1, fs. 262v–69v. Regimento régio (12 October 1502), AHM 17 (1973), dcc. 203, 356; I, 98–98v; AHM (1973), doc. 258, 429–31.

52. ARM, RGCMF, vol. 1, fs. 262v–69v; *regimento* in AHM 17 (1973), no. 203, 356; vol. 1, f. 98–98v; *carta régia*, no. 258, 429–31. The term "açúcar quebrado" sometimes refers to what was called in the Caribbean muscovado sugar.

53. ARM, RGCMF, 1, fs. 34v, 36v; AHM 16 (1973), no. 145, 241–42; AHM (1973), 1, fs. 107–7v; AHM (1973), 1, no. 284, 451–52. Contrary to Manuela Marrero, "De la esclavitud en Tenerife," *Revista de história* 100 (1952): 434, slaves were linked to the sugar harvest and there is reference to at least one sugar master in Telde. See Manuel Lobo Cabrera, *Esclavos indios en Canarias* (Madrid, 1983); AHM 3 (1933): 154–59; and ARM, *Capelas*, caixa 118, no. 4, doc. 684.

54. Moquet, *Voyages*, bk. 1, 50, cited by Vitorino Magalháes Godinho, *Os descobrimentos e a economia mundial*, 4 vols. (Lisbon: Ed. Presença, 1983), 4:201; ARM, *Misericórdia de Funchal*, no. 684, fol. 539.

55. Lobo Cabrera, *Esclavitud en las Canarias Orientales*, 233–35; Manuel Lobo Cabrera, *Los libertos en la sociedad Canaria del siglo XVI* (Madrid: Consejo Superior de Investigaciones Científicas and Tenerife: Instituto de Estudios Canarios, 1983), 51, 61.

56. Rivero Suárez, *El azúcar*, 43–93.

57. M. Coello Gómez et al., *Protocolos de Alonso Gutiérrez (1522–1525)* (Santa Cruz de Tenerife: Cabildo Insular de Tenerife and Instituto de Estudios Canarios, 1980), 178 n. 333.

58. Alberto Vieira, *O comércio inter-insular nos séculos XV e XVI* (Funchal: Centro de Estudos de História do Atlântico, 1987), 57.

59. Vitorino Magalhães Godinho, "Preços e conjuntura do século XV ao XIX," *Dicionário de história de Portugal* (Lisbon: Iniciativas Editoriais, 1971), 4:488–516; José Gentil da Silva, "Echanges et Troc: l'Example des Canaries au debut du XVI siècle," *Annales* 165 (1961): 1004–11; Manuel Lobo Cabrera, *Monedas pesos y medidas en Canarias en el siglo XVI* (Las Palmas: Cabildo Insular de Gran Canaria, 1989), 10–13; Rivero Suárez, *El azúcar*, 147–48.

60. Fernando Jasmins Pereira, *Estudos sobre história de Madeira* (Funchal: Centro de Estudos de História do Atlântico, 1991), 232–34.

61. AHM 15 (1972): 64.

62. AHM 15 (1972): 46, 229, 313, 318, 372–80; Frutuoso, *Livro primeiro*, 113; Armando de Castro, "O sistema monetário," in *História de Portugal*, ed. José Hermano Saraiva, 6 vols. (Lisbon: Alfa, 1983), 3:236–38; Lobo Cabrera, *El comercio canario-europeo*, 117.

63. Lobo Cabrera, *El comercio canario-europeo*, 116; Pereira, *Estudios*, 219–24.

64. Magalhães Godinho, *Os descobrimentos*, 87; ARM, Câmara municipal do Funchal, *registro geral* 1, fs. 1–1v, letter on the sugar trade (Alcochete, 14 July 1469); AHM 15 (1972): 45–49.

65. ARM, RGCMF, 1, fs. 308v–309 (Sintra, 7–8 August 1508), published in AHM 18 (1973): 503–4; Alvaro Rodrigues de Azevedo, "Notas," in *Saudades da terra* (Funchal, 1873), 501.

66. See Lobo Cabrera, *El comercio canario-europeo*, 7.

67. Mauro, *Le Portugal*, 225; ARM, RGCMF, 1, fs. 5v–6; AHM 15, 57 (Funchal, 1972), 57; fol. 148–148v; AHM, 15, 68 (Funchal: Boletim do Arquivo Distrital do Funchal, 1972); Henrique Gama Barros, *História da administração pública em Portugal nos séculos XII a XV*, 2d ed., 11 vols. (Lisbon: Liv. Sa da Costa, 1945–54), 10:152–53; Rau and Macedo, *O açúcar na Madeira*, 26 n. 27; *Monumenta henricina*, 15 vols. (Coimbra, 1960–74), 15:87–89; ARM, CMF, no. 1298, f. 37; f. 68; f. 87v; ARM, RGCMF, 1, fs. 292–93; ANTT, Gavetas XV-5-8, summarized in *As Gavetas do Torre do Tombo*, 12 vols. (Lisbon, 1960–1977), 4:169–70.

68. Gama Barros, *História*, 10:155; Fernando Jasmins Pereira, *Alguns elementos para o estudo da história da Madeira* (Funchal: Centro de Estudos de História do Atlântico, 1991), 139–62; ARM, RGCMF, 1, f. 262v, 291v.–292, in AHM 17 (1973): 350–58, 369. See also Rodrigues de Azevedo, "Anotações," in *Saudades*, 681–82.

69. See his correspondence in Alberto Vieira, ed., *O público e o privado na história da Madeira*, 2 vols. (Funchal: Centro de Estudos de História do Atlântico, 1996–98).

70. My comments here are based on a broad range of archival sources. See also Lobo Cabrera, *El comercio canario-europeo*, 19; Manuela Marrero Rodrigues, "Los genoveses en la colonización de Tenerife," *Revista de historia canaria* 16 (1950); Aznar Vallejo, *Documentos canarios en el registro de sello*, 196; Augustín Guimerá Ravina, "El repartimiento de Daute (Tenerife, 1498–1529)," in *III CHCA* (1980), 1:127–28; and Fernando Clavijo Hernández, *Protocolos de Hernán Guerra (1510–1511)* (Santa Cruz de Tenerife: Cabildo Insular/Instituto de Estudios Canarios, 1980), 39–40.

71. G. Camacho y Pérez Galdós, *La hacienda de los principes* (La Laguna: Imprenta Curbelo, 1943), 524. This author notes eighty-eight Genoese merchants of whom eighty-one (82 percent) were *vecinos*. In my review of printed sources, I only found fifty-four Genoese, of which 29 percent were *vecinos*.

72. M. Marrero Rodrigues, *Los mercadores flamengos* (n.p.), 601–9.

73. Giles Hana, a Flemish merchant and *vecino* of Tenerife, married Francisca de Carminatis, daughter of the Lombard merchant Juan Jácome de Carminatis, who himself was married to the daughter of Jaime Joven, a Catalan merchant and *vecino* of the island. Flemish merchant Juan de Xembrens married Ana de Betancor, daughter of Guillén de Betancor. See Marrero Rodrigues, "Los genoveses," 611–14.

74. Marrero Rodrigues, *Los mercadores*, 351 n. 177; José Peraza de Ayala, "Historia de la casa de Monteverde," in Francisco Fernández de Béthencourt, *Nobiliário de Canarias* (La

Laguna: Régulo Pérez, 1959), 2:491–579; Manuel Lobo Cabrera, "Los vecinos de Las Palmas y sus viajes de pesqueria," *III CHCA* (1980), 2:471; Guilherme Camacho y Pérez Galdós, "El cultivo de la caña de azúcar," *AEA* 7 (1961): 33–34; Manuela Marrero Rodrigues, "Una sociedad para comerciar en Castilla, Canarias y Flandres en la primera mitad del siglo XVI," *III CHCA* (1980), 1:161. See also various articles by J. M. Madurell Marimon: "Notas sobre el antiguo comercio," *AEA* 3 (1957): 563–92; "El antiguo comercio," *AEA* 7 (1961): 71–74; and "Miscellanea de documentos historicos atlânticos," *AEA* 25 (1979): 224–25, 235–38.

75. Elisa Torres Santana, *El comercio de las Canarias Orientales en tiempo de Felipe III* (Las Palmas: Cabildo do Insular de Gran Canaria, 1991), 304–8.

76. See Pereira, *Livro de contas da ilha da Madeira, 1504–1537*, 2 vols. (Funchal: Centro de Estudos de História do Atlântico, 1989).

77. David Ferreira de Gouvea, "O açúcar e a economia madeirense (1420–1550): Consumo de excedentes," *Islenha* 8 (1991): 11–22.

78. Lobo Cabrera, *El Comercio canario-europeo*, 113–14; Rivero Suárez, *El azúcar en Tenerife*, 147–48.

79. Joel Serrão, "Nota sobre o comércio do açúcar entre Viana do Castelo e o Funchal," *Revista de Economia* 3 (1950): 209–12; Virginia Rau, *A exploração e o comércio do sal em Setúbal: Estudo de história económica* (Lisbon: n.p., 1951); ARM, RGCMF, 1, f. 301–301v, published in AHM 17 (1973): 453–54; Domenico Geoffré, *Documenti sulle Relazioni fra Genova ed il Portogallo del 1493 al 1539* (Rome, 1961); Madurell Marimón, "Notas," 486–87, 493–94, 497–99.

80. *Acuerdos del cabildo de Tenerife* (La Laguna: Instituto de Estudios Canarios, 1948) 1, 83, no. 447 (26 March 1505).

81. The island received manufactured products, especially textiles from Antwerp, Ghent, Holland, and Rouen, and these were traded for money and sugar by Genoese and Flemish merchants, such as Bernardino Anehesi, Jerónimo Lerca, Lamberto Broque, Sebastian Búron, and Jerónimo Fránquez. See Eddy Stols, "Les Canaries et l'expansion coloniales des Pays-Bas Méridionaux," *IV CHCA* (2000), 1:908; and Manuel Lobo Cabrera, "El comercio entre Gràn Canaria," 32–33.

82. Magalhães Godinho, *Os descobrimentos*, 4:98.

83. José Gonsalves Salvador, *Cristãos Novos e o comércio no atlântico meridional* (São Paulo: Pioneira/MEC, 1978), 247.

84. For example, in January 1596, the town councillors prohibited António Mendes from unloading the sugar of Balthazar Dias. Three years later, he was obliged to reship a cargo of Bahian sugar without unloading any of it. See ARM, RGCMF, t. 3, f. 44v; ARM, RGCMF, documentos avulsos, caixa 4, no. 504, fs. 12v–13v refers to the prohibitions of 1591, 1597, 1601; ARM, CMF, no. 1312, fs. 7–8v; no. 1313, fs. 20–23.

85. CMF, no. 396, fs. 75v–76; ARM, RGCMF, t. 9, fs. 29v–30v (10 June 1664).

86. Torres Santana, *El comercio de las Canarias Orientales*, 300.

87. Sidney Greenfield, "Madeira and the Beginnings of New World Sugar Cane Cultivation and Plantation Slavery," in *Comparative Perspectives on Slavery in New World Plantation Societies*, ed. Vera Rubin and Arthur Tuden (New York: Academy of Sciences, 1977); Charles Verlinden, "Les origines coloniales de la civilization atlantique: Antecedents et types de

structure," *Journal of World History* 4 (1953): 378–98. On slavery, see Alberto Vieira, *Os escravos no arquipélago da Madeira: Séculos XV a XVII* (Funchal: Centro de Estudos de História do Atlântico, 1991).

88. Gloria Daz Padilla y José Miguel Rodríguez Yanes, *El señorio en las Canaria Ocidentales* (Santa Cruz de Tenerife: Cabildo de Insular de El Hierro, 1990), 316.

89. José Pérez Vidal, "Canarias, el azúcar, los dulces y las conservas," *II jornadas de estudios canarios-américa* (1981): 176–79.

90. ARM, CMF, *vereaçoes* 1527, f. 23v.

91. ARM, DA, no. 66 (8 February 1528).

92. ARM, RGCMF, 1, f. 372v.

93. ARM, JRC, fs. 391–96, Testament of 11 September 1599. In 1579 (ARM, *Misericórdia do Funchal*, no. 71, fs. 114–215) Gonçalo Ribeiro refers to being in debt to Manuel Luís, *Mestre de açúcar* "who is now in Pernambuco." See José António Gonçalves de Mello, *João Fernandes Vieira: Mestre de campo do terço da infantaria de Pernambuco*, 2 vols. (Recife: Universidad do Recife, 1956), 2:201–67.

CHAPTER FOUR

The Sugar Economy of Española in the Sixteenth Century

Genaro Rodríguez Morel

The sugar economy of Española was born in the wake of the collapse of gold mining, the principal source of the original conquistadors' wealth.[1] This decline in mining, which took place at the same time as the disappearance of the indigenous population, accelerated rapid social and economic changes in the island's society. The struggles for control of the political and economic resources of the island that accompanied these changes created the context in which the sugar industry was established.

From the beginning, gold mining was an economic activity that seemed to lack any long-term rationale. Neither those in political control of the island nor those who held Indians as dependents in grants of labor or *encomiendas*, and were thus the principal figures in the society, looked beyond the immediate glow cast by the much-desired metal. This lack of foresight was brutally evident in the way in which the colonists treated the natives of the island; a treatment so unjust and unreasonable that in two decades the indigenous population was almost completely eliminated.[2]

The colonists considered the calamitous decline of mining and of the indigenous population a serious blow. In part, the development of the agrarian economy resulted as a response to those declines. The colonists from the first days of settlement had practiced agriculture, but it had been directed toward local consumption. Although the mining economy was already showing signs of decline in the first decade of the sixteenth century, it was in the 1510s that a new agricultural system centered on the production of sugarcane and the "plantation" took form.

It is important to emphasize that as the mining economy weakened, and before sugar was firmly established, the island's colonists had tried to establish other lucrative enterprises. The most profitable of these was the import and export of horses. Brought first from Castile to be used in the takeover of the island, horses were later exported for the conquest of Mexico and Peru.[3] This trade permitted a number of individuals to increase their personal fortunes that were then invested in the sugar industry. The horse trade served as a stepping-stone to the sugar estate.[4]

The economic model of the plantation, which grew with the development of the sugar economy, intensified social differentiation among the white population of the island. The fact that the building of a sugar mill or *ingenio* required considerable economic resources essentially prohibited the Spanish population of middling resources from participating in this activity. This division reinforced the predominance of the island's upper class while it diminished the purchasing power of the rest of the Spanish population. One result of these economic rivalries was emigration. Sectors of the population, which had received the fewest benefits, or had not gained great advantages from the distribution of wealth, were the most likely to move from the island.

Those who could not leave were forced to seek other means of subsistence, and many concentrated in the city of Santo Domingo, the neurological center of the economy, where numerous small shops of shoemakers, tailors, carpenters, and masons flourished. Parallel to this small-time artisan activity was small-scale agriculture in maize, yuca (manioc), and vegetables along with the raising of chickens and livestock to serve the needs of the island's white population.[5] These products were sold primarily in the island's main cities of Santo Domingo, La Vega, Santiago, and Puerto Plata and were exported to neighboring islands such as San Juan (Puerto Rico) and Cuba.

The existence of this system of small-scale agriculture and trade supported the establishment of the sugar economy. The system allowed those who could to invest their capital in the building of *ingenios* while others participated in the export economy through small-scale local commerce that provided food to the white population as well as clothing and supplies to the slaves.

The Introduction of Sugarcane

The first sugarcanes were probably brought to Española from the island of Madeira, which, as we have seen in chapter 3, had a long experience with this plant. Columbus had lived on Madeira and later made his last stop there before his second voyage to America. The cane cuttings he brought from Madeira were planted in La Isabela, and although they took root, they did not have a great impact on the island. In fact, we do not know if the colonists ate those canes during the economic crisis that the island suffered in those first years. In any case, Columbus seems to have already had the development of a sugar economy in mind at that time, but his disputes with the political elites that accompanied him made this project difficult from the outset.

The fact that the first sugarcanes took root indicated that the climatic conditions and the soil were appropriate for its cultivation.[6] Nevertheless, what was lacking for the next eight years was a political situation that could make the expansion of the sugar economy possible. In 1501, sugarcane was introduced again, this time by Pedro de Atienza, *vecino* (resident with legal rights) of La Concepción who, with Miguel de Ballester, another *vecino* of that town, actually extracted juice from the cane, although they did not succeed in producing crystallized sugar.[7] The following years between 1501 and 1506 were particularly important for the establishment of sugarcane on the island. In this period there were various experiments, like that of Alonso Gutiérrez de Aguilón, to produce sugar in La Concepción, which was the center of the mining economy and, because of its proximity to Santo Domingo, also the demographic center of the colony.[8] By 1511 Miguel de Ballester and others resident in La Concepción were making sugar for local consumption.[9]

The slow growth of the sugar economy can be attributed to the lack of financial resources among the producers and thus to the absence of the tools and specialists needed to produce the sugar. Only when this situation was improved did the sugar economy begin to expand, in the second decade of the sixteenth century around La Concepción.[10] By 1515 a few *ingenios* with the appropriate technology were in operation, due primarily to the importation of specialists from the Canary Islands by the physician Gonzalo de Vellosa.[11]

The general disbursement of Indians carried out in 1514 by Lic. Rodrigo de Albuquerque accelerated the economic growth of the island.[12] This *repartimiento* had various objectives. First, it was designed to reassign the indigenous population to those colonists who were closest to the treasurer Miguel de Pasamonte, and thus it was against the interests of the friends of the Viceroy Diego Colon.[13] Second, it was intended to thin out the population of La Concepción and by doing so to stimulate the repopulation of certain other places, which had been abandoned by their old settlers because they presented few opportunities for wealth. As a result of the *repartimiento* many of the persons who had settled in La Concepción were forced to change their residence in order to derive any benefit from the assignment of Indians. Pedro de Atienza, for example, had to change his residence and affiliation (*vecindad*) to Santo Domingo. Similarly, Alonso Gutiérrez de Aguilón and Hernando Gorjón established residence in Azua, where they set up their *ingenios*. Licenciado Pedro Vázquez received some Indians in Puerto Plata on the condition that he move there and establish a sugar mill in the area.[14]

This repopulation process was cut short, however, by a smallpox epidemic in 1518 that devastated the island's indigenous population. The political leadership of the colony, seeing that the disbursement of the indigenous population would provide little stimulus to the Spanish settlers, turned to other incentives such as the granting of pastures, water, lands, livestock, and money with which to build sugar mills. It was precisely during this period that armadas were sent out again to capture natives from the "useless islands" (Bahamas), and the taking of Caribs from the Lesser Antilles increased. Although indigenous slaving was intense in this period, it was not very profitable, especially since the colonists were also receiving licenses to acquire black slaves at this time.

The help received by the colonists allowed a large number of people to initiate the construction of sugar mills. It was this set of circumstances in the 1510s that stimulated the growth of the sugar economy, and with it the plantation system. From this moment forward the struggle of groups in conflict on the island revolved around the acquisition of those economic factors that permitted their growth as a class.

Incentives and Benefits for the Building of Sugar Mills

The decline of mining and the economic crisis on the island contributed to the massive emigration to other conquests, particularly New Spain. In order to stop this migration, the Crown had to support for the first time, although somewhat timidly, the efforts of the encomenderos to change from the mining economy to the new economic activity of sugar. The Crown had its reasons, hoping to slow the deterioration of the island's economy, while the encomenderos sought new ways to reproduce themselves as a class.

By the second decade of the century, the encomenderos had achieved a certain coherence and political autonomy as a class, although not in terms of economic independence.[15] The Crown wished to maintain the island as a springboard for further conquests and the encomenderos needed the economic help of the state to achieve their social ambitions. The first royal assistance designed to stimulate the sugar economy took place during the administration of the Jeronymite friars (after 1516), but the limited nature of these efforts suggest that they were more a response to the pressures and demands of the encomendero elite than a policy of the state.

It was only after the Jeronymite friars began to actively support the sugar producers that the state's interest became clear. In March 1518, a license was granted to Cristóbal de Tapia to bring ten sugar masters and other specialists in sugar making to the island without having to pay any taxes.[16] In December of the same year, a royal order (*real cédula*) was sent to Lic. Rodrigo de Figueroa, indicating that all those who wished to stay on the island and were willing to build *ingenios* could receive funds from the Royal Treasury.[17] This was the first time that state funds were specifically committed to the construction of *ingenios*. Other measures followed. Like the privilege given to Cristóbal de Tapia, specialists from the Canary Islands could be brought in without having to pay any taxes, and Rome was requested to reduce the normal tithe ($\frac{1}{10}$) on sugar from the Indies with an impost of only 3 percent ($\frac{1}{30}$).[18] In addition, superior magistrates and in particular the *Veedor*, Cristóbal de Tapia, was authorized to distribute lands and water rights to those residents wishing to build mills.[19] Earlier grants made by the governors of the island for such purposes were also confirmed even though they had lacked the authority to make such concessions.[20] These state measures, except in a few particular cases, simply recognized and legalized existing practices.

It was not until the 1520s that state measures really began to transform the sugar economy. For the first time residents on the island were permitted to cast or smelt the copper essential for the kettles in the sugar-making process. This was a tremendous relief for them since it eliminated the cost of bringing that metal from Castile and, at the same time, it implied a loss to Crown revenues from the taxes that the imported copper had generated.[21] In June 1521, the Crown ordered treasurer Miguel de Pasamonte to take up to 6,000 pesos from the royal treasury for the purpose of distributing these funds to those willing to build *ingenios*, a measure stimulated by Antonio Serrano, the general counsel (*procurador general*) for the island and one of the principal representatives of the elite.[22]

It is important to point out that royal funds were not given to everyone who asked for them, but only to certain individuals who could meet the conditions of the loan. Those wishing to obtain these funds had to present guarantees that they had sufficient collateral.[23] Borrowers also had to commit themselves to repayment in no more than two years at the same time that they were also committed to build the new *ingenio* in the same period. Failure to do so resulted in the loss of their properties to the Crown.

Beyond the direct financial assistance provided to the colonists, the Crown also abolished taxes on the major imports, especially those needed for the production of sugar.[24] From this point in 1520 forward, tools and equipment, much of it produced outside of Spain, were imported.[25] Despite these measures, the encomenderos sought further loans and more time to complete the mill-building projects in progress.

In August 1520, another concession was made to help the sugar producers. A royal decree was sent to Treasurer Miguel de Pasamonte ordering further loans totaling 4,500 pesos with the same restrictions and guarantees as before. Table 4.1 presents a list of the recipients and the amounts received. Many of those who received these loans were not able to build their *ingenios* in the time allotted and sought extensions. For example, Lic. Alonso Fernández de las Varas sought a one-year extension, while Juan de Orihuela, who had begun an *ingenio* in the town of Azua, asked for two years.[26] Another case was that of Lic. Antonio Serrano, who had received land, water rights, and Indian laborers, and who in 1520 had still not completed his mill. Claiming that the delay was caused by the death of his Indians in the smallpox epidemic of 1518, he sought permission to build a horse-driven *trapiche* instead of the larger water-powered *ingenio*, which he had promised; the *trapiche* was completed in 1521.[27]

One of the most renowned cases of these extensions was that of Lucas Váz-quez de Ayllón who in 1523 had still not completed the *ingenio* he had begun in Azua. He argued that he had been unable to complete the mill because of his services to the Crown in New Spain. In response to his request for an extension, he received a royal decree ordering that "You do not vacate or consent to be unoccupied the said lands and rights that you were given. For the present We extend the said period and We order that you begin [to operate the mill] after Christmas day on the first day of spring."[28]

Beyond those mentioned above, there were others who had received money, land, water rights, and Indians, but never constructed a mill and sold off what they had received. Estéban de Pasamonte, nephew of the powerful Treasurer Miguel de Pasamonte, was denounced for selling certain Indians who had been given to him to construct an *ingenio*. He had apparently sold them without hesitation for a thousand ducats to Francisco Orejón, a *vecino* of La Vega.[29] The cases of Serrano, Ayllón, and Pasamonte were clearly exceptional in that these men formed part of the social and economic elite of the colony, a fact that explains why it was difficult to take measures against them.

But in general, it seems clear that the colonists were disposed to build *ingenios* because the loans given in 1519 were paid on time by the creditors, especially those who already had some experience in building mills. This was also the situation of Hernando de Berrio, Gonzalo Guzmán, Pedro de Valenzuela, Her-nando de Carvajal, and Diego Franco, among others, who paid off their loans.[30] This was also the case of Hernando Gorjón of Azua, who in December 1521 had

TABLE 4.1. Beneficiaries of Royal Loans for the Construction of *Ingenios* on Hispaniola

Beneficiary	Pesos (in Gold)	Indians[a]	Livestock /Land	Place
"Certain residents"	500	60	200 cows	Bonao
Francisco Orejón and Alonso Román	500	100		La Vega
García de la Barrera	500	150		La Vega
Juan Carillo and Pedro López de la Mesa	500	Yes		La Vega
Esteban de Pasamonte		Yes		Santo Domingo
Miguel de Pasamonte	400	100		Santo Domingo
Bachiller Moreno	500	100		San Juan de la Maguana
Pedro Alonso and companions	300	100		Santo Domingo
Hernando Gorjón	400			Azua
Lope de Bardeci	400			Santo Domingo
Hernando de Berrio	400			Santo Domingo
Diego Caballero	400			Santo Domingo
Antonio Serrano	600			Santo Domingo
Hernando de Carvajal	400			Santo Domingo
Gonzalo de Guzmán	400			Santo Domingo
Diego Franco	200			Santo Domingo
Pedro de Valenzuela	250	Yes		Santa María del Puerto
Francisco Cerón	500	Yes		San Juan de la Maguana
Judges and officials of Santo Domingo	400			Santo Domingo
Juan Freyre	400	Yes		La Sabana
Pedro Vázquez and Diego de Morales	500	Yes		Puerto Plata
Diego de Moguer	450	Yes		Puerto Plata
Fernando de Toval	500		Cows	San Juan de la Maguana
Francisco Tostado	400			Santo Domingo
Alonso Gutiérrez de Aguilón	400			Azua
Gonzalo de Vellosa	400			Azua
Lucas Vázquez de Ayllón			Land	Azua
Francisco Ceballos				Puerto Plata
Cristóbal de Tapia				Santo Domingo
Juan de Orihuela	400		Land	Azua

[a] "Yes" indicates an award was made but numbers are unknown.

his debt paid off, the money later returned to him to help in the building of a school and a hospital that he had begun.[31] Clearly, despite the early difficulties, the sugar industry was expanding, and while the claim of Rodrigo de Figueroa around 1520 that more than forty mills had begun construction seems exaggerated, the scale suggested is indicative of expansion.[32]

We know that by this date the mill of Lic. Garcia de Barred and Pedro Alonso in San Juan de Mauna was in operation, as was that of Lucas Vázquez de Ayllón and Francisco Cembalos in Puerto Plata, those in Azua owned by Hernando Gorjón and another by Alonso Gutiérrez de Aguilón, and around Santo Domingo a mill owned by the brothers Tapia and another by Gonzalo de Vellosa.[33] This is notable because even in the decade of the 1540s there were fewer than forty mills in operation. The president of the *audiencia* (appellate court) of Santo Domingo, Alonso López Cerrato, claimed that when he arrived in 1544 only ten mills were in operation and that by 1548 there were thirty-four functioning, but we must be cautious here too and take into account his own desire to demonstrate his efficiency.[34]

As a consequence of the royal loans and grants for the construction of *ingenios*, the colonial administration faced a crisis. The flexibility with which the loans were made and the laxity shown toward the collection of the debts allowed the colonists to go for decades without paying. This was due, in part, to the fact that the principal debtors were also the representatives of royal authority. In 1544 when Cerrato was named president of the *audiencia* and a general accounting was made of the treasurers Miguel, Juan, and Estéban Pasamonte, as well as of the factor Juan de Ampiés, a large portion of the money was still unpaid, totaling in official calculations some 50,000 castellanos.[35]

This situation of fiscal disorder, the malfunctioning of public institutions, and the rising costs of the *ingenios* created conditions that attracted the investment of foreign capital in the island's sugar sector. These investments coincided with the accession to power in Spain of the Emperor Charles V, who began to implement a new economic policy that favored Flemish, Genoese, and German interest groups. Among these, the Genoese, although they already had a certain presence on the island, were the most dynamic, and their financial experience contributed even further to their success. They began associating themselves with local entrepreneurs, forming partnerships for the building of mills.[36] One of the most important of the Genoese investors was Melchior Centurión who, as early as 1520, had allied with other Genoese and Spaniards resident on the island to build a mill.

Not only did the Genoese invest directly in the construction of mills, they also financed loans for the sugar business. Families such as the Centurión, Vivaldo,

Justinián, Grimaldi, Castellón, Forne, and Basinana, among others, invested in the island's production and commerce. The profitability of their loans was assured by the high rate of interest they charged and by the expansion that the sugar trade began to experience. As the sugar economy expanded in this period, thanks in part to the level of foreign investments, it also became increasingly encumbered with debt, not only to the Genoese but to investors and merchants in Seville, particularly to the Jorge family, one of that city's principal merchant consortiums.

Table 4.2 presents a listing of both *ingenios* and *trapiches* constructed on Española prior to the decade of the 1580s.

Investment Arrangements and the Costs of Ingenio Construction

Although the costs of construction and *ingenio* or a *trapiche* were considerable, they were not as high as has been claimed, at least during the beginning of the island's sugar economy. According to Gonzalo Fernández de Oviedo, one of the few authors who have provided concrete figures, the building of an *ingenio* required 10–15,000 ducats, but this seems to be exaggerated.[37]

According to the statements made by those who actually built mills in this period, the costs were not so high. Let us take, for example, the declarations made by Pedro Vázquez, the *alcalde ordinario* of Santo Domingo and his partner Diego de Morales, councilman (*regidor*) of the town of Santiago, who formed a partnership (*sociedad*) in 1519 to build a water-powered mill. They stated that their costs, including the purchase of the mill wheel, axles, presses, copper, and furnaces, did not exceed 800 castellanos and that included in this price was the purchase of 5,000 molds for preparing sugar as well as the construction of a thatch-roof purging shed. The sum also included the purchase of seven *suertes* of cane as well as a group of Indians. Pedro Vázquez stated, "Your Majesty knows that land and water cost the *ingenio* nothing, they are given to whoever asks for them. What is necessary to build an *ingenio* and work the land is people and provisions [*comida*]. I put people on the land as well as 400 pesos that I invested and with the effort that he [Diego de Morales] made and with the provisions and people that he had, the *ingenio* was finished with very little cost and I spent almost nothing in cash."[38]

Evidently state-sponsored assistance and concessions had lowered the costs of establishing an *ingenio*, but the rapid growth of the industry began to produce an inflationary trend and to affect the price of land, especially in urban areas and for those lands closest to the rivers. Moreover, since the level of sugar production depended on the size of the operation, many sugar mill owners began to build

TABLE 4.2. Sugar Mills Built on Española before 1590

Owner	Mill	Type	Location
Juan de Villoria	Samate	I	Higüey
	Santi Espiritus	I	Cazuy
Trejo brothers	La Magdalena	I	Azua
Almirante		I	Río Ibuaca
Benito de Astorga		I	Río Ibuaca
Pedro Serrano and Francisco del Prado	San Antón del Valle Hermoso	I	Santo Domingo
Alvaro Caballero		I	Santo Domingo
Pedro Vázquez		I	Haina
Francisco de Tapia		I	Haina
Cristóbal Lebrón	Arbol Gordo	I	Buenaventura
Juan de Ampiés		I	Nigua
Miguel de Pasamonte		I	Nigua
Francisco Tostado		I	Nigua
Francisco de Tapia		I	Nigua
Diego Caballero		I	Nigua
Lope de Bardeci		I	Nizao
Alonso Dávila		I	Nizao
Miguel de Pasamonte		I	Nizao
Alonso Gutiérrez de Aguilón		I	Nizao
Diego Caballero	Zepizepi	I	Ocoa
Hernando Gorjón	Santiago de la Paz	I	Azua
Alonso Zuazo	La Veracruz de Ocoa	I	Ocoa
Hernán Velázquez	Santiago	I	Santo Domingo
	San Cristóbal	I	Santo Domingo
Lucas Vázquez de Ayllón; Francisco Ceballos	San Marcos	I	Puerto Plata
Alonso Zuazo		I	San Juan de la Managua
Juan de Soderin	Santi Espiritus	I/T	Azua
Juan de Villoria	Sanate	I	Azua
Antonio Meléndez		T	Santo Domingo
Alvaro Caballero	San Sebastián	I	Haina
Bautista Justián		T	Santo Domingo
Gómez Hernández		T	Santo Domingo
Alvaro Caballero		T	Santo Domingo
Alvaro Caballero	La Concepción	T	Haina
Alvaro Caballero	San Cristóbal	T	Nigua

TABLE 4.2. *continued*

Owner	Mill	Type	Location
Cristóbal Colón		T	Santo Domingo
Diego Caballero		I	Santo Domingo
Alonso de Peralta		I	Azua
Ruy Díaz Caballero		I	Azua
Garcia de Escalante		I	
Doña Inés de Fuentes		I	
Doña Leonor de Tapia		I	
Señora de Astorga		I	
Diego de Aguilar		I	
Baltasar García		I	
Melchor de Torres	La Trinidad	I	San Juan de la Maguana
Melchor de Torres	Santa Bárbola	I	Higüey
Melchor de Torres		I	Azua
Hernán Sánche Alemán		I	
Lope de Bardeci		I	
Lorenzo Solano		I	
Diego de Herrera		I	Haina
Melchor de Torres		T	
Lic. Estévez		I	
Hernando de Hoyo		I	
Tomás Justinián		I	
Pedo Vázquez de Ayllón		I	
Diego de Guzmán		I	
Martín García		I	
Juan Caballero de Bazán		I	
Francisco Caballero		I	
Doña Catalina de Velázquez		I	
Juan Bautista de Berrio		T	Azua
Juan Bautista Gómez		T	
Antonio Meléndez		T	
García de Aguilar		T	
Juan Mosquera		I	
Juan de Vadillo		I	
Don Luis Colón		I	

Note: I = *ingenio*; T = *trapiche*.

large constructions on their properties. By the 1520s a dam seventy meters long, one of the largest of its time, was constructed to provide waterpower on an *ingenio* near Santo Domingo on the Isabela River. The wealthy Castilian merchant Benito de Astorga, who spent 12,000 castellanos and some seven years in its construction, due in part to the technical mistakes made in the building of the dam, began this mill in 1525.[39] Diego Caballero reported having spent some 15,000 ducats on his *ingenio* San Cristóbal on the Nigua River, but this outlay included the construction of sixty stone houses on the property as well as the purchase and seeding of extensive lands for the feeding of the slaves.[40] We do not know the amount spent by Lic. Estévez, an *oídor* of the *audiencia* of Santo Domingo, in the construction of his mill, but it included a stone structure one hundred meters long by ten meters wide that was also used to warehouse the sugar, as well as a stone building sixty by fifteen meters that housed the slaves and also served as the milling house. Such buildings indicate the increasing size of the island's *ingenios* and the rising level of expenditure in their construction.[41]

The cost of *ingenio* building seems to have increased as the sugar industry grew and consolidated itself. The price of land, cattle, and slave labor rose sharply and by the end of the second half of the sixteenth century, there was a race to purchase land. Undoubtedly, it was at this point that the great colonial latifundio began to take form.[42]

The colonists had from the outset of settlement sought to acquire great extensions of land, especially those near Santo Domingo and those best for the growing of sugar, but the Crown had limited them. Once it realized the interest of the sugar planters in taking lands, the Crown had ordered that "they be given only the lands and waters that they might need, giving to each that which he can make use of."[43] The indebtedness of the sugar planters grew in proportion to the growth of the industry, although they often exaggerated their debts in order to gain further grants and aid from the state or to have the state excuse them from their debts. They were to some extent successful. In 1529 a royal provision prohibited creditors from foreclosing on debts incurred in building *ingenios* by sugar planters in both Española and Puerto Rico.[44]

Even when the costs of *ingenio* construction had not been high, many colonists had created partnerships (*compañías*) to meet the expenses and to limit the risks involved. The first to form such an association for the manufacture of sugar were Miguel de Ballester and Alonso Gutiérrez de Aguilón, who did so in 1503. Another such group was formed in 1516 by Gonzalo de Vellosa, the Cristóbal brothers, and Francisco de Tapia. The practice became common among sugar producers, especially after 1519 as the building of mills intensified.[45] One of the characteristics of these *compañías* was their limited duration, often considerably

The earliest representation of a sugar mill in the New World shows various stages of production, from the cutting to the process of clarification and the filling of the forms. In the distance is a waterwheel probably driving a two-cylinder press in the shed; in the foreground is an edge runner press. From Theodor de Bry, Americae, pars quinta (Frankfurt, 1595). Courtesy of the John Carter Brown Library at Brown University.

shorter than the lives of those who entered into them. Usually one of the partners ended up with the whole *ingenio*. Such was the case in the association formed by Gonzalo de Vellosa and the Tapias in which after a short period, the brothers ended up with the entire estate. This process was due to the early investors' insecurity and lack of confidence in this incipient mercantile enterprise. Those with more limited resources tended to sell out their interests and were unable to wait to see the full result of their investments.

Ingenios *and* Trapiches: *Production Units and Their Technology*

There were two types of sugar mills. Las Casas and Oviedo, the first authors to mention the two types of sugar mills, agree in their definitions and in the fact that water-powered *ingenios* were more productive than the animal-driven *trapiches*. Both types utilized the same basic system of wheels, gears, and presses, but the equipment of the *ingenios* being larger demanded a greater motive force and thus the need for water power, while the smaller *trapiches* could be moved by horses or ox teams. Las Casas related that the first *trapiche* was built by Gonzalo de Vellosa, who in 1516 made "what is called a *trapiche*, a mill that is moved by horses where the sugarcane is pressed and from which they take the sweet juice that makes the sugar."[46] I believe that it was this same Vellosa who, after having constructed the *trapiche*, modified the part that moved the gears of the mill for something larger and more powerful. This is probably the meaning of the royal order sent in 1518 to Lic. Rodrigo de Figueroa, who was at that time *juez de residencia de Española*, which congratulated Vellosa for "having invented an *ingenio* to make sugar."[47] This invention must have been a great innovation, since according to the residents of Santo Domingo it spread so rapidly across the island. I believe this to be the *ingenio* of vertical cylinders, for after that date all the mills constructed made use of this system. In reality, unlike the wooden presses used by Aguilón and Ballester at La Concepción de la Vega, which mashed the canes, both the *trapiche* and the *ingenio* with the new mechanism squeezed the canes between the vertical cylinders and extracted almost all the juice from them. After Vellosa's invention no other change or innovation was made in the milling system of the *ingenios*.[48]

The water-powered *ingenios* being larger productive units required more capital and tended to be constructed by wealthier colonists, although the various grants and benefits offered by the Crown inclined many persons to attempt the construction of mills.[49] While the milling system remained the same through the sixteenth and seventeenth centuries, some innovations were made in other aspects of sugar production. The wealthy hacendado Hernando Gorjón, using Vellosa's hydraulic system, adapted it to an undershot mill in which the water passed underneath the wheel instead of being conveyed to it overhead. Many, particularly the wealthiest mill owners, such as Alonso Zuazo, Francisco de Tapia, Diego Caballero, and Catalina Velázquez, quickly adopted this system. Gorjón's system eliminated some of the most common risks and accidents of the former method and also the need to construct large tanks and aqueducts to bring the water to the mill wheel.[50]

Another change came a few decades after Gorjón's innovation. This was the idea of Francisco de Acosta, who designed a new system for the arrangement of the kettles that contemporaries said "has not been used before and that will use only half the firewood that has been used until now to produce the same [amount] of sugar with the same effect, and one saves the costs that the others have."[51] Although we do not know the exact nature of this invention, we know it was used by a great number of the island's sugar producers.[52]

Early planter residence on Española. Now known as ingenio Engombe, *it was probably called Santa Ana in the sixteenth century when it was first constructed. It was the property of the powerful Justinián family, the Genoese, who were among the first on the island to trade in sugar. Photograph by Stuart B. Schwartz.*

It seems clear that the technical changes in the island's sugar industry significantly increased production, but not all the producers were, for economic reasons, able to employ the new technologies. The lack of capital forced some sugar planters to sell their properties while others chose to scale them down to more modest and less expensive *trapiches*.

The second half of the sixteenth century witnessed a growth in these smaller units. One of the advantages, which the animal-powered *trapiches* offered, was their ability to operate throughout the year, even in times of drought, unlike the *ingenios*, which depended on the natural current of the rivers. The *trapiches* only had to stop due to accidents or shortage of cane or firewood while the complicated system of water control at the *ingenios* made them subject to problems such as damage to the holding tanks and aqueducts or breakage of the waterwheel.[53] It was not uncommon for owners of *ingenios* to also own a *trapiche*, which could continue to operate in case of accidents or damage at the waterwheel. This was the case of the *contador* Alvaro Caballero, the wealthy planter Melchior de Torres, and Alonso Zuazo, all of who could be counted among the principal producers on the island.

Levels of Productivity

Sugar production depended on three fundamental factors. First, the equipment of the mill, including the kettles, teaches, and sugar forms, as well as the mill itself, was a key element. Here we can also include the specialists and technicians needed for its operation, especially the *maestros de azúcar* who managed the operation. Second, the land, its value depending on its location and its qualities, was, of course, essential. Third, the slaves and the resources available for their subsistence made the operation possible.

The technicians who worked on the sugar estates were relatively few in number. The first were apparently brought from Madeira and the Canary Islands in the second decade of the sixteenth century. The skills of sugar masters, kettle men, and purgers were especially appreciated, but because the early planters lacked the resources to maintain a strict division of labor, it was common to find the specialists, especially the sugar masters, performing other tasks at the mills.[54] Slaves sometimes acquired the necessary skills to do the specialized jobs, and some became sugar masters, but this practice was not generalized. Despite the proliferation of slaves in the later sixteenth century, whites continued to occupy these crucial positions.[55]

Sugar mills employing specialists had certain advantages. With them, they produced more and better sugar and were even able to make good sugar from the *remieles* or skimmings of the kettles, which resulted from the clarification process. Although the specialists were paid substantial sums for the work they performed, they also shared some of the uncertainty of the process since they were responsible for and had to guarantee the quality of the sugar produced, just as had been the practice in the Canaries.[56] Sugar planters who used slaves for these tasks had no recourse if the harvest was lost or the sugar of poor quality except to punish them, something that was rarely done.

The location, extent, and quality of the land determined a sugar estate's possibilities. *Ingenios* had to be built on lands near rivers to take advantage of waterpower for the mills and for irrigation of the cane fields. Those mills near the port of Santo Domingo, the principal center of export, were also advantaged because of the ease of transport. *Ingenios* were concentrated on the rivers Haina, Nizao, Nigua, and Quiabon. Sugar specialists believed that the best land for sugar on the island were found in the San Juan valley and that these lands produced the island's best grade of sugar.[57] Although the Crown at first had granted lands to anyone willing to build a sugar mill, as the industry took form this policy changed under the pressure of sugar planters who demanded large grants of the best lands for themselves.

Access to firewood was another consideration related to the problem of land. The mills needed enormous quantities of fuel in the heating and clarification process, and thus access to forests provided considerable savings to sugar producers. In the mid-sixteenth century it was calculated that to process one *tarea* (a days worth) of land planted in cane, twenty to twenty-five carts of firewood were required, each cart with between sixty and seventy *arrobas* of fuel, or 739 kilograms of fuel per day. It was also estimated that each cart of fuel was equivalent to forty *arrobas* (about 450 kilograms). A well-managed mill could process six *suertes* of cane in a year, for which it would need about 4,000 cartloads of firewood. This situation caused considerable deforestation as well as speculation in the sale of firewood and in the price of lands close to forested areas. Those who could not buy firewood sent their slaves to cut it. Of twenty slaves employed in this activity, only half actually cut the wood; the rest transported it to the mill. While the price of firewood was not great, the cost of getting it was a constant burden on the sugar producers.

All the factors we have outlined so far influenced production in one way or another, and we can ask how they affected productivity. To answer this question, however, we first must examine levels of production on Española in the sixteenth century, but this task is not easy given the incomplete nature of the historical record. Thus we will have to make some estimates based on the general figures that are available.

The chroniclers Gonzalo Fernández de Oviedo and Bartolomé de Las Casas offer some observations about sugar production. Oviedo stated that the annual production of a large *ingenio* was some 5,000 *arrobas*, while a *trapiche* produced about half that amount.[58] Las Casas, with less precision, stated that a water-powered *ingenio* could produce the same amount of sugar as three *trapiches*.[59] I believe both statements are true, although they have to be explained. The sugar mills produced various types of sugar. Oviedo did not specify the type of sugar he was referring to. I believe his reference to 5,000 *arrobas* was only to white sugar or those sugars on which the tithe was paid.

To better understand production levels we must review the system of cultivation. Lands were prepared by weeding, burning the vegetative cover, and cleaning. Teams of oxen then plowed the land, the number of *rejas* (shares) of the plows contributing to the success of the planting. Most plows on the island were equipped with four *rejas*, but plows with five or six were not unknown.[60] This was followed by the planting of the cane pieces in the furrows and their covering with soil; a task generally done by slaves known as *sembradores* (seeders). Once the cane had begun to put out its first shoots, it was weeded (*aporcada*) two or three times before it had become strong. This had the effect of lessening infestation by rats, which often presented a considerable threat to the planted cane.

Once the canes had begun to grow, the planters waited until they were *en sazón*, ready to be cut. This depended greatly on the quality of the land and on the treatment that the cane received from the *cañaveros*, who were responsible for the cultivation process. According to specialists of the time, the best lands were situated in the valleys of San Juan de la Maguana, Puerto Plata, La Yaguana, Azua, and some of the areas around the city of Santo Domingo. Here the cane could be cut in ten months. Elsewhere, longer periods were necessary.[61]

The time when the cane was actually cut also depended on a variety of factors. Generally, water-powered *ingenios* had to wait until the rainy season to take advantage of the river currents. Wealthier planters who could build holding tanks were able to mill cane throughout the year. Those who milled cane in the rainy season suffered the disadvantage of lower sucrose content in the cane.[62] Most planters milled the cane from December to February, the months of heavy rainfall.[63]

The question of production levels is also complicated by the lack of sources and their often fragmentary or contradictory nature. The few register books (*libros de cuentas*) that survive are incomplete and thus estimates must depend on scattered observations. Although these come from people linked to the industry and thus knowledgeable about it, they must be used carefully because these men were also not disinterested observers. Thus I have used two measures to estimate production: (1) an areal estimate of production per *suerte* of planted cane, and (2) the number of sugar loaves produced in relation to the kettles. The similarity of the results from these two measures provides some confidence in my estimates.

According to the register books, a *suerte* of cane of 6,500 montones, harvested and milled in its season at a large *ingenio*, would yield about 1,200 *arrobas* or about 13.6 tons of white sugar. Conditions and capacity of the mill would determine the length of time needed to mill this cane. Juan de Palencia, mayordomo of the Columbus mill, said a week was necessary to mill a *suerte*, but Juan del Valle, another mayordomo at the same mill, said a month was needed.[64] Both, however, agreed on the amount of sugar that could be produced from an area this size. The milling year lasted about six months, during which about 180 to 200 *tareas* or approximately six *suertes* could be milled. The merchant Anton de Torres, *vecino* of Santo Domingo, stated in 1535 that a well-endowed *ingenio* could mill 200 *tareas* a year, from which we can infer that a mill in the best conditions would produce 8,000 *arrobas* of white sugar.[65] This, of course, was optimum production. Francisco Gómez, manager of Lope de Bardeci's *ingenio*, claimed that the island's best *ingenios* could make over 5,000 *arrobas* a year of white and *quebrado* sugar without counting lower grades (*espumas* and *panelas*) and that animal-powered *trapiches* were capable of making 3,000 *arrobas* of white sugar plus the lower grades.[66] These figures fall into the range suggested by Oviedo.[67]

Following Las Casas, we assume that an *ingenio* could produce three times as much sugar as a *trapiche*. Other contemporary sources from the industry put *trapiche* production at about 4,000 *arrobas* a year.[68] This calculation was supported by Alonso de Monsalve, who claimed that for every 1,000 *arrobas* of white sugar, an *ingenio* would produce 3,000 to 4,000 *arrobas* of lower grades. We have an annual average of over 24,000 *arrobas* a year, at least in the best managed and equipped mills.[69]

We can also establish production levels by another method. Sixteenth-century registers of *ingenio* operations indicate that a *suerte* of cane produced 270 kettles of cane juice (*melaza*), from which 2,200 loaves of sugar, each weighing 1.8–2 *arrobas* could be made, or approximately 4,000 *arrobas*. *Trapiches* would be able to extract 210 kettles of juice from a *suerte* or 1,680 loaves, an amount equivalent to 3,360 *arrobas*. If a *trapiche* produced one-third the sugar of an *ingenio* or two *suertes* a year, it would have an annual production of 6,000–7,000 *arrobas*.

The maximum levels of production on the island were reached in the second half of the sixteenth century at a moment when there were about twenty-five mills operating and the island was exporting to Seville about 100,000 *arrobas* a year, not counting sugar marketed or exported directly by the producers, contraband, and sugar sold locally, all of which multiplied the real levels of production.[70] By 1589 export had fallen to only 20 percent of former levels, but this did not necessarily indicate a fall in production as much as a rise in contraband carried out by the Dutch, English, French, and others.

Productive Forces and Levels of Development

The growth of slave labor on Española during the first decades of the sixteenth century resulted from two fundamental factors: the large numbers of indigenous slaves introduced from other regions and islands, and the avalanche of Africans. The first group replaced the indigenous inhabitants of the island, and the Africans eventually came to represent a new social synthesis there. Although it is difficult to speak of the precise numbers of Indian slaves brought to the island during the height of this trade, between 1514 and 1519, a period marked by a smallpox (*viruelas*) epidemic on the island in 1518, probably a minimum of 10,000 slaves were introduced. We know that between 1514 and 1524, eighty registered slaving expeditions (*armadas*) sailed from Santo Domingo, to say nothing of those that escaped official notice.[71] The imported workers, along with the native inhabitants, served as the basis for the early growth of the industry and continued to predominate into the middle of the 1520s, far outnumbering the few Africans who began to appear as workers.[72] During the 1530s, when the indigenous population dropped by about 90 percent, there were nineteen *inge-nios* and a few *trapiches* operating on the island. At this time the slave population did not exceed 2,500, of which 200 were Indians.

The slave trade presented a series of problems, not the least of which were the conflicts between Sevillian merchants who wished to control the transoceanic trade and the merchants and planters on the island who objected to their disadvantages in competition with the Europeans. While blacks had been present from the early stages of colonization, only at the close of the 1520s did they become common, as small-scale production with indigenous laborers was replaced by the intensive use of slave labor. By 1540 about 15,000 Africans had arrived, most of them for use on sugar estates, although some were also employed in mining, truck farming, and domestic service, especially in Santo Domingo, where about 2,000 worked.[73]

The size of slave holdings varied considerably. Although an average of 100 slaves per mill was common, there were *ingenios* with over 300 slaves and some slave owners held over 1,000 slaves, divided among various properties. Melchior de Torres in 1577 had over 1,000 slaves divided among his three mills with his *ingenio* Santa Barbola alone holding 370. In addition, he had other slaves on his ranches and in domestic service. Alvaro Caballero, one of the most powerful planters, had over five hundred slaves on his sugar properties, 200 on "La Concepción de Nuestra Señora" and 150 each on "San Cristóbal" and "San Miguel de la Jagua." He held other slaves on his other farms and enterprises. These owners were, of course, exceptional, but the potential size of a large sugar operation is underlined by these figures, which signal the full establishment of the plantation regime of production.[74]

The increasing use of slaves on the sugar plantations caused a rise in slave prices. During the decade of the 1520s the price of an African male slave oscillated between forty and fifty *castellanos*, while that of an Indian slave was ten to fifteen, although some were priced at twenty *castellanos*.[75] In the second third of the sixteenth century as the sugar economy matured, the price of black slaves doubled and at times reached as much as 200 *castellanos*. The principal sugar planters reacted by complaining against the extortion and control of the labor supply by the Sevillian merchants and traders.[76] In turn, the merchants claimed they were forced to raise prices because of the devaluation of the currency in this period, which during the 1530s lost 10 percent of its value.[77] Because of a shortage of specie, planters preferred to pay for the slaves in sugar, which often led to even wider profit margins in the merchant's favor who sometimes asked for 130 *arrobas* of this committed or *lealdado* sugar.[78]

The plantation system reinvigorated the declining fortunes of the island. This was especially the case in those sectors of the economy most turned toward Europe, such as the production of sugar and hides. Over an expanded network of roads that linked the cane fields to the island's principal port, these goods moved to the city of Santo Domingo, and from there to Spain. The rise in this activity produced an increase in those agents who promoted and who benefited from the sugar economy; slave traders, merchants, and suppliers of Spanish products. It especially benefited small merchants and the muleteers, carters, and porters responsible for transport from the fields and on the docks. Whereas before ships from Spain had returned empty or with products of little value, the rise of the sugar economy coincided with a rise in sugar prices in Europe.[79] These conditions moved the municipal council of Santo Domingo to seek permission to sell the island's sugar directly to Flanders and the Canaries and bypass the Seville monopoly and the perceived stranglehold of that city's merchants.[80] The struggle between merchants and producers and the growing market for sugar in Europe now stimulated the local economy, which had been moribund since the decline in the pearl fishery and the expeditions to capture Indians. The rise of the plantation economy changed all this and now made Santo Domingo and its merchant community the center of activity in the region. The sugar elite was able to appropriate the already existing fleet of ships that had been used previously for other activities, and with them the sugar planters were able to export not only sugar and hides to Europe, but *casabe* (manioc bread), salt, corn, and cotton to other parts of the region.[81] Meanwhile, they continued to import wines, oil, flour, and clothing.

Sevillian merchants complained bitterly to the Council of the Indies and to royal officials against the control of the island's trade exercised by local merchants. They argued, for example, that shipmasters from the island were unqualified to sail across the Atlantic since they were uncertified by the Council of the Indies. This was an attempt to forestall an island-based competition to the Sevillian monopoly.[82] The Andalusian merchants responded by raising the price of the goods shipped to the island, producing margins of profit of over 100 percent. The town council of Santo Domingo reacted by imposing import taxes especially on wheat, which was now forced to pay a *castellano de oro* for each *arroba* unloaded in Santo Domingo.[83] To this challenge the Sevillians threatened to send neither ships nor goods to the island. Their representative, Juan de Loaysa, asked the Crown to insure that their goods could continue arriving in the island without imposts or intervention from the municipal council of Santo Domingo.[84] The island merchants answered that their Andalusian counterparts were unsatisfied with buying the local products for a song and selling them in Europe for excessive prices, but that on the irreplaceable foodstuffs they imported, they were not well satisfied to double their gain and even sought to increase it one hundredfold.[85] Gonzalo Fernández de Oviedo came to the island's defense. He accused the peninsular merchants of having destroyed the island and the city of Santo Domingo: "and when wine and bread were plentiful in Spain is when the merchants claimed in the Indies that it was scarce, and they did not hesitate to sell their goods and supplies at the highest prices. Because nothing arrives there except that which they have registered, and they know their costs and do not introduce a pipe of wine or of flour beyond what they see fit, there we never escape from need and hunger nor do they fail to make all they want."[86]

Of course the arguments on both sides were self-interested and must be evaluated with care. The Sevillian traders wished to hold their monopoly while the Santo Domingo merchants also sought to sell their products locally and faced serious competition from the better quality of Spanish imported goods. This competition had a negative effect on the local white population, the major consumer of the imported goods. The Andalusian merchants reduced the number of ships sailing to the island and also put restrictions on the sale of goods beyond the island, a measure that also affected local merchants who resold goods for gold and silver in Tierra Firme, Río de la Hacha, Cartagena, Puerto Rico, Cuba, and elsewhere. While the Crown responded at times to the demands of the Sevillian merchants, it also opened the island's trade to ships sailing directly from the Canaries, at least until pressure from the Sevillians stopped this trade in the mid-1540s.[87]

In this struggle local merchants were disadvantaged. First, the Sevillian merchants were able to determine the conditions of trade and the island's economy because it remained so dependent on one product, and sugar offered few alternatives in competition with the variety of goods arriving from Spain. Then too, the Sevillian merchants had an upper hand since the sugar economy was also dependent on capital from the metropolis.

How did this commercial crisis affect the sugar economy? First, we must recognize that if this situation had a negative effect on the export of the island's principal products. The people who suffered the most were not the principal producers, but rather middling and small-scale settlers resident on the island. As we have seen, the owners of the large *ingenios*, men such as Alonso Zuazo, Hernando Gorjón, the Tapia brothers, and the *contador* Alvaro Caballero, also owned the main means of transportation—ships, caravels, and galleons. More-over, few of them sold their products directly to Spain. Official registers indicate that a large proportion of the island exports were handled by Spanish inter-mediaries, usually small-scale merchants or *tratantes* who had taken up resi-dence on the island and who often sold the sugar well above the regulated price. The majority of the island products were, in fact, carried in privately owned ships or smuggled out as contraband.

Crisis and Decadence of Sugar

In general terms the decline of the sugar industry was not an isolated fact, but part of a systemic crisis. The plantation sector began to decline gradually and then fell sharply by the end of the century, but by the 1550s some owners of *ingenios*, unable to maintain large operations, had converted their mills to mod-est *trapiches*.[88] Even though the most powerful sectors of the island sought to change this situation or modify its effects, little could be achieved within the limitations of the slave system, and most of all, because the principal problems were external to the island itself. Attempts at diversification like the cultivation of ginger, which was introduced in the 1550s, produced few positive results.[89] Market conditions and shipping problems impeded its success. Moreover, its cultivation had broad social implications since it did not call for large capital outlays or large slave forces. It permitted a broad spectrum of producers to operate and may have stimulated the growth of a peasantry on the island for the first time, but in any case its cultivation was short-lived. Local investors seeking ways of salvaging the economy turned to other activities such as a trade in hides. Without any control by the producers, this commerce led to the indiscriminate slaughter of cattle, including cows that had just given birth and those about to deliver. The result was a meat shortage in the urban butcher shops, while in the countryside the meat was fed to the pigs. The Audiencia was forced to intervene in an attempt to stop these practices.[90]

The continuing and worsening crisis of the island's economy limited the ability of the residents to make purchases or to restock their labor supply on the sugar estates. The cost of a slave in the second half of the sixteenth century rose to 350 pesos de oro, a prohibitive price for most producers. This caused a reconsideration of plans to capture and enslave Caribs as an alternate to the increasingly expensive Africans, who were not only costly to acquire, but who also carried a tax of thirty ducats each upon arrival on the island.[91] The shortage of labor on the sugar estates became acute, and some mills operated with only half the needed labor force. This was reflected in the fall of production. Sugar planters were also confronted by shortage and the rising costs of subsistence crops such as maize, cassava, and cazabe. Planters faced the choice of permitting slaves time to grow their own food or intensifying slave activities in sugar production and thus hoping to create the margin of profit needed to buy provisions. The latter strategy depended on the existence of a class of small-scale suppliers who provided foodstuffs to the cities. This dependence was disruptive to the social organization originally conceived by the Crown, which had envisioned each sugar mill as a self-sufficient unit and presumably a class of small-scale producers who would provide for the urban areas. Complaints and conflicts developed between competing urban consumers and the sugar planters, a situation made even worse at the end of the 1560s by an infestation of worms that attacked the maize and cassava crops. Hunger set off a desertion of the countryside. The president of the Audiencia remarked that if there had not been a prohibition on leaving the island without permission, the island would have been abandoned. The archbishop of Santo Domingo noted that places that had held a population of 500 *vecinos* now had no more than thirty inhabitants.[92]

The crisis was also made worse by other conditions both internal and external to the island. The low value of the island's currency moved merchants to stop importing goods or to charge exorbitant prices. They also raised the costs of shipping, especially on sugar and hides, and those planters without their own ships were sometimes forced to pay fourteen to fifteen ducats for each crate of sugar and seven *reales* for the export of each hide. Some merchant firms charged as much as half the value of the cargo as a shipping fee and found other means to extract exorbitant profits.[93]

In addition, there was now increased competition in the European market for sugar with the rise of Brazilian production and that of Granada and the Levantine coast of Spain. About the competition from Brazil little could be done, but the Audiencia of Santo Domingo petitioned Phillip II in 1584 to protect the sugar producers of Española by prohibiting the sale in Andalusia of sugar made in Granada and Almería. As a result of such pressures, the *cabildo* of Seville removed all duties on sugar and other goods coming from the island, and eventually the shipping taxes were cut in half.[94] These measures did have some effect and there were years when the tax yield exceeded 20,000 *castellanos*, enough to cover the salaries of the royal functionaries and provide for repair of the fortifications, but in the long term these measures could not stem the decline of the island's economy.[95]

Clearly the fall of the sugar economy was not an isolated fact, but part of a general crisis of the colonial system of the island. Since sugar was the principal activity on Española by this time, its fate determined that of the island's economy. Had the local and international economic contexts been more positive, the plantation system associated with sugar would have remained in place much longer on the island.

NOTES

Abbreviations

AGI   Archivo General de Indias (Seville)
AHMS   Archivo Historico Municipal de Sevilla

1. Roberto Cassá, *Historia social y económica de la Republica Dominicana* (Santo Domingo: Alfa y Omega, 1987), 1, 65.

2. Pierre Chaunu, *Sevilla y América en el siglo XVI* (Sevilla: SEVPEN, 1983), 74–80.

3. The first horses were brought by Governor Nicolás de Ovando in 1502. In his fleet of that year he brought fifty-nine horses and six oxen. AGI, Contratación 3250.

4. See Justo L. del Río, *Caballos y équidos españoles en la conquista y colonización de América (siglo XVI)* (Seville, 1992), 110–40.

5. Small-scale agriculture had been initiated by Ovando during his rule, although he maintained a monopoly over these activities. See Genaro Rodríguez Morel, "Poder y luchas políticas en la Española, 1502–1514" (unpublished).

6. According to Pedro Martir de Anglería, the canes planted at La Isabela grew rapidly in the first two weeks. See *Las decadas del Nuevo Mundo* (Madrid: Polifemo, 1989), *decada primera*, 30.

7. Bartolomé de las Casas, *Historia de las Indias* (Santo Domingo, 1985), lib. 3, cap. 129, 273–74.

8. Mervyn Ratekin, "The Early Sugar Industry in Española," *Hispanic American Historical Review* 34, no. 1 (1954): 4.

9. José Pérez Vidal, "Catalanes y valencianos en la propagación de la industria azucarera," in *América y la España del siglo XVI*, ed. Francisco de Solano and Fermin del Pino, 2 vols. (Madrid: CSIC, 1983), 2:305–14.

10. Estéban Mira Caballos, "La economia de la Española a través de las cuentas del tesorero Cristóbal de Santa Clara, 1505–1507," *Ibero Amerikanischit Archiv* 24, nos. 3–4 (1998).

11. Lic. Vellosa was among the first doctors and surgeons of the city of Santo Domingo. He arrived in Ovando's fleet and he received a grant of 100 Indians for his services in 1511. Later, in the *Repartimiento de Albuquerque* of 1514, he received an encomienda of thirty-one Indians and their chieftain. It is likely that his knowledge and experience with sugar was owed to his relationship to his wife's family. His wife, Luisa de Betancourt, was from one of the traditional sugar families of the Canary Islands and Madeira. On the first sugar technicians to arrive on Española, see Fernando Ortiz, *Los primeros técnicos azucareros de América* (Havana), 1–21.

12. *Licenciado* was the title used by a person holding a degree, often in law, roughly equivalent to a master's degree in this period.

13. Rodríguez Morel, "Poder y luchas políticas."

14. Charges made against Lic. Pedro Vazquez in the investigation (*residencia*) made by Alonso Zuazo in 1519. See AGI, Justicia 45, fls. 227–39.

15. Rodriguez Morel, "Poder y luchas políticas,"

16. *Real cédula* sent to Cristóbal de Tapia, Valladolid, 2 March 1518, AGI, indiferente general 419, libro 7, fs. 40v–41.

17. Instructions given to Lic. Rodrigo de Figueroa, Zaragoza, 9 December 1518, AGI, indiferente general 419, libro 7, fs. 146v–49. See also Irene A. Wright, "The Commencement of the Cane Sugar Industry in America," *American Historical Review* 21, no. 4 (1916): 757–58.

18. *Real cédula* to the governor of the Canary Islands, Barcelona, 16 August 1519, AGI, indiferente general 420, libro 8, 249–50; *Real cédula* to don Luis de Carroz, ambassador to Rome, Barcelona, 14 September 1519, AGI, indiferente general 420, libro 8, fs. 139–40. This document includes a letter of the king to the Pope.

19. *Real cédula*, Barcelona, 23 September 1519, AGI, indiferente general 420, libro 8, fs. 145–46.

20. *Real cédula* sent to Lic. Antonio Serrano, *regidor* of the city of Santo Domingo (Barcelona, 27 October 1519), AGI, Indiferente general 419, libro 8, fs. 153v–54.

21. *Real provisión* sent to D. Diego Colón, Burgos, 11 April 1521, AGI, indiferente general 420, libro 8, f. 285; Ratekin, "Early Sugar Industry," 11.

22. *Real cédula* sent to Miguel de Pasamonte, Valladolid, 9 June 1520, AGI, indiferente general 420, libro 8, fs. 236v–37. A note on currency: In this period a *castellano* was worth 8 *reales* of silver while a *ducado* or ducat equaled 11 *reales*.

23. There are cases of persons without sufficient collateral who had to seek a bondsman to provide security in the case of failure to repay the debt. This was the case of Diego Franco, who sought out Juan de León as his bondsman for a loan of 200 pesos.

24. *Real cédula* to the royal officials of Española, Valladolid, 9 June 1520, AGI, indiferente general 420, libro 8, fs. 236v–37.

25. Alonso Fernández de las Varas was among the first to benefit from this license. He imported copper tools from Flanders for the mill he had set up.

26. *Real cédula* to Diego Colón, Burgos, 6 September 1520, AGI, indiferente general 420, libro 8, f. 319; *Real cédula* to Juan de Orihuela, Burgos, 6 September 1520, AGI, indiferente general 420, libro 8, f. 320.

27. He had received the original aid from the Jeronymite friars. See also *Real cédula*, La Coruña, 17 May 1520, AGI, indiferente general 420, libro 8, fs. 251v–52v.

28. *Real cédula*, Valladolid, 26 June 1523, AGI, indiferente general, libro 9, f. 151.

29. *Real provisión* issued to Lic. Cristóbal Lebrón, juéz de residencia of Española, Granada, 20 June 1526, AGI, indiferente general 421, libro 11, fs. 24–3l.

30. Ibid.

31. AGI, Contaduria 1050, f. 243.

32. Rodrigo de Figueroa to Charles I, Santo Domingo, 14 November 1520, AGI, Patronato 174, ramo 19.

33. The Crown's protection of the nascent sugar industry extended to Seville, where a royal provision of 1525 ordered the town council to charge no taxes on sugar arriving from Española. See AHMS, Carpeta 24, n. 15, Madrid, 11 January 1525.

34. López Cerrato to the king, Santo Domingo, 10 January 1548, AGI, Santo Domingo 77, ramo 3, doc. 115.

35. Cerrato to the king, 12 September 1544, AGI, Santo Domingo 77, ramo 5, doc. 134.

36. On this subject, see Ramón Carande, *Carlos V y sus banqueros*, 3 vols. (Barcelona: Crítica, 2000).

37. *Historia general y natural de las Indias* (Madrid, 1992), 1, chap. 8, 106–7.

38. Contract made between Pedro Vázquez and Diego de Morales, Santo Domingo, 22 November 1519, AGI, Justicia 45, fs. 149–53.

39. Bento de Astorga to the king, Santo Domingo, 13 May 1532, AGI, Santo Domingo 77, ramo 3, doc. 66. See also Justo L. del Río Moreno, *Los inicios de la agricultura europea en el Nuevo Mundo, 1492–1542* (Seville, 1991), 346–47. On the expensive dam built by Esteban de Pasamonte, see AGI, Santo Domingo 74, ramo 1, doc. 4; and Wright, "Commencement of the Cane Sugar Industry," 760–63.

40. *Real cédula* to Audiencia of Santo Domingo, Valladolid, 26 February 1538, AGI, Santo Domingo 868, libro 1, fs. 113–14.

41. Estévez to the king, Santo Domingo, 4 July 1563, AGI, Santo Domingo 71, libro 1.

42. Genaro Rodríguez Morel, "La formación de la propriedad territorial y los grandes latifundios en Española. siglo XVI" (unpublished); Francisco Moscoso, "Propriedad y pastos comunes el la Española, 1541–1550," paper presented at the Seventh Dominican Congress of History, Santo Domingo, 1995.

43. "Solo se les dieron las tierras y aguas que hubieren menester dando a cada cual . . . lo que justamente bastaba," AGI, 23 September 1519, indiferente general 420, libro 8, fs. 145–46.

44. AGI, Patronato 275, Toledo, 15 January 1529, ramo 1; AGI, Toledo, 15 January 1529, Justicia 21, n. 1, ramo 2.

45. We can mention those formed in 1519: Juan de Villoria, Isabel de Campusano, Melchior Centurión, and Esteban Justinián, Lic. Pedro Vázquez, and Diego de Morales; in 1520, Orduño Ordóñez and Alonso Gutiérrez de Aguilón; in 1521, Rodrigo de Figueroa and Hernando de Burende, Rodrigo de Figueroa and Juan de León. See Sidney Mintz, "Imagen y realidad en el paisaje antillano de plantaciones," *Paisajes del azúcar: Actas del quinto seminário internacional sobre la caña de azúcar, Motril, 20–24 de sptiembre 1993* (Granada: Diputación, 1995); and AGI, Justicia 45, fs. 149–58.

46. "Alcanzó a hacer uno que llaman trapiche, que es el molino o ingenio que se trae con caballos donde las cañas se estrujan o exprimen y se le saca el zumo melifluo del que se hace azúcar" (Las Casas, *Historia*, 272). I believe that the invention of the *trapiche* was earlier since by 1515 Gorjón as well as Aguilón had mills working in the area around Azua.

47. "Por haber inventado un yngenio para hacer açucar," *Real cédula*, 29 October 1518, AGI, indiferente general 419, libro 7, fl. 130.

48. I do not believe that the island's *ingenios* ever made use of millstones to press the cane.

49. Frank Moya Pons, *La Española en el siglo XVI, 1493–1520* (Santiago: Universidad Católica, Madre y Maestra 1978), 261.

50. Genaro Rodríguez Morel, "Cartas privadas de Hernando Gorjón," *Anuario de estudios americanos* 52, no. 2 (1995). The costs of Benito de Astorga had run so high because after two years constructing an aqueduct and tank, due to an error in its construction, the water did not rise high enough to make the system function properly. See AGI, Santo Domingo 77, doc. 53, ramo 3.

51. *Real cédula*, Madrid, 5 February 1562, AGI, indiferente general 426, libro 25, fs. 156v–57v. See also Fernando Ortiz, *Los primeros técnicos azucaceros de América* (Havana, 1955), 16.

52. Although Fernando Ortiz mentions the claim of Gabriel Lozada that he had invented a new machine to make mills, there is no evidence that this mechanism was associated with *ingenios* to produce sugar. See Ortiz, *Los primeros técnicos*, 16; and AGI, indiferente general 425, libro 23, fs. 215v–16.

53. Rodríguez Morel, "Esclavitud y vida social," 96.

54. Given the high cost of employing these technicians, few sugar mills could afford them. See the testimony of the sugar master at the *trapiche* of Pedro de Meléndez, Rodrigo Alvarez de Lorca, in AGI, Justicia 983, n. 3, 1574.

55. Among the Canarian technicians in this period, we can mention Juan Sanchez Trucha, *vecino* of Azua, Francisco Gómez, Juan Sánchez Morcillo, *vecino* of Tenerife, Pedro Dominguez, sugar master on the properties of Juan de Soderín, sugar master Rodrigo Alvarez de Lorca, and Juan Luis, sugar master on the *ingenio* of Luis Colón.

56. On sugar in the Canary Islands, see Manuel Lobo Cabrera, "Esclavirud y azúcar en Canarias," *Actas del seminario internacional: Escravos com e sem açúcar* (Funchal: Centro de Estudos Atlánticos, 1996); and Guillermo Camacho Pérez Galdós, "El cultivo de la caña de azúcar y la industria azucarera en Gran Canaria (1510–1535)," *Anuario de estudios atlánticos* 7 (1961): 11–71.

57. Rodríguez Morel, "Esclavitud y vida rural," 94.

58. Oviedo, *Historia*, 1, lib. 4, cap. 8, 107.

59. Las Casas, *Historia*, 274.

60. Rodríguez Morel, *Esclavitud y vida rural*, 6.

61. In the lands of Columbus fourteen or fifteen months were necessary. There were some planters, especially wealthier ones, who could afford to allow the cane to stand longer, allowing the sucrose to concentrate and thus produce more sugar in a single harvest.

62. According to sugar masters and *cañaveros*, cane cut in the dry season yielded more sugar.

63. This information comes from Pedro de Balboa, sugar master on one of the mills of the contador Diego Caballero. He also worked at one of the admiral's mills in 1535; see AGI, Justicia 14, n. 2, item 2.

64. Ibid., Libro de cuenta de Juan del Valle, AGI, Justicia 14, n. 1.

65. AGI, Justicia 14, n. 1.

66. Some planters preferred to produce lower grades despite their lesser value because they were not taxed as heavily. For every twenty-five *arrobas* of white sugar, one *arroba* was paid as a tithe and this caused producers to lower their output of white sugar, especially since the cost of producing lower grades was also less. The issue is further complicated by refining opportunities. At first, there were mechanisms for refining lower grades in Seville, but by

mid-century the *cabildo* of Seville had prohibited their refining in Andalusia. See the relevant documentation from the *cabildo* of Seville on the refining and sale of sugar in AHMS, sec. 1, fol. 39, n. 73. See also Justo L. del Río Moreno, "Refinería del azúcar en Sevilla: Siglos XVI y XVII," *La caña de azúcar en tiempos de los grandes descubrimientos, 1450–1550: Actas del primer seminario internacional," Motril, 25–28 de septiembre 1989* (Motril: Junta de Andalucía and Ayuntamiento de Motril, 1990).

67. AGI, Justicia 983, n. 2.

68. For example, the *trapiche* "La Magdalena" in the Azua area, owned by Alonso de Peralta, choirmaster (chantre) of the cathedral of Santo Domingo, and his nephew Alonso de Heredia, produced 4,000 *arrobas* a year, according to its register. This mill was administered by Damián de Peralta, Alonso's brother; see AGI, Justicia 112, ramo 4, n. 2. The same amount was produced by the *trapiche* "San Sebastián" of Alvaro Caballero, located on the banks of the Haina River.

69. Some members of the municipal council of Santo Domingo stated that the island's *ingenios* produced over 20,000 *arrobas* a year. See the letter from the king to the *procuradores* Alonso de Encina and don Antonio Manriquez, 25 March 1573, AGI, Santo Domingo 73, ramo 2, doc. 59a.

70. Melchior de Castro to king, Santo Domingo, 25 June 1543, AGI, Santo Domingo 77, ramo 5, doc. 130. The wealthy hacendado Alvaro Caballero sold his sugar within the area of the audiencia of Santo Domingo rather than selling it to Europe.

71. These numbers are suggested by Enrique Otte, *Las perlas del Caribe* (Caracas, 1977), and more recently by Esteban Mira Caballos, *El indio antillano* (Seville: Muñoz Moya, 1997).

72. In this period on the *ingenio* "Santi Espiritus," owned by a group of Genoese and Juan de Villoria, there were eighty-five workers, of which only a third were blacks. See Rodríguez Morel, "Esclavitud y vida rural," 92.

73. Information offered by the merchant, Juan Medina de Villavicencio, to the Board of Trade, Seville, 17 June 1548, AGI, Santo Domingo 2687.

74. Genaro Rodríguez Morel, "Dos plantaciones azucareras en la Española en el siglo XVI: Un análisis comparativo."

75. See the list of prices of Indians carried in the armada of Captain Diego de Ocampo in 1521, AGI, Contaduria 1050, n. 1, f. 245.

76. Cabildo of Santo Domingo to king, 31 May 1537, AGI, Santo Domingo 73, ramo 1, doc. 19.

77. See the exhaustive study by Cipriano de Utrera, *La moneda provincial de la Española* (Santo Domingo, 1951).

78. Lealdado sugar sold for five pesos per pound in Seville. See Cabildo of Santo Domingo to king, 10 October 1552, AGI, Santo Domingo 73, ramo 11, doc. 67.

79. See Earl J. Hamilton, *American Treasure and the Price Revolution* (Cambridge, Mass.: Harvard University Press, 1934).

80. The letter of 19 August 1530 was signed by a number of town councilors, such as Miguel de Pasamonte, Diego Caballero, Martin de Landa, and others, who were also sugar planters; see AGI, Santo Domingo 73, ramo 1, doc. 1.

81. On the owners of the privately owned ships, see Otte, *Las perlas del Caribe*, 102–3. On the fate of the caravels that came with Ovando's fleet, see AGI, Contratación 3250.

82. On this matter the officials of the Casa de Contratacción received a royal order informing them that the island-based pilots and masters had not been examined and certified for these positions. The document refers to the masters employed by Francisco Ceballos and Bartolomé de Monesterio, both of whom were *vecinos* in Puerto Plata; see AGI, indiferente general 1964, Madrid, 26 March 1545, lib. 10, fs. 16–16v.

83. Cabildo of Santo Domingo to king, 1 December 1531, AGI, Santo Domingo 71, lib. 2, f. 130.

84. AGI, Justicia 973, ramo 1, n. 4.

85. Ibid.

86. Rodríguez Morel, "Esclavitud y vida rural," 106.

87. Pressure from merchants from the Sevillian quarter of Triana had an impact on the Canarian trade with Santo Domingo; see AGI, Santo Domingo 49, ramo 3, f. 127.

88. Pedro Sánchez de Angulo for the Cabildo of Santo Domingo, 19 June 1558, AGI, Santo Domingo 71, libro 1, f. 170–170v.

89. Justo L. del Río Moreno and Lorenzo E. López y Sebastián, "El jenjibre: Historia de un monocultivo caribeño del siglo XVI," *Revista complutense de historia de América*, 18 (1992).

90. Lic. Estevez to Crown, AGI, Santo Domingo 49, ramo 3, doc. 147.

91. Memorial of the Audiencia of Santo Domingo, 10 January 1572, AGI, Santo Domingo 50, ramo 1, doc. 35.

92. Andrés de Carvajal to king, Santo Domingo, 8 May 1569, AGI, Santo Domingo 71, libro 2I, f. 408.

93. See Eufemio Lorenzo Sanz, *El comercio de España con América en la época de Felipe II*, 2 vols. (Valladolid: Diputación Provincial, 1986), 1:289–98; and Ruth Pike, *Enterprise and Adventure: The Genoese in Seville and the Opening of the New World* (Ithaca: Cornell University Press, 1966).

94. On Brazil, see Stuart B. Schwartz, *Sugar Plantations and the Formation of Brazilian Society: Bahia, 1550–1830* (Cambridge, Mass.: Cambridge University Press, 1985). On sugar in the Spanish Levante, see José Pérez Vidal, *La cultura de la caña de azúcar en el levante español* (Madrid: Consejo Superior de Investigaciones Científicas, 1973).

95. AGI, Santo Domingo 71, libro 2, f. 130.

CHAPTER FIVE

Sugar and Slavery in Early Colonial Cuba

Alejandro de la Fuente

According to a popular Cuban saying, without sugar there is no country ("sin azúcar no hay país"). What the adage does not state is that sugar was produced in the island well before any sociopolitical entity resembling a "country" existed. It is frequently forgotten that Cuba's spectacular rise to a prime world producer of sugarcane in the early nineteenth century was based not only on a favorable market conjuncture, due to the destruction of Haiti's productive capacity, but also on a long local tradition of production of high-quality sugars. According to British sources, by the early eighteenth century the island made "the best sugars in the West Indies."[1]

For the most part, modern historians have ignored this long productive tradition. We have very limited knowledge of sugar and slavery during the early colonial period. With a few notable exceptions, students of colonial Cuba have concentrated their research efforts on the period that corresponds to the rise and expansion of the slave-based, export-oriented plantation complex that developed on the island at the end of the eighteenth century.[2] If only by default, modern historiography has contributed to reproducing the old vision that Cuba's pre-plantation history is, in fact, prehistory. It is a vision that subordinates the very existence of the colony to its role as a supplier of sugar to the North Atlantic markets and that blends Cuba's own history with that of the system into which the island was inserted.

This lack of scholarship cannot be attributed to a couple of factors: scarcity of sources, and difficult access to existing sources. The early colonial period has been overlooked mainly because Cuba's postrevolutionary historiography has focused on the "precedents" of the revolution itself, none of which can be found in the first two centuries of colonial history.[3] The lack of serious empirical research about this period has led, in turn, to sweeping characterizations about slavery and sugar in colonial Cuba. Numerous authors have assumed that the plantation model typifies slavery in general, regardless of chronological, geographical, or socioeconomic factors.[4] It is as if sugar production generated, per se, a given set of fixed social and productive relations.

During the seventeenth century sugar was produced in Cuba under circumstances that were vastly different from those found in nineteenth-century plantations. Several institutional and commercial constraints prevented the early manufactures from becoming plantations, in the sense of productive units serving a highly competitive international market and driven by a permanent search for efficiency.[5] Rather than specialized industrial units, seventeenth-century mills were basically self-sufficient agricultural concerns that manufactured sugar in an artisan-like manner with a limited number of slaves.

A limited and irregular supply of slaves was one of the leading factors that explains Cuba's late entrance into sugar production. Despite the relatively early conquest by Spaniards in 1511, concrete evidence about sugar production on the island cannot be found until the 1590s. By this time, Havana had become a crucial maritime center of the Spanish empire and played an active role in the Atlantic world in the making, which was, to no small degree, a world of slaves and sugar. The challenge, however, is to explain not only why the production of sugar did not start in Cuba until the late sixteenth century but also to explain why a century later the island remained a modest producer. In contrast to Brazil or the British West Indies, sugar did not become the colony's cash crop in the seventeenth century. If official records are to be trusted, by the second half of the century other commercial crops, particularly tobacco, successfully competed with sugar in terms of value, while traditional products, such as wood and hides, still represented a large share of the island's total exports. It would take another century for sugar to be queen and displace all other economic activities.[6]

By looking at the early stages of sugar manufacturing in Cuba, this chapter seeks to fill a significant gap in the historiography of sugar and slavery in the Americas. The only other study devoted specifically to this subject is now over eighty years old.[7] After a brief discussion of the numerous attempts made to build sugar mills during the sixteenth century and the obstacles placed by the Spanish colonial system on the growth of this economic activity during the 1600s, this essay explores the main technological and productive features of these units, including their use of the slave labor force.

The Expansion of Sugar Production

Following the example of Española, where sugar production started in the late 1510s, the first European settlers of Cuba sought to build sugar mills since at least

1523. A *real cédula* of 1526 asserted that some mills were being constructed around Santiago, but later evidence demonstrates that none had been actually built. Indeed, when a royal official requested a license to build a mill in 1534, he assured the Crown that his would be "the first on this island." The same year Governor Gonzalo de Guzmán asserted that conditions were not propitious for sugar to prosper in the colony.[8]

The governor was right. By the mid-1530s the very existence of the colony was threatened not only by the disastrous decline of its indigenous population but also by its decreasing production of gold. Most of the initial European settlers had abandoned the colony, moving to more promising territories to the west and the south. Hernando de Soto's expedition to Florida in 1538 further contributed to the European exodus from Cuba. It has been estimated that between 1519 and 1539 about 90 percent of the indigenous population perished, while the total number of *vecinos* (heads of households) on the island declined from more than 1,000 to less than 200, an 80 percent drop.[9] African slaves had been imported since the 1510s, but their number did not offset the loss of the indigenous labor force. Thus, at the same time that sugar production was being introduced in Puerto Rico and was expanding in Española, Cuba lacked even the minimal conditions to initiate the construction of costly sugar mills.

Vecinos were fully aware that royal support was crucial to overcome these adverse conditions. Their initial attempts to build sugar mills on the island were preceded by requests for loans and licenses to import duty-free African slaves. Such petitions were directed to the Crown in 1523, 1525, 1532, 1534, and 1550, always with negative results. Although the colonial governor affirmed that a number of *maestros* from Española were building a mill (*trapiche*) in 1547, no other evidence corroborates this assertion.[10]

Furthermore, despite royal efforts to the contrary, by the mid-sixteenth century the colony had reached its lowest demographic point and faced a real threat of total European depopulation. The initial colonial model, based on gold and the *encomienda* system, had collapsed. There were neither mines nor Indians. An initial cycle of agricultural prosperity generated by expeditions that used the island as a provisioning base ended as soon as the new colonies were able to provide for their own food and animals. After the conquest of Mexico, the Spanish settlements, most of which had been established on the southern coast, facing the continental areas of the empire, became suddenly obsolete.

But this same process ultimately explains the recovery of Cuba's initial importance within the empire. The discovery of the Gulf Stream and the organization of the fleet system turned the northwest of the island into an area of strategic importance in the Spanish transatlantic system of communications and trade.

FIGURE 5.1. Cuban Exports to Seville, 1560–1699

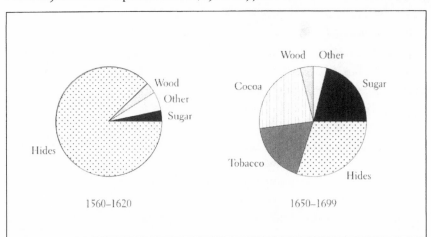

Sources: 1560–1620: Chaunu, *Seville et l'Atlantique*; 1650–1699: Garcia Fuentes, *El comercio español con América*.

While the initial history of the colony blends with that of the east and the south, around Santiago de Cuba, after the 1550s the colony's economic and political center moved to Havana, in the north and the west.[11] Due to its unique role as the meeting point for the returning fleets, by the end of the century Havana had become an important trade and military post with a growing tertiary sector and an expanding agricultural sector. The city was also one of the most important shipbuilding centers in the early Atlantic, with seven shipyards in operation. Between 1570 and 1610 the number of *vecinos* increased ten-fold, while the slave population experienced a similar growth. It has been estimated that between 1570 and 1620 Havana was the fastest growing urban center in the Americas.[12]

It was in this environment, around 1595, that sugar production was initiated in Havana. In contrast to Española, where sugar had replaced gold as the main export product after mining activities began to decline, what the remaining settlers of Cuba needed in the mid-sixteenth century was an economic activity that required neither abundant slaves nor capital. Ranching fulfilled both re-quirements. The island had excellent natural conditions for the animals to reproduce, no initial outlay of capital was needed, and the production of hides involved a limited number of workers or slaves. During the second half of the sixteenth century hides became the colony's main export product (see figure 5.1). According to official figures, between 1560 and 1620 Cuba was, after Nueva España, Spain's main supplier of hides (see table 5.1).

The cattle economy and the commercial opportunities generated by Havana's

TABLE 5.1. Origin of Selected Imports in Seville: Percentage Distribution of Their Value, 1560–1620 and 1650–1700

Origin	Hides		Sugar		Wood	Tobacco
	I	II	I	II	II	II
Nueva España	62.6	3.6	7.7	1.3	58.0	0.9
Tierra Firme	2.8	5.3	1.0	29.0	2.3	2.2
Santo Domingo	11.0	31.0	74.0	0.8	0.4	7.0
Puerto Rico	1.8	2.4	14.0	5.0	—	0.1
Cuba	13.4	11.8	1.4	63.0	38.0	57.3

Sources: Hugette Chaunu and Pierre Chaunu, *Seville et l'Atlantique (1504–1650)*, 8 vols. (Paris: SEVPEN, 1955–59); Lutgardo García Fuentes, *El comercio español con América, 1650–1700* (Seville: Diputación Provincial, 1980).
Note: I = 1560–1620; II = 1650–1700.

maritime activities allowed a number of *vecinos* to accumulate the necessary capital to initiate the construction of sugar mills. Even before the first mills were constructed in the 1590s, some *melado* (sugarcane syrup) and low-grade sugars were being manually produced for local consumption. At least since the 1580s, sugarcane fields appear in Havana's notarial records as part of the inventories of the *estancias*, small farms that grew food for the local market. Some of these fields, the governor asserted in 1597, had been cultivated for decades. In 1593 Havana was already exporting *melado* "from the island's harvest" to Florida, and by 1597 more than 3,000 *arrobas* (thirty-four tons) of low-grade sugars had been exported to Spain, Cartagena de Indias and Campeche, Mexico.[13]

Several factors facilitated the establishment of the first sugar mills around Havana in the mid-1590s. The prosperous cattle-raising economy provided food for the slaves, as well as power for the animal-driven mills. The construction of Havana's *zanja* (aqueduct), which brought water from the river of La Chorrera (currently Almendares) to the city, allowed for the irrigation of agricultural lands and the construction of water-powered mills. Five of the nine well-known mills of the initial seventeen that benefited from a 1602 royal loan were built on La Zanja or La Chorrera. Moreover, the construction of two major forts—El Morro and La Punta—during the last decade of the sixteenth century attracted not only abundant currency and slaves to the city but also a large number of skilled artisans who had the ability to build sugar mills. Gonzalo de la Rocha, a master carpenter who went to the city in 1597 to work in the construction of forts, built two sugar mills, one in 1598 and another in 1599.[14] In Santiago, the royal administrator of the copper mines, Captain Francisco Sánchez de Moya, constructed

MAP 5.1. Sugar Regions of Cuba and Española

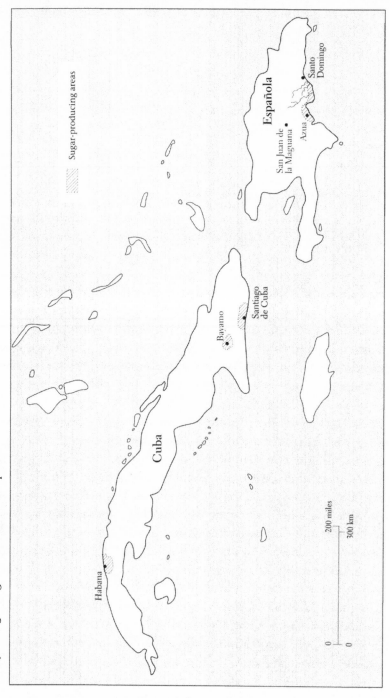

Sugar-producing areas

Española

Santo
Domingo

Azua

San Juan de
la Maguana

Santiago
de Cuba

Bayamo

Cuba

Habana

200 miles

300 km

two mills using "the carpenters, laborers, slaves" and even animals owned by the king to operate the mines.[15] Royal regulations to the contrary notwithstanding, a military foundry established around 1597 produced cauldrons, kettles, spare parts, and various copper utensils for the mills. The master smelter, Francisco Ballesteros, was being investigated in 1603 because of his use of the military foundry for private purposes, including the production of "a few kettles for the sugar mills." Indeed, the inventories of some mills included copper utensils made on the island.[16]

In addition to these local conditions, three equally favorable regional and international factors contributed to the establishment of sugar manufactures in the island. First, the concession of the slave-trade monopoly to the Portuguese merchant Pedro Gómez Reinel in 1595 increased the supply of African slaves to colonial territories. Havana, which was included among the possible destinations for the slave cargoes, benefited from this trade. Between 1595 and 1600 four cargoes entered the city directly from Africa, with a total of about eight hundred slaves. At least sixty more were received from Cartagena de Indias, the great slave port of Spanish America under the Portuguese *asientos*.[17]

Second, prices were high. According to Earl Hamilton, between 1511 and 1599 the price of an *arroba* of sugar in Andalusia went from 265 *maravedís* (eight *reales*) to 2,384 *maravedís* (seventy *reales*).[18] Last but not the least, other Caribbean producers, particularly Española and Puerto Rico, had entered a phase of decline and were facing great difficulties. Displaced from the official trade routes, the sugar produced on these islands had become increasingly expensive. Other Atlantic producers, such as Madeira and the Canary Islands, were also in decline.[19] Sugar manufacturing had expanded into Mexico and Peru, but their production was destined mainly for local markets.[20]

The initial group of *señores de ingenio* (sugar mill owners) attempted to maximize these propitious conditions even further through various concessions from the Crown. With the support of Governor Juan Maldonado Barnuevo (1594–1602), who later became a sugar mill owner, they obtained in 1595 a *real cédula* granting Cuban producers the same privileges and immunities given to mill owners in Española in 1529 and 1534.[21] Basically, this meant that *ingenios* were exempted from liquidation because of debts. Also, with Maldonado's support, these *vecinos* asked and obtained a royal loan of 40,000 *ducados* (440,000 *reales*) to build sugar mills. The loan was granted in 1600 as a lump sum to be distributed by the governor. It was actually disbursed in 1602, under Governor Pedro de Valdés, among seventeen prominent *vecinos* of the city.[22]

Concerning other requests, the *señores de ingenio* were less successful. In 1604 and 1606 Havana's town council instructed its representative in the Spanish court to petition the king for additional concessions to stimulate the recently established sugar business on the island.[23] In both cases the requests revolved around one crucial issue: trading. The councilmen claimed that by the time the fleets returned to Spain the harvest had not ended and asked for the right to export Havana's sugar outside the fleets, in the so-called *navíos sueltos*. They also requested the elimination of all export and import duties, which amounted in all to 10 percent of the product's final value. Equally unsuccessful was another petition for a loan of 80,000 *ducados* by a group of fourteen *vecinos* not included among the beneficiaries of the 1602 loan. In their commission to Juan Gutiérrez del Rayo, to whom they promised to pay 2,000 *ducados* if he succeeded in obtaining the loan, the petitioners asserted that many of them had sugarcane fields and were even building some *ingenios* on their own.[24]

According to these petitions and other related documents, there were between twenty and twenty-five mills in Havana around 1610. These were small units, with a capacity of production that rarely exceeded 1,000 *arrobas* (11.4 tons) per year. Between 1603 and 1610 the city exported legally 57,000 *arrobas* (648 tons) of sugar to Seville, an annual average of 6,300 *arrobas* (72 tons). If we assume that about two-thirds of total production was exported to Spain, Havana's output must have amounted to about 10,000 *arrobas* per year, that is, around 500 annual *arrobas* per mill. Although this is a comparatively low figure, it is probably close to the real one.[25] The governor of Santiago asserted in 1617 that each animal-powered mill produced, if it was well supplied, a maximum of 800 *arrobas* per year.[26] However, many mills were, in fact, chronically undersupplied. In the 1670s, the best harvest of *ingenio* San Miguel, for which we have specific production figures, amounted to 900 *arrobas*; between 1671 and 1675 its annual average production was 550 *arrobas*.[27]

This estimate is corroborated also by the amounts of sugar registered to Seville by some individual producers in the early years of the century. Juan Maldonado "el mozo" (the young one), co-owner of the water-powered mill San Diego, for instance, exported an average of 410 *arrobas* per year between 1606 and 1609. If this accounted for two-thirds of his mill's production, then the total annual output was around 615 *arrobas*. Pedro de Oñate, owner of a smaller animal-powered mill, exported an annual average of 320 *arrobas* between 1604 and 1606 (its total output would have been 480 *arrobas* per year).[28]

A second important center of sugar production developed in Eastern Cuba, around Bayamo and Santiago, roughly by the same time that the first mills were being constructed in Havana, if not earlier. There is evidence that Bayamo was producing some sugar and *melado* by 1585.[29] Due to its isolation from the official trade routes, Bayamo became the main center of contraband in the island,

FIGURE 5.2. Petitions of Land for Sugar Mills, Havana, 1590–1700

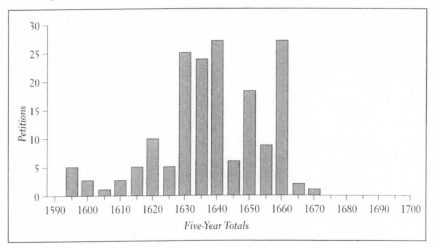

Source: ACAHO, 1590–1700.

prompting Governor Gabriel de Luján to assert in 1583 that exports of hides from the east had dwindled because the French took them all.[30] The problem persisted into the early 1600s, when Governor Valdés claimed that more than 40,000 hides were annually smuggled out of Bayamo and other eastern towns through French, British, Dutch, and Portuguese ships.[31] Bayamo's contraband-based growth was so impressive that by 1570 it was Cuba's most populous urban center, "the best town in the island," according to the bishop.[32] A protected interior enclave connected to the sea through the Cauto River, the city could trade peacefully with the enemies of Spain with little risk of a military attack. In addition to hides, by the early seventeenth century Bayamo exported sugar, wood, cocoa, and indigo. By 1617 there were twenty-seven sugar mills operating in the region, sixteen in Santiago and eleven in Bayamo. These mills produced about 22,000 *arrobas* per year, some of which was legally exported to Cartagena de Indias.[33]

Although we know little about the evolution of sugar manufacturing in the island, it is clear that production expanded during the seventeenth century.[34] This growth was based on the multiplication of mills, rather than on the expansion of the productive capacity of individual units. Havana's town council received 109 petitions of land to build mills in the first half of the century. According to these petitions, the process of expansion went on through the early 1660s (see figure 5.2). Juan Diez de la Calle asserted that by mid-century the city exported 60,000 *arrobas* (682 tons) of sugar per year, a six-fold increase compared to the first decade of the century. Indicative of this growth was also the accusation that the royal officials allowed the illegal exportation of 80,000 *arrobas* of sugar from Havana, between 1635 and 1640.[35] In Santiago, the governor reported that in 1628 the number of mills was also increasing, an assertion confirmed by other sources. A mid-seventeenth-century description reported that there were "many sugar mills" in the region and that their product was exported to Cartagena de Indias.[36]

This process of expansion, however, came to an abrupt end during the 1660s. Although the business survived, the post-1660s conjuncture is marked by a number of negative factors. Like other commercial crops, sugar production was particularly sensitive to changing market conditions and heavily dependent on a stable system of communications and trade. But this is, precisely, a prime indicator of the negative conjuncture: during the second half of the century the fleet system virtually collapsed. The number of ships registered from Havana to Seville in the 1690s amounted to only 52 percent of the ships returning in the 1650s. Compared to the first decade of the century the decline was a dramatic 8 percent.[37] Furthermore, this contraction of the official trade took place at a time in which contraband sales probably decreased as well, due to the expansion of production on the British sugar islands, which had become the largest world producers by the end of the century.[38] Instead of sugar, what eastern Cuba was exporting to the West Indies at that time was a growing number of live animals that were used to power their mills and feed slaves.

The crucial slave trade also experienced difficulties. Until 1640, during the so-called Portuguese period of the slave trade, the number of slaves entering Havana legally seems to have been close to adequate.[39] Using the notarial records and other local sources, I have identified twenty-four different ships importing slaves into the city, between 1600 and 1639. Conversely, not a single cargo has been found for the 1640s. In fact, with the revolt of Portugal, Spain had to confront the fact that the main suppliers of slaves—Dutch, British, Portuguese— were all enemies. The trade was largely reorganized around the Caribbean repositories of Curaçao, Barbados, and Jamaica, but the number of slaves entering Cuba clearly declined. Between 1640 and 1650 the average price of healthy, prime-age (eighteen to thirty years old) African slaves in Havana increased 25 percent. They remained at this level until 1680, when prices began to decline slowly.[40] Starting in 1649, several epidemics made slaves even scarcer.[41]

Adding to these hardships was a growing fiscal pressure, which tended to make Cuban sugars less competitive. As early as 1638 the Consejo de Indias estimated that taxes added thirty-six *reales* to the price of each *arroba* of sugar. That is, sugar export prices doubled because of these high duties. But in 1670, a new regulation ordered that export duties (2.5 percent) should be paid for by the producers, rather than by merchants, passing on to mill owners part of the growing fiscal pressure. Havana's town council protested against this measure, explaining that such duties would make sugar production unprofitable in the island.[42]

In fact, the tax could not have come at a worst moment for Cuban producers because sugar prices collapsed in the 1670s. Between 1600 and the 1660s average prices in the Havana local market had been fairly stable, varying between thirty and thirty-six *reales* per *arroba* of white sugar and between twenty and twenty-two *reales* for *quebrado*, darker sugar. After 1670 the average price declined to twenty-four and sixteen *reales*, respectively.[43] A sugar merchant resident in the city asserted that before the peace with Portugal (1670) sugar was sold at "four pesos and even more" (thirty-two *reales*), whereas at that time (1690) it was quoted at twenty-four *reales*. The peace had facilitated the importation of Brazilian cheaper sugars, which, according to the Consulado de Sevilla, had displaced those produced in Havana. Thus the city council's request to the king was that no sugars "from Brazil, Virginia, Jamaica, Curaçao and Barbados" be admitted into the Spanish market.[44] With low prices, higher duties on production, growing competition, and inadequate communications and supply of slaves, sugar manufacturing in the island could barely survive. In the last two decades of the century numerous mills were demolished, their land used for other productive purposes. In the area of Cojímar alone, to the east of Havana, five mills were dismantled between 1680 and 1692.[45]

Sugar production did not disappear despite this adverse conjuncture. A representative of Havana's town council asserted in 1687 that one-third of the mills had been destroyed during the crisis. Yet there were still about seventy mills operating in the area.[46] Although sugar had not prospered to the point of displacing other commercial crops, during the seventeenth century it had become one of the island's leading export products (see figure 5.1). In the process, Cuba had also become the largest supplier of the Spanish legal sugar market. Española, the main source of Spain's sugar imports during the sixteenth century, had retreated into the cattle economy and was, after the 1650s, the main exporter of hides to the peninsula, replacing Cuba (see table 5.1). The island had gone through a process that was roughly the opposite of Cuba's, where the cattle economy had served to fill the space between the gold and sugar cycles. In Española, it was sugar that filled the gap between the end of the gold cycle and the expansion of the cattle-based economy.

Sugar also affected the demographic composition of the population. Not only did the number of slaves increase due to the expansion of sugar production, but the age and sex structures of the slave population resident in the sugar producing

TABLE 5.2. Beneficiaries of Royal Loans for the Construction of *Ingenios* in Havana, 1602

Beneficiary	Type	Mill
Antonio de Ribera	*Ingenio*	Nuestra Señora del Rosario
Juan Maldonado el mozo	*Ingenio*	San Diego
Hernán Manrrique de Rojas	*Trapiche*	Santa Cruz
Diego Ochoa de la Vega	*Ingenio*	Santa María de Palma
Pedro Suárez de Gamboa	*Trapiche*	Matanzas
Antonio Matos da Gama	*Ingenio*	—
Martín Calvo de la Puerta	*Ingenio*	Santiago
Melchor Casas	*Ingenio*	Tres Reyes
Ginés de Orta Yuste	*Ingenio*	Nuestra Señora del Rosario
Sebastián Fernández Pacheco	*Ingenio*	San Sebastián
Pedro de Oñate	*Ingenio-trapiche*	La Candelaria
Benito Rodríguez	*Trapiche*	San Miguel
Hernán Rodríguez Tabares	*Ingenio-trapiche*	—
Baltasar de Rojas	*Ingenio-trapiche*	San Juan
Silvestre Morta	*Trapiche*	San Miguel
Hernando de Espinar	*Trapiche*	San Antonio
Lucas de Rojas	*Trapiche*	Santa Cruz

Source: Archivo General de Indias, Santo Domingo, leg. 116.

areas were significantly different from those of the urban area. Whereas in the urban area of Havana slaves represented 29 percent of the population by 1691, in the sugar-producing area of Jesús del Monte they were the majority of the population, 59 percent. And whereas children made up 14 percent of the slave population in the city, in Jesús del Monte they were less than 1 percent.[47] These contrasts anticipated trends that the plantation complex would reinforce later.

The Pioneers: Señores de Ingenio

The most salient feature of the first group of mills and their owners is their heterogeneity, both in terms of the size of the units and the social background of the *señores de ingenios*. According to the information contained in the 1602 loan, the number of slaves mortgaged by each mill owner ranged from two to twenty-eight. Few of the mills were water-powered, but most used animal traction. Some were already producing, but others were still under construction. Con-

Slaves	Amount (Ducados)
16	4,400
26	3,500
14	3,000
28	3,000
28	3,000
11	2,500
11	2,500
16	2,500
22	2,500
19	2,500
11	2,500
7	2,500
10	2,000
5	1,000
6	600
12	500
2	500

cerning the *señores de ingenios*, some belonged to the local landed elite, with roots that went back to the origins of the town while others were newcomers who had used royal service to get into the sugar business; a few seem to have come from rather humble backgrounds.

Although we do not have information on all the seventeen beneficiaries of the loan, we have some data on most of them (see table 5.2). The reconstruction of family links during this period is considerably difficult because descendants did not always keep their father's last name. Identification is further complicated by the fact that sometimes there are two individuals with the same name living in the city, not necessarily related. Moreover, unless there is a petition for some royal favor or appointment by a member of the family, which usually generated a genealogical dossier proving that they were old Christians without blood stains, it is impossible to track the origins of these families back to Europe.[48]

At least seven of the seventeen beneficiaries of the loan were part of or were linked through marriage to the local landed elite. Hernán Manrique de Rojas,

FIGURE 5.3. The Initial Sugar Elite

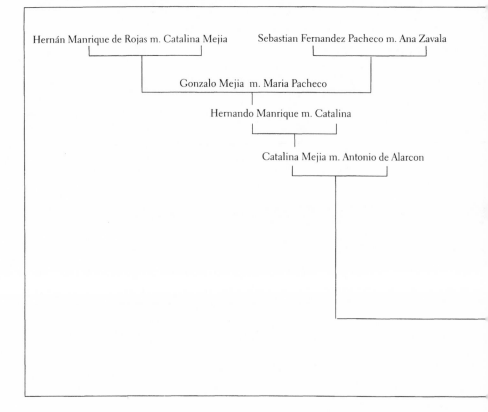

Baltasar de Rojas, and Lucas de Rojas were all members of the Rojas clan, one of the most powerful family groups in sixteenth-century Havana. Lucas and Baltasar were brothers, sons of Alonso de Rojas, who went to Havana in the 1540s with his brother Diego de Soto (or Sotolongo) and an uncle, Juan de Soto. In the 1550s Alonso de Rojas represented Havana's *cabildo* (town council) before the Audiencia de Santo Domingo. He was elected *regidor* in 1564, 1568, 1570, and 1576. Between 1569 and 1585, he was also elected *alcalde* four times. In the process, Alonso de Rojas accumulated numerous lands, including the *hatos* (cattle lands) San Felipe y Santiago, Las Cruces, and San Francisco de las Vegas. In all, he received seven concessions of agricultural and ranching land from the *cabildo* between 1559 and 1590. Using his family connections in Madrid, in 1573 he obtained a royal grant to occupy and exploit for ten years what he had described in his petition to the king as "the small" Isla de Pinos, the second largest island of the Cuban archipelago. It is possible that he was given possession of other smaller islands, for seventeenth-century navigation manuals refer

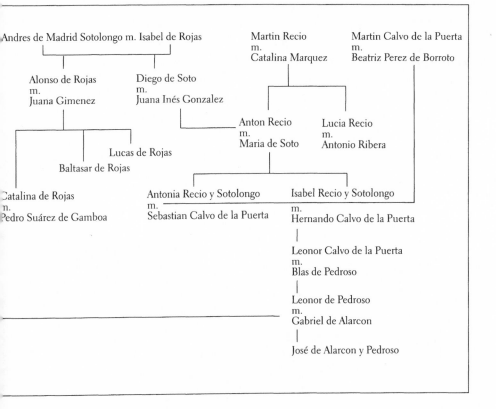

Andres de Madrid Sotolongo m. Isabel de Rojas

Alonso de Rojas
m.
Juana Gimenez

Diego de Soto
m.
Juana Inés Gonzalez

Martin Recio
m.
Catalina Marquez

Martin Calvo de la Puerta
m.
Beatriz Perez de Borroto

Anton Recio
m.
Maria de Soto

Lucia Recio
m.
Antonio Ribera

Lucas de Rojas

Baltasar de Rojas

Catalina de Rojas
m.
Pedro Suárez de Gamboa

Antonia Recio y Sotolongo
m.
Sebastian Calvo de la Puerta

Isabel Recio y Sotolongo
m.
Hernando Calvo de la Puerta

Leonor Calvo de la Puerta
m.
Blas de Pedroso

Leonor de Pedroso
m.
Gabriel de Alarcon

José de Alarcon y Pedroso

to some "keys of Alonso Rojas" in the northwest of the island.[49] Lucas and Baltasar, in turn, added a few extra lands to the family holdings and were elected twice to the *cabildo*. Baltasar de Rojas was an officer in the prestigious cavalry company of the city[50] and owned the *corrales* (pig-farm lands) Río Hondo and Espíritu Santo, which he had received from the *cabildo* in 1578 and 1598. Lucas de Rojas, elected *alcalde* in 1610, had also received a parcel of cattle land in 1603.

Linked directly to the family was a third member of the first group of *señores de ingenio*: Pedro Suárez de Gamboa, son of Alonso Suárez de Toledo and Inés de Gamboa, and husband of Catalina de Rojas, daughter of Alonso de Rojas and sister of Baltasar and Lucas (see figure 5.3). Suárez de Gamboa was the captain of the city's cavalry company, the one in which his brother-in-law Baltasar de Rojas was an officer. Both of his parents could legitimately claim to be members of the local landed elite. Suárez de Toledo had been part of the royal treasury, *regidor* and *alcalde*. He was the owner of three *corrales* in the area of Matanzas, which he had obtained from the *cabildo* in 1564, 1568 and 1570. His wife, Inés de

Gamboa, had also received several concessions of land in the same area, including two *hatos* (Canímar and Macurijes) and the *corral* Puerto Escondido. As a result, their son Pedro Suárez de Gamboa came to own an enormous extension of land—much larger than the current city of Matanzas—including some properties strategically located by the bay, one of the best in Cuba's northern coast. Because of its relative isolation, Matanzas Bay was known to be a center of contraband and other illegal activities, from which the family profited. In 1581 Suárez de Toledo was under investigation because one of his properties served as a provisioning base for enemy ships.[51] Suárez de Toledo himself was not found guilty, but several of his employees were. It was in this relatively isolated enclave that Pedro Suárez de Gamboa built his sugar mill.

The other member of the Rojas clan promoting a mill was Hernán Manrique de Rojas, one of the wealthiest and most enterprising characters in Havana's sixteenth-century society.[52] Manrique had been in Havana since at least the 1560s and had performed numerous services to the Crown. In 1564, for instance, Governor Diego de Mazariegos named him captain of a twenty-five-man force that went to Santa Helena, Florida, to gather military intelligence about French activities in the region. Upon his return, Manrique de Rojas brought with him a marble column bearing the arms of the French royal house, a prisoner, and detailed information about the coast, its harbors, and their location. Twenty years later, in 1586, he played a prominent role in organizing the defense of the city against Francis Drake, whose forces had attacked Santo Domingo and Cartagena de Indias. Governor Gabriel de Luján referred to him as "a person of experience and a very good soldier."[53] Other royal and honorific positions to which Manrique was appointed included governor of Jamaica (ca. 1570), "protector" of the Indians relocated to the town of Guanabacoa (1577), patron of the Royal Hospital (1575), and *alcalde* (1603).[54]

His economic activities were equally varied. There is evidence that at least part of his fortune—and certainly that of his brother Gómez de Rojas—had originated in contraband and other illegal operations. His years as governor of Jamaica—one of the most active smuggling areas in the whole Caribbean—had given him the opportunity to profit directly from contraband. According to some testimonies, he had taken "large quantities of money and hides" from the island. Hernán Manrique was accused also in 1565 before the Consejo de Indias for trading with Portuguese smugglers in Cuba, from whom he received slaves in exchange for hides. Twenty years later, he owed money to the Crown, but when the royal officials attempted to collect the debt he left the city and went to *tierra adentro*, that is, to the Cuban interior.[55] Equally suggestive of his involvement in contraband is the purchase in 1590 of the *hatos* of Isla de Pinos—those given originally to Alonso de Rojas—for 5,500 *ducados* (60,500 *reales*). Because of its isolation, Isla de Pinos was an ideal place to provision the numerous foreign ships that sailed through the Caribbean as well as to engage in other illegal operations.

But some of his economic enterprises were legal. In 1583 he bought the license to operate exclusively the copper mines of Santiago de Cuba. Using his power and prestige, in the last decade of the century, Hernán Manrique obtained from the *cabildo* the collection and administration of the *sisa*, a local tax on wine and meat used to pay for the construction of La Zanja, Havana's aqueduct. Manrique also obtained the contract to build the aqueduct, for which he had received 10,000 *ducados* (110,000 *reales*).[56]

Hernán Manrique was related to another *señor de ingenio*: Sebastián Fernández Pacheco. Gonzalo Mejía, son of Manrique and Catalina Mejía, married María Pacheco, one of the two daughters of Fernández Pacheco and Ana Zabala (or Zavala) (see figure 5.3). Fernández Pacheco was originally from San Miguel, one of the Azores, where some members of his family still lived at the end of the century. He was not a member of the traditional landed elite; rather, it seems that he had made his fortune as a merchant. Between 1594 and 1602 he participated in a trading partnership with Melchor López, a merchant resident in Garachico (Tenerife, Canary Islands) who shipped him wine, tar, and other products.[57] Fernández Pacheco was elected *alcalde* in 1604, but in his case wealth and sugar preceded his entrance into the exclusive circle of the town council, not vice versa.

Three generations later, these two families became linked to the descendants and relatives of still two others of the original *señores de ingenio*: Antonio de Ribera and Martín Calvo de la Puerta.[58] In 1591 Ribera married a member of the Recio clan, one of the most powerful family groups in Havana. His wife Lucia Recio Márquez was daughter of Catalina Márquez and Martín Recio, who, together with his brother Antón, had established the family in Havana. Their local influence becomes evident in the twenty-six concessions of land that the Recios received from the *cabildo*, between 1550 and 1610. When Antón Recio (son of Martín) requested a *hato* in 1576, half the *cabildo* had to excuse themselves because they were his relatives. In some years, such as 1601, two of the *regidores* or *alcaldes* were members of the family.

Antón Recio, brother of Lucía Recio (wife of Antonio de Ribera), was related to sugar mill owner Martín Calvo de la Puerta, a distinguished member of the elite and founder of one of the most prominent families in the colony. Calvo de la Puerta was a notary, an *alcalde* in 1602 and 1608, petitioner of the *cabildo* in 1597, and steward of the town council in 1605. A *real cédula* of 1583 had ordered the colonial governor to employ him in the royal service "according to his quality and ability."[59] His local influence was further guaranteed through a marriage with Beatriz Pérez de Borroto, daughter of notary and *alcalde* Francisco Pérez de Borroto.

The link between Calvo de la Puerta and the Recios came through two of his sons, Hernando and Sebastián, who married two daughters of Antón Recio, brother of Lucía (the wife of Antonio de Ribera). A granddaughter of one of these unions—between Hernando Calvo and Isabel Recio—married a great-great-grandson of Hernán Manrique de Rojas and Sebastián Fernández Pacheco (see figure 5.3).

But there is more. Antón Recio was also related, through his wife, María de Sotolongo, to sugar mill owners Baltasar de Rojas, Lucas de Rojas, and Pedro Suárez de Gamboa, husband of Catalina de Rojas. María de Sotolongo was daughter of Diego de Soto (or Sotolongo), brother of Alonso de Rojas and uncle of Lucas, Baltasar and Catalina. That is, they were cousins of Antón Recio's wife and second cousins of their daughters Isabel and Antonia, married to two of the sons of Calvo de la Puerta. What this genealogical reconstruction shows is that seven out of the initial seventeen *señores de ingenio* were—or would soon be— somehow related through marriage and family links: Antonio de Ribera, Martín Calvo de la Puerta, Hernán Manrique de Rojas, Sebastián Fernández Pacheco, the brothers Lucas and Baltasar de Rojas, and their brother in-law-Pedro Suárez de Gamboa.

At least two of the beneficiaries of the 1602 loan were latecomers who had gotten into the sugar business through royal service. The most obvious case is that of Juan Maldonado "el mozo," nephew, not son, as it is usually stated, of Governor Juan Maldonado Barnuevo, with whom he shared the property of the mill.[60] Juan Maldonado "el mozo" was captain in Havana's garrison and was elected *alcalde* in 1608. By this time he had strengthened his ties to the local elite through a marriage with Maria Bohorquez (1604), probably a daughter of Maria Millán de Bohorquez and notary Juan Bautista Guilisasti.[61]

A similar case was that of Diego Ochoa de la Vega, who was commissioned by the Audiencia de Santo Domingo in 1593 to perform a "visit" to the island—a legal institution equivalent to an inspection tour. In 1600 Ochoa de la Vega was accountant of the royal treasury. He was elected *regidor* in 1597 and was, together with Gómez de Rojas Manrique (the brother of Hernán Manrique de Rojas), the author of the first ordinances to fight against maronage ever approved in the island (see below).

We know little about the other *señores de ingenio*. In a headcount made in 1582 for military purposes, public notary Melchor Casas is enumerated among the most important *vecinos*, those qualified as "trustworthy." He married Juana de Inestrosa in 1609, probably a descendant of Juan de Inestrosa, treasurer and accountant of the royal treasury and *regidor* of the *cabildo*. Juan de Inestrosa was the son of Manuel de Rojas, governor of the island in 1524 and 1531–35. The identification of Casas is further complicated because there was a second Melchor Casas in Havana, a merchant, who died around 1592.[62] Casas and Inestrosa initiated a relatively important family in the city, where several of their descendants were "commissaries" of the Holy Inquisition during the 1670s.[63]

Some of the initial *señores de ingenio* came from less distinguished social backgrounds, an indication that Havana's thriving mercantile economy opened some opportunities for social mobility. In the 1582 list, two of the sugar mill owners appear among those "who live by working": Hernán Rodríguez Tabares and Ginés de Orta (or Dorta) Yuste. The latter, listed as water supplier (*aguador*), had obtained the *cabildo*'s contract to carry water to the town in 1576. Rodríguez Tabares is named without any specific trade. Wealth and sugar opened the doors of the *cabildo* to these lower *vecinos*, although it is noteworthy that neither Orta nor Rodríguez Tabares were ever elected or *regidor* in the 1585–1610 period. Orta Yuste was elected *alcalde* de la Santa Hermandad in 1600, the body in charge of persecuting runaway slaves. Rodríguez Tabares had been appointed to this position before (1599) and he had also been the steward of the town council in 1597 and 1598.

Antonio Matos de Gama (or da Gama) not only lived from his work but was also a latecomer. The earliest reference found about him in the local records corresponds to 1591, when he bought an *estancia* where he later built his mill. He never received a concession of land from the *cabildo*. What seems to have opened the doors of the exclusive sugar business to him was his trade: Matos de Gama was a *maestro de azúcar* (sugar master) from Madeira.[64]

The Pioneers: Sugar Mills

If this first group of *señores of ingenio* was diverse in terms of their origin, so were their mills. The few mills actually producing sugar in the 1590s, before the 40,000-*ducado* loan was received, were rather modest units manufacturing low-grade sugars and *melado*. It is not by chance that when these mills were sold, they were not designated as either *ingenios* or *trapiches*, but rather as agricultural units that had incorporated some machinery to process sugarcane. Thus, in 1591, Pedro de Carvajal, a Sevillian resident in the city, sold half of his "*estancia* with a *trapiche*," not vice versa. Similarly, Pedro de Oñate owned, in 1595, an "*estancia* with a *trapiche* where *melado* is made."[65] What seems to be a terminological question is, in fact, a reflection of production realities. These units were not yet defined by the presence of the processing machinery.

Much smaller than the cattle-oriented *hatos* and *corrales*, the *estancias* were the most dynamic agricultural units in Havana during the second half of the sixteenth century. This dynamism was based primarily on their economic purpose: the production of staples for the local market and the fleets. Alonso de Cáceres y Ovando, a lawyer from the Audiencia de Santo Domingo who wrote Havana's first local ordinances (1574), asserted that the main purpose of the estancias was to produce "bread," that is, to grow yuca (cassava) and to produce cassava bread (*pan casabe*).[66] These units surrounded the city like an agricultural belt, with easy access to the local market. Three-quarters of the *estancias* grew cassava and bananas, one-third produced corn and fruits, and nearly half included some domestic animals as well.[67]

It was in these units that sugarcane was first grown and that the first mills were built. Twenty-one percent of all *estancias* included cane among their products in the 1578–1610 period. An estancia became a sugar mill when the unit started producing sugar, not only *melado*. This required, in addition to the extracting machinery (the mill itself), separate installations to boil the syrup (boiling house) and purify sugar (purging house). It was then designated and sold as either an *ingenio* or a *trapiche*.

These designations have been the source of much confusion among historians. Following the writings of Bartolomé de las Casas and Gonzalo Fernández de Oviedo, both of whom described sugar production in Española, Cuban historians have asserted that water-powered mills were called *ingenios*, whereas *trapiches* were smaller, animal-driven units. Using this distinction, Fernando Ortiz claims that the immunities and privileges granted by the Crown to sugar mill owners were given to those who built *ingenios*, not *trapiches*.[68]

Contemporary documents do refer to *ingenios* and *trapiches* as different types of mill. The 1602 loan covered eight *ingenios*, six *trapiches*, and three *ingenios-trapiches*. The distinction between these units, however, was not based on the type of power source. We know, for instance, that the *trapiche* Santa Cruz, built by Manrique de Rojas in la Chorrera, was water-powered and that the *ingenios* Nuestra Señora del Rosario and Los Tres Reyes, of Antonio de Ribera and Melchor Casas, were animal-driven.[69]

The difference between *ingenios* and *trapiches* was thus based on different criteria. When Governor Maldonado reported to the Crown how he would distribute the royal loan, he suggested that those building *ingenios de agua* (water-driven mills) should get 8,000 *ducados*; those who built *ingenios de caballo de rueda grande voladora* (horse-driven mills with a large overshot wheel) should receive half this amount, whereas those building *trapiches pequeños* (small mills) should not get more than 2,000 or 3,000 *ducados*. Maldonado calls *ingenios* not only those driven by water, but also by animals.[70]

This, however, does not elucidate the sort of technological and productive features that differentiated *ingenios* from *trapiches*, beyond the fact that the latter were smaller, less expensive units. The coexistence of different technologies and the changing nature of terms further complicates the problem. In the early years of sugar production in Cuba, it seems that *ingenio* referred to units using either the old Mediterranean technology of a heavy millstone rolling over small pieces of cane, or the horizontal mill of two rollers that was also used at this time in Brazil.[71] Because of their inefficiency, these *ingenios* always required a comple-mentary press (*prensa*) to further extract juice from the cane. In 1603, the *ingenio* Nuestra Señora del Rosario used a two-roller mill and two wood presses "full of stone, so they are heavier." When this *ingenio* was built around 1598 its owner described it as a "two-roller mill with its overshot wheel to mill cane such as those of Motril and Salobreña."[72] In other words, this mill replicated the technology used in southern Spain to manufacture sugar.

Which types of mills were designated as *trapiches*? One possibility is that they were smaller traditional mills, without a wheel, powered by the slaves them-selves. Small hand presses had been used to produce sugar in Madeira since the fifteenth century and were also used in Mexico.[73] In his report to the Crown, Maldonado specified that those building small *trapiches* would receive money to help them with "the blacks and the copper for kettles." In contrast to the *ingenios*, he makes no reference to any other power source. The other possibility is that these *trapiches* referred to the new, three-roller vertical mills that spread out throughout Brazil, the Spanish colonies, and the West Indies during the seventeenth century. The problem is that these new mills first appeared, be-tween 1608 and 1613, in Brazil, and were probably introduced by a Spanish priest from Peru.[74] Although the available data is not conclusive, it seems that this type of mill was used in Cuba before these dates. As early as 1606, sugar mill owner Hernando de Espinar included in his will "the *trapiche* San Antonio with three new rollers." Three years later Juana Rodríguez owned an "*ingenio* with two small rollers and a large one broken."[75] Also noteworthy is that in 1617 the governor of Santiago de Cuba reported that sugar production had prospered in the area because "they have invented small *ingenios* of three rollers called *trapiches*."[76] If this technology was first introduced and applied in Brazil, it spread to Cuba remarkably fast. It is of course possible that the island received the technology directly from Peru through the fleet of Tierra Firme, but the puzzling question is that these vertical mills were frequently referred to as "of

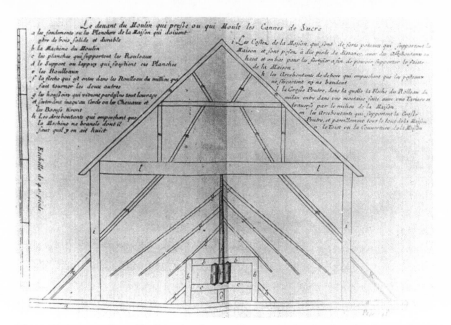

The vertical three-roller mill. This plan became the standard technology for the pressing of sugarcane in the seventeenth century. This arrangement was apparently both more effective and less expensive to operate than previous systems, and it spread rapidly to all the Atlantic sugar-producing regions. Its origins, however, remain controversial. From Richard Ligon, A True and Exact History of the Island of Barbados *(London, 1657), as reprinted in* Recueil des divers voyages faites en Afrique et en l'Amerique *(Paris, 1674).*

new type" (*a la nueva usanza*) or as "of Brazilian type" (*a usanza del Brasil*), an indication that the vertical mill did get into Cuba from the Portuguese colony.[77] After this initial moment, however, the term *ingenio* was used to designate the sugar mill as a whole, whereas *trapiche* referred specifically to the machinery used to extract liquid from the cane.

These early mills were transitional in more than a technological sense. The *ingenios* were changing also in terms of their land extension, the nature of the labor force used, and the sources of their supplies, particularly copper for the boiling kettles and the potter's clay for *hormas*, forms or pots, needed to purge sugar.

Since most of the *ingenios* were built on *estancias*, their availability of land was limited. The average size of an estancia during the period from 1578 to 1610 was about 2.5 *caballerías* (eighty-three acres). Only part of this land could be used to grow cane, for most mills had to devote significant portions of their land for firewood and food production. To buy firewood added greatly to the expenses of the *ingenio*,[78] especially since Havana's town council attempted to restrict the

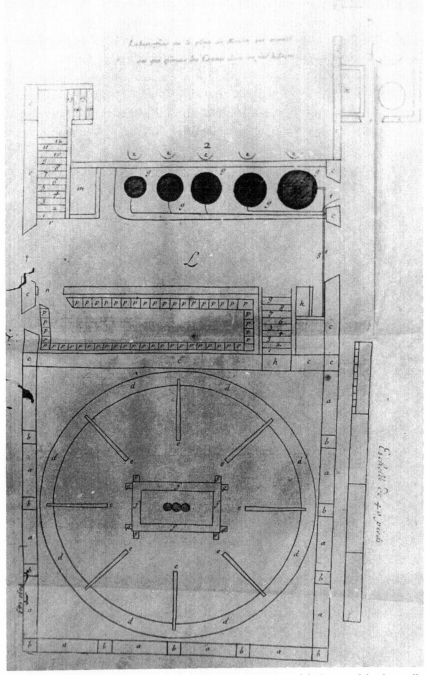

The mill and the kettle house. This plan shows an overhead view of the layout of the three-roller mill (right) and the series of kettles and pans (left) where the juice of the cane was heated and clarified before being placed in forms, where the liquid crystallized into sugar. From Richard Ligon, A True and Exact History of the Island of Barbados *(London, 1657).*

production of firewood in the area peripheral to the city for defensive purposes and to protect local construction and shipyard activities.[79] Therefore, several *señores de ingenio* incorporated additional parcels of land to their properties in the early 1600s, either through *cabildo* concessions or through purchase. *Ingenio* San Antonio de Padua, originally established on three *caballerías* of land (100 acres), had incorporated three *estancias* by 1608, with a total of ten additional *caballerías* (333 acres) of land. The Maldonados acquired the *"estancia* and *trapiche"* of a Manuel Pérez and added it to their mill in 1603.[80] In 1601 and 1603, Hernando de Espinar and Baltasar de Rojas obtained from the *cabildo* parcels of land to incorporate into their respective mills. Hernán Rodríguez Tabares did the same in 1608, when he requested a piece of land adjacent to his *ingenio* "to cut firewood for his mill." In order to cut costs, several *señores de ingenio* also built sawmills in their units. These mills were then capable of producing their own sugar-packing boxes and the different wood parts needed to operate the *ingenio*. The slaves employed in the sawmill of the *ingenio* Nuestra Señora del Rosario, for instance, produced boxes, boards for the purging house, and oxcarts in 1603.[81]

Land was also needed to rotate crops and produce food for self-consumption and, if possible, for the local market. Information about the precise amount of land used to grow sugarcane is extremely scarce and we can only advance very rough estimates about its size. In the second half of the seventeenth century, more than 60 percent of the total sugar mill land was reserved for the production of firewood. That is, all the mill installations—including notably the *casas de molienda, calderas y purga* (mill, boiling, and purging houses)—the *cañaverales* (sugarcane fields) and the other crops occupied less than 40 percent of the total land, a proportion well below those used by sugar mills in the British West Indies during the same period or in Cuba during the plantation era.[82] In a further step to avoid market uncertainty, most mills included other crops among their products. Corn was grown to feed the animals; large *platanales* (banana fields) supplied one of the main staples in the slave diet: "It is their sustenance instead of bread and they do not get anything else but meat," a sugar mill owner declared in 1603. Additionally, sugar mills grew rice, beans, pumpkins, and other vegetables. Despite its limited demand and commercial value, some mills produced tobacco for slave consumption. Cassava was also grown, but at least part of it was destined to produce "bread" for the local market.[83]

Sugar mill owners reacted to these initial limitations of land and sugarcane acreage in yet another way. Some of them milled canes grown by others, a system similar to the Brazilian *lavradores de cana*. Although it is not possible to establish how prevalent this practice was in the early years of sugar manufactur-

TABLE 5.3. Contracts with *Maestros de Azúcar* in Havana, 1599–1608

Year	Señor de Ingenio	Maestro	Payment (*Reales*)	Time
1599	Sebastián Fernández Pacheco	Antonio de Salazar	2/form	2 years
1600	Antonio de Ribera	Pedro González	1,000/year	2 years
1601	Hernán Manrique de Rojas	Amador Rodríguez	26/*tarea*	1 year
1603	Domingo de Viera	Antonio Veloso	1,100/year	2 years
1603	Juan Maldonado	Nicolás Hernández	44/*tarea*	1 year
1603	Luis Hernández	Nicolás Hernández	1,100/year	1 year
1604	Juan Maldonado	Nicolás Hernández	55/*tarea*	2 years
1605	Lucas de Rojas	Miguel de Estrada	800/year	3 years
1607	Juan de las Cabezas Altamirano	Miguel de Estrada	1,100/year	1 year
1608	Rafael Sanz	Alonso Gómez	1,100/year	1 year

Source: Archivo Nacional de Cuba, Protocolos Notariales de la Habana, Escribanía Regueira, 1599–1608.

ing in Cuba, it is clear that, contrary to Brazil, these sugarcane farmers did not constitute an essential part of the sugar economy in the island.[84] In a 1598 report to the Crown, those planning to build sugar mills asserted that the manufacture would benefit the poor *vecinos* because they would have the opportunity to grow cane "and take them to be milled at the *ingenios*" of the wealthy *vecinos*. Sugar mill owner Luis Hernández, for instance, milled cane from the *estancia* of Jusepe Rodríguez in 1604; similarly, Manuel Baez, a Portuguese, supplied cane to the mill of Juan Pérez de Oporto in 1610. According to sugar mill owner Antonio de Ribera—who explicitly instructed his administrator not to mill outside cane—it was customary for growers to provide some "slave help" during the harvest period and to pay *señores de ingenio* half the sugar and *melado* produced.[85] Although this system never disappeared completely, the trend was for mills to grow their own cane and to become self-sufficient units. Furthermore, with the introduction of the three-roller vertical *trapiches*, sugar technology became significantly cheaper, allowing some of the initial farmers to build their own mills.

The initial *señores de ingenio* had difficulties, also, in securing a stable supply of labor force with the qualifications needed to produce white, purged sugar. The first *maestros de azúcar* (sugar masters) who worked in Cuban mills were white, well-paid artisans usually hired for one or two years (see table 5.3). Two examples indicate how scarce these skilled workers were. In 1600 sugar mill owner Martín Calvo de la Puerta commissioned a resident of La Palma to hire in the Canary Islands a "sugar master to come to this city to serve the sugar of my mill." Juan Maldonado granted a similar commission in 1602.[86] In 1603 Maldonado and Luis Hernández shared the services of maestro Nicolás Hernández, who became responsible for production in both mills. One of these maestros, Pedro González, ended up co-owning two *ingenios* in Havana.[87]

As with the land, the *señores de ingenio* attempted to eliminate this expense and their dependency on hired labor by having their own slaves trained into the trade. As table 5.3 shows, the sugar masters' labor was expensive—more expensive, in fact, than that of a physician in charge of the mill's slaves.[88] The sugar masters' yearly salary was similar to that of a soldier or a low officer in the local garrison and double that of a nonqualified rural worker.[89] Moreover, the masters customarily received, in addition, free housing and food. Some mill owners hired *maestros de azúcar* only on condition that they teach one of their slaves "how to make sugar and purge it, so he can make a living as a sugar master."[90] By the second half of the seventeenth century slaves performed this task almost exclusively.

Equally problematic was guaranteeing the supply of copper utensils and purging forms. Those promoting the sugar business in the 1590s attempted initially to import these supplies from Portugal, alleging that they were the best available and could not be produced satisfactorily in the island. In 1597 they informed the Crown that to initiate production they lacked "two main things: copper kettles and earthenware forms because they are not available on the island, and no one knows how to make them." In fact, the *señores de ingenios* were attempting to circumvent Seville's trading monopoly by importing these supplies directly from Portugal "in one or two ships . . . without having to go to Seville to declare" the merchandise. The same year they contracted the acquisition of copper utensils and 50,000 forms from Aveiro, Portugal with merchant Juan Rodríguez Quintero, asserting that these were widely used by all other producers, including those in the Canary Islands, Madeira, São Tomé, and Brazil.[91]

Both needs were soon met through local resources. By the early 1600s sugar mills were using locally produced forms, and new pottery works were being established in the sugar-producing areas. Moreover, in a further move toward self-sufficiency, some *ingenios* began producing their own forms and trained slaves into the trade.[92] When the *señores de ingenio* rented their potteries to independent masters they included a yearly supply of free sugar forms as part of the price. Fueled by sugar demand, by mid-century some of these pottery works had grown into units employing as many as eight slaves.[93]

A similar solution was applied to the supply of cauldrons, kettles and the many other copper utensils required by each mill. The copper mines of Santiago del Prado, in eastern Cuba, provided abundant raw material; the military foundry

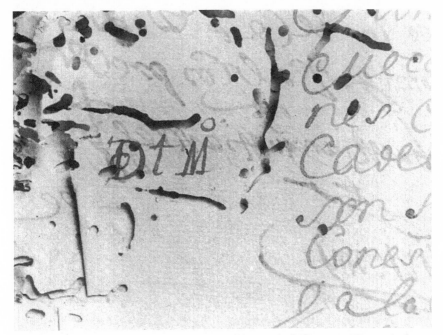

Identification mark placed on sugar crates shipped from Cuba by Domingo Montero in the nao *Nuestra Señora de Regla in the Tierra Firme fleet of 1653. From the Archivo Nacional de Cuba, Escribania Requeira, 1653. Photograph by Alejandro de la Fuente.*

established in Havana with royal money processed it, so in practice the Crown was supporting the nascent sugar manufactures in still another way. Furthermore, in 1630 the king authorized each *señor de ingenio* to buy up to fifty *quintales* (approximately 5,000 pounds) of copper from the Santiago del Prado mines, eliminating one of the legal barriers that prevented the growth of sugar manufacturing in the island.[94]

Making the mills self-sufficient in as many ways as possible—including cane, supplies such as forms, firewood and packing boxes, food, and labor—represented an effort to minimize their vulnerability to unpredictable market conditions and maximize protection against the chronic lack of liquid capital that afflicted sugar mill owners. Several of the first mills seem to have run into great financial difficulties to maintain operations and pay their debts (above all the 1602 loan).[95] Some *señores de ingenio* mortgaged their units to obtain fresh capital, frequently from the church or local merchants, an early indication of the dominant position enjoyed by commercial capital in the colonial setting. Cuba's sugar history during the whole colonial period is the history of subordination of local producers to merchants and other lenders. As early as 1610 several

TABLE 5.4. Sugar Mills in Havana, 1600–1615

Year	Owner	Mill	Type	Price (*Ducados*)
1601	Melchor Casas	Los Tres Reyes	A	10,000
1602	Juan Maldonado	San Diego	W	—
1602	Ginés de Orta Yuste	El Rosario	A	—
1603	Antonio de Ribera	N. Sra. del Rosario	A	—
1606	Hernando de Espinar	San Antonio	A	—
1606	Luis Hernández	—	A	—
1607	P. Suárez de Gamboa	—	A	13,000
1607	A. Matos da Gama	San Francisco	A	7,000
1608	Domingo de Viera	San Antonio	A	3,800
1608	Gaspar de Salazar	La Trinidad	A	3,000
1608	Juan Mordazo	—	A	—
1615	Juan del Poyo	San Antonio	A	9,759

Source: Archivo Nacional de Cuba, Protocolos Notariales de la Habana, Escribanía Regueira, 1601–1615.
Note: A = animal-driven; W = water-driven.

clergymen and merchants had been able to take control of some of the mills. Antonio de Ribera borrowed 300 *ducados* (3,300 *reales*) from the monastery of Santo Domingo in 1601, which he guaranteed with a mortgage on his mill Nuestra Señora del Rosario and for which he paid 21.4 *ducados* per year (a 7.1 percent rate). Gaspar de Salazar, Havana's parish priest, bought half of the mill San Juan from Baltasar de Rojas in 1606; he owned the other half already. The bishop himself, Juan de las Cabezas Altamirano, owned a mill by 1607.[96] Merchant Francisco López de Piedra, who had served as guarantor of three *señores de ingenio* on occasion of the 1602 royal loan, had acquired control of the mill originally owned by Antonio Matos de Gama by 1607. He had also taken control of Melchor Casas's Los Tres Reyes, which he had dismantled.[97] Another merchant, Enrique Méndez de Noroña, a Portuguese who had come to the city around 1597, bought the mill that had belonged to Lucas de Rojas.[98] In 1608 sugar mill owners Domingo de Viera and his wife Juana Núñez sold their mill San Antonio de Padua to another merchant, Juan del Poyo Valenzuela, who sold it off in 1615.[99]

Several sugar mill owners responded to capital scarcity by selling portions of their mills to others, by creating "companies" to build and to administer mills, and by paying debts through participation in the ownership of the *ingenio*.

	Animals		
Slaves	Horses	Mules	Forms
14	8	8	—
31	—	—	—
24	—	—	—
14	8	8	700
14	7	4	600
27	2	9	1,000
16	8	—	1,200
9	8	5	—
4	7	1	500
3	9	—	—
5	9	—	—
10	16	8	445

Martín Calvo de la Puerta sold half of his *ingenio* Santiago to sugar master Pedro González as early as 1598 and the latter also bought one-fourth of Melchor Casas's mill three years later. Hernán Manrique de Rojas and Melchor Casas agreed to build jointly a water-driven sugar mill in 1600, as did Juan Pérez and Pedro González Cordero in 1603.[100] Similarly, the mill San Diego was co-owned by Juan Maldonado and his uncle, former Governor Juan Maldonado Barnuevo, despite the fact that only the former appeared as beneficiary of the 1602 loan.

The proliferation of these partnerships indicates that the sugar business required a relatively large outlay of capital. The scant information available suggests that the total market value of these initial mills varied widely, from some 3,000 *ducados* (33,000 *reales*) to perhaps as much as 20,000 *ducados* (220,000 *reales*), depending on the size of the unit, the technology used and, above all, the number of slaves it possessed (see table 5.4). Each slave added, on average, about 300 *ducados* to the total value of the unit. The two least valuable mills included in table 5.4, those of Viera and Salazar, had only three or four slaves. Conversely, those with total value of 10,000 *ducados* or more had three times as many slaves. The *ingenio* of the Maldonados, with thirty-one slaves (the largest identified during this period) was probably worth about 20,000 *ducados*, if not more. The significance of slaves in the total value of the *ingenio* was reflected in the value of the mill San Antonio, which was sold for 3,800 *ducados* in 1608, when it only had four slaves. Seven years later, the same mill—then owned by merchant Juan del Poyo Valenzuela—was sold for an amount that more than doubled the original price. In the process, however, the new owner had added six new slaves to the *dotación* (the mill's slaves), which alone represented about one-third of the price increase.

Since slaves were an expensive, indispensable, and difficult-to-replace element of the production process, the *señores de ingenio* had a vested interest in providing for their basic needs. Those in the mill Nuestra Señora del Rosario received two sets of clothing per year, one made of *cañamazo* (a low-quality heavy linen), and the other made of *jerga* (a thick, coarse woolen cloth that was reputed to be warmer for the winter period). They also received a blanket every year, usually imported from Nueva España. As mentioned above, their diet was composed mainly of bananas and meat (pork and turtle), but it is likely that the slaves grew some other staples to supplement their diets in their own *conucos*. Some *vecinos* took on the production of salted turtle meat for the slaves. Poultry meat was reserved for those who fell ill.[101]

Sugar mill owners insisted as well that their slaves receive religious indoctrination. Every night, after supper, they were supposed to pray and to learn "the Christian doctrine." A priest was to come to the mill during Lent to give mass and to administer confession to the slaves.[102] Indeed, it was frequent for slaves during this period to be baptized and to get access to the sacrament of the Christian marriage. In 38 percent of all legal marriages registered in Havana's parish between 1585 and 1644, at least one of the spouses was a slave.[103] The sexual composition of the mills' slave populations was extremely unbalanced—with masculinity rates as high as 533—but opportunities for physical mobility of both rural and urban slaves facilitated contacts. Those in the sugar mills were allowed to visit the city during religious festivities, while female urban slaves seemed to have been able to visit the mills quite frequently. Besides, the church supported the slaves's right to marry and, to the extent that this did not interrupt the production process and served to pacify the labor force, it was accepted and perhaps even encouraged by the slave owners.

In normal circumstances, however, slaves were strictly prohibited from leaving the mill, particularly during the *zafra* (the harvest period), which began in late December, after Christmas, and extended through July, during the rainy season. Runaway slaves from the mill Nuestra Señora del Rosario were to be "punished in their body," not only for their own correction, but also to serve as "an example for the others."

Slave owners reacted to the rise of sugar manufacturing and the growing numbers of slaves entering the island with the elaboration of a repressive apparatus unlike anything the colony had witnessed before. Starting in 1599 the town council of Havana frequently discussed the need to repress maronage, given "the large number of runaway slaves" in the area. The *cabildo* called for an open meeting with *vecinos* and approved in 1600 its first slave code, designed precisely for the repression of maronage.[104] This *Ordenanzas de cimarrones* established a gradation of corporal punishments that began with fifty lashes for first-time offenders, two hundred lashes in public for second-time offenders, and death for those who bore arms or headed a runaway gang. Owners of runaway slaves were asked to sell them outside the colony. The town council subsequently approved even harsher punishments, condemning first-time offenders to two hundred lashes in public, and second-time runaways with the same, plus mutilation of both ears. In order to facilitate their identification, in 1610 the *cabildo* ordered mutilation of the nose for runaway slaves.[105] Once confronted with the reality of a growing slave population, slave owners did not hesitate to use their legislative powers to control their labor force in the most brutal way.

These initial mills foreshadowed a number of features that would become the norm for the typical seventeenth-century mill. Cuba's sugar growth was based on the proliferation of small units, rather than on the expansion of the productive capacity of each *ingenio*. Operating under uncertain—at times, adverse—market conditions, the sugar mills incorporated even more land, diversified their production, became as self-sufficient as possible, and were able to exploit their scarce and valuable slave labor force only in a limited way. The final section of this chapter summarizes these findings by looking at the structure of sugar mills during the 1640–1700 period.

Conclusions: Seventeenth-Century Mills

Compared to the *ingenios* built around Havana during the early years of sugar manufacturing on the island, seventeenth-century mills were much larger units, but only in terms of their land acreage.[106] The basics of the productive process were still the same. The three-roller vertical *trapiche* became almost universal; the few attempts made to improve milling techniques were not successful.[107] The auxiliary *prensas*, typical of the *ingenios* based on the traditional technology, had virtually disappeared by mid-century. The *trapiches* incorporated a growing number of metal parts to enhance their durability, first of bronze, then of iron, but remained a minor part of the total sugar investment.[108] Toward the end of the century (1670–1700) the average price of a *trapiche* was 500 pesos

(4,000 *reales*), which amounted to only 2 to 5 percent of the total value of the mill. Like their predecessors, seventeenth-century mills were agricultural units that processed a limited amount of cane for an equally restricted market.

By the 1640–1700 period, 75 percent of mills had 15 *caballerías* (500 acres) of land or more, a size approached by some of the earlier mills through the incorporation of additional *estancias* into the unit. The typical mill during the second half of the century occupied between twenty and thirty *caballerías* (666 to 999 acres) and a few had an even larger acreage. Conversely, other indicators had remained basically stagnant. The average number of forms, for instance, actually declined to some 300 or 400 per mill. The size of the *dotación* remained unchanged. The average mill had only about sixteen slaves. More than half of all *ingenios* had even fewer slaves and were forced to hire additional labor during the harvest, which added greatly to the expenses of the mill. In the 1670s it was estimated that a slave-owner had to spend from fifteen to twenty *reales* per month to sustain a slave. Half that amount (ten to thirteen *reales*) went into food. Clothing accounted for only two *reales*, given that the slave received only one set of canvas clothing per year, worth twenty *reales*. Three extra *reales* were calculated as medical expenses—the assumption was that each slave would need medical services only once a year. Finally, two annual visits of a priest added 0.3 *reales* per month to the maintenance of each slave.[109]

Hired labor was much more expensive. The *jornal* (rent) paid for a hired slave amounted to seventy-two *reales* per month (three daily, for twenty-four days of work). Even considering the amortization of the slave price among the monthly expenses, it was cheaper to buy than to rent their labor. Poorly stocked mills had to devote as much as 40 percent of all their expenses to pay for rented slaves. Free laborers' salaries were even higher. An overseer received between thirty and thirty-five pesos (240 to 280 *reales*) per month, including two daily *reales* as food allowance.[110] With the monthly salary of a free worker it was possible to sustain a slave for a whole year. Still, most mills were forced to hire occasional free workers, mainly to repair the *trapiche*, fix the cauldrons and kettles, and perform other specialized tasks.[111]

The lack of an adequate supply of cheap slaves is reflected not only in the small size of the *dotaciones* and the use of a hired labor force, but also in the age structure of the sugar slave population. Despite the fact that at least 89 percent of the mills' slaves were Africans who had entered the island at a young age—the imports' average age was 18.5 years—the mean age of sugar slaves was astonishingly high: 41 years. This was true even among those without any specific trade or qualification, 49 percent of whom were 40 years or older. Since learning a trade was a long empirical process, the average age of skilled slaves tended to be higher, depending on the complexity of the trade. On the one hand, sugar and purging masters displayed the highest average ages, 50 and 46, respectively. At the other end, the slaves in charge of milling the cane and the sawyers were substantially younger (35 and 33 years), but these activities did not require a long training process, although they were more physically intense. Milling slaves who did not perform their tasks at the speed required could get one of their hands trapped between the rollers of the *trapiche*, resulting in their permanent disability.

Whereas the aging of the slave mill population reflects primarily the inadequacy of the slave supply, it shows also that in these mills the productive life of the slave was longer than usually assessed. The age-structure of the slaves working in Cuba's nineteenth-century plantations was significantly different. Ninety percent of the plantation slaves were grouped in highly productive ages (15 to 40 years old),[112] compared to only 54 percent in seventeenth-century mills. Similarly, although the average price of healthy African male slaves declined by half between the ages of 30 and 50 in the plantations, in the late 1600s this level of depreciation was not reached until the slaves were 60 or 65 years old.[113] The slaves listed with a particular skill did not reach their prime age until they were 35, at which point they were valued over 400 pesos (3,200 *reales*). The unskilled slaves' prime age was 25 and they were valued, on average, 16 percent less than those with specific qualifications. This price gap was particularly wide between the ages of 40 and 55, when skilled slaves, still productive, had managed to master the secrets of their trade.

If seventeenth-century mills were able to produce with this aging slave population, it is because, in contrast to sugar plantations, they were agricultural units crafting a high-quality product for a protected market. Whereas plantations maximized their resources to produce a cash crop at the lowest possible cost, seventeenth-century mills diversified as much as possible to avoid the market and to enjoy, as Jean-Pierre Berthe puts it, the relative security of a closed economy.[114] According to Moreno Fraginals, "until the mid-eighteenth century, the sugar mill was an agricultural institution with an initial capitalization dominated by the value of cane lands, reserves of standing wood-fuel, oxen and their pastures, food plots for slave and employee maintenance, and typically agricultural implements. . . . It was in fact a small center for processing an agricultural product, the sowing, care, cutting, and transportation of which occupied a high percentage of the labor force."[115] The "industrial" installations of these *ingenios* (mill, boiling, and purging houses) represented less than one-fifth of its total value (19 percent). Conversely, the land (26 percent), slaves (27 percent) and crops (10 percent, including sugarcane, bananas, and others) amounted to more than 60 percent of the total investment. It is symptomatic that during difficult times, it made economic sense for mills to rent part of their land.

It is, then, imprecise at best to characterize sugar production—and slavery, more generally—in colonial Cuba using the plantation as a model. Sugar and slavery coexisted in the island under very different circumstances during the long seventeenth century. It was not until a century later that, in Roland T. Ely's words,[116] her majesty sugar became queen and the plantation complex began shaping Cuban society at large.

NOTES

Abbreviations

ACAHO	Actas Capitulares del Ayuntamiento de la Habana: Actas Originales
ACAHT	Actas Capitulares del Ayuntamiento de la Habana, Trasuntadas
AGI	Archivo General de Indias (Seville)
AHN	Archivo Histórico Nacional, Madrid
ANC	Archivo Nacional de Cuba (Havana)
BM	British Museum (London)
BN	Biblioteca Nacional (Madrid)
ER	Escribanía Regueira
PNH	Protocolos Notariales de la Habana
RAH	Real Academia de la Historia (Madrid)

1. John Atkins, *A Voyage to Guinea, Brasil and the West Indies* (London: Ward and Chandler, 1737), 223.

2. Historiography about early colonial Cuba is scant, particularly on sugar and slavery. These issues are covered in some general studies, the best of which is that of Leví Marrero, *Cuba: Economía y sociedad*, 15 vols. (San Juan and Madrid: Playor S.A., 1974–92). Also useful are the following works: Isabelo Macías, *Cuba en la primera mitad del siglo XVII* (Seville: Escuela de Estudios Hispanoamericanos de Sevilla, 1978); Francisco Castillo Meléndez, *La defensa de la isla de Cuba en la segunda mitad del siglo XVII* (Seville: Diputación Provincial, 1986); Arturo Sorhegui and Alejandro de la Fuente, "El surgimiento de la sociedad criolla" and "La organización de la sociedad criolla (1608–1699)," in *Historia de Cuba: La colonia, evolución socioeconómica y formación nacional*, ed. Instituto de Historia de Cuba (Havana: Editora Política, 1994); and three classic works of Irene A. Wright: *Historia documentada de San Cristóbal de la Habana en el siglo XVI*, 2 vols. (Havana: El Siglo XX, 1927); *Historia documentada de la Habana en la primera mitad del siglo XVII* (Havana: El Siglo XX, 1930); and *Santiago de Cuba and Its District (1607–1640)* (Madrid: Felipe Peña Cruz, 1918). Concerning sugar production specifically, see de la Fuente, "Los ingenios de azúcar en la Habana del siglo XVII (1640–1700): estructura y mano de obra," *Revista de Historia Económ-*

ica 9, no. 1 (1991); and Castillo Meléndez, "Un año en la vida de un ingenio cubano (1655–1656)," *Anuario de Estudios Americanos* 39 (1982). For the early eighteenth century, see Mercedes García, "Ingenios habaneros del siglo XVIII," *Arbor* 547–48 (1991).

3. For overviews of this historiography, see Louis A. Pérez Jr., *Essays on Cuban History: Historiography and Research* (Gainesville: University Press of Florida, 1995); and Jorge Ibarra, "Historiografía y revolución," *Temas* 1 (1995).

4. For examples of these sweeping characterizations, see Jesús Guanche, *Procesos etnoculturales en Cuba* (Havana: Letras Cubanas, 1983), 224; Julio Le Riverend, *Selección de lecturas de historia de Cuba* (Havana: Editora Política, 1984), 85; Sergio Aguirre, *Historia de Cuba*, 3 vols. (Havana: Editorial Nacional de Cuba, 1966), 1:66; and Calixto Masó y Velázquez, *Historia de Cuba* (Miami: Ediciones Universal, 1976), 66.

5. I have borrowed this meaning from the classic study of Eric R. Wolff and Sidney W. Mintz, "Haciendas y plantaciones en Mesoamérica y las Antillas," in *Haciendas, latifundios y plantaciones en América Latina*, ed. Enrique Florescano (Mexico City: Siglo Veintiuno, 1978). For examples of authors referring to seventeenth-century sugar mills in Cuba as plantations, see Castillo, "Un año en la vida de un ingenio," 463; and Eduardo Torres-Cuevas and Eusebio Reyes, *Esclavitud y sociedad: Notas y documentos para la historia de la esclavitud negra en Cuba* (Havana: Editorial de Ciencias Sociales, 1986), 39.

6. The process of the rise and consolidation of the plantation complex has been masterfully studied by Manuel Moreno Fraginals, *El ingenio: Complejo económico social cubano del azúcar*, 3 vols. (Havana: Editorial de Ciencias Sociales, 1978). Important contributions to this subject include Laird Bergad, *Cuban Rural Society in the Nineteenth Century: The Social and Economic History of Monoculture in Matanzas* (Princeton: Princeton University Press, 1990); Franklin W. Knight, *Slave Society in Cuba during the Nineteenth Century* (Madison: University of Wisconsin Press, 1970); and Pablo Tornero Tinajero, *Crecimiento económico y transformaciones sociales: esclavos, hacendados y comerciantes en la Cuba colonial (1760–1840)* (Madrid: Ministerio de Trabajo y Seguridad Social, 1996).

7. Irene A. Wright, "El establecimiento de la industria azucarera en Cuba," *La reforma social* (1916).

8. *Real cédula* (13 February 1523), ANC, Academia de la Historia, leg. 50, no. 329; Libros generalísimos de reales órdenes, año 1526, ANC, Academia de la Historia, leg. 80, no. 7, f. 2; RAH, *Colección de documentos inéditos relativos al descubrimiento, conquista y organización de las antiguas posesiones españolas de ultramar* (segunda serie), 25 vols. (Madrid: Establecimiento Tipográfico "Sucesores de Rivadeneyra," 1885–1932), 4:359; Ramón de la Sagra, *Historia física, política y natural de la isla de Cuba*, 12 vols. (Paris: Arthus Bertrand, 1839–56), app. 83, 2:39. A good summary of the earliest effort to promote sugar manufacturing in the island is provided by Marrero, *Cuba: Economía y sociedad*, 2:311–12.

9. Juan Pérez de la Riva, "Desaparición de la población indígena cubana," *Revista de la Universidad de la Habana* 196–97 (1972); Alejandro de la Fuente, "Población y crecimiento en Cuba (siglos XVI y XVII): Un estudio regional," *European Review of Latin American and Caribbean Studies* 55 (1993).

10. For a summary of these petitions, see my "Introducción al estudio de la trata en Cuba: Siglos XVI y XVII," *Santiago* 61 (1986). See also RAH, *Colección de documentos inéditos* (segunda serie), 4:301.

11. Pierre Chaunu, *Sevilla y América, siglos XVI y XVII* (Seville: Universidad de Sevilla, 1983), 87.

12. This process of growth is studied by Alejandro de la Fuente, César García del Pino, and Bernardo Iglesias Delgado, "Havana and the Fleet System: Trade and Growth in the Periphery of the Spanish Empire, 1550–1610," *Colonial Latin American Review* 5, no. 1 (1996).

13. AGI, Contaduría, leg. 1089; Juan Maldonado Barnuevo to the king (Havana, 12 August 1598), AGI, Santo Domingo, leg. 116.

14. De la Rocha's passage to the city is recorded in Luis Romera Iruela and María del C. Galbis Díez, *Catálogo de pasajeros a indias durante los siglos XVI, XVII y XVIII: Volumen 7 (1586–1599).* (Madrid: Ministerio de Cultura, 1980–1986), 668. See also the contracts to build sugar mills in ANC, PNH, ER 1598, f. 70v; 1599, f. 580.

15. Causa seguida a Juan de Eguiluz (11 February 1634), AGI, Santo Domingo, leg. 104.

16. *Real cédula* (6 September 1603) and Valdés to the king (Havana, 12 July 1604), AGI, Santo Domingo, leg. 100, ramo 1. About the foundry's activities, see Macías, *Cuba en la primera mitad del siglo XVII.*

17. Diego de Encinas, *Cedulario indiano*, 4 vols. (Madrid: Cultura Hispánica, 1946), 4:401; Alejandro de la Fuente, "El mercado esclavista habanero, 1580–1699: Las armazones de esclavos," *Revista de Indias* 189 (1990): 376–77.

18. Quoted by Marrero, *Cuba: Economía y sociedad*, 2:320.

19. Genaro Rodríguez Morel, "Esclavitud y vida rural en las plantaciones azucareras de Santo Domingo, siglo XVI," *Anuario de Estudios Americanos* 49 (1992); Frank Moya Pons, "Azúcar, negros y sociedad en la Española en el siglo XVI," *Revista EME EME, Estudios Dominicanos* 1, no. 4 (1973); Rodríguez Morel, "Slavery and Sugar Plantation in Puerto Rico, XVI Century," in *Slaves With or Without Sugar*, ed. Alberto Vieira (Coimbra: Centro de Estudos Atlânticos, 1996); T. Bentley Duncan, *Atlantic Islands: Madeira, the Azores and the Cape Verdes in the Seventeenth-Century Commerce and Navigation* (Chicago: University of Chicago Press, 1972), 31–37; Manuel Lobo Cabrera, *La esclavitud en las Canarias orientales en el siglo XVI (negros, moros y moriscos)* (Gran Canaria: Excmo. Cabildo Insular, 1982), 232.

20. Ward Barret, *The Sugar Hacienda of the Marqueses del Valle* (Minneapolis: University of Minnesota Press, 1970), 4; François Chevalier, *Land and Society in Colonial Mexico* (Berkeley: University of California Press, 1963), 78; Colin A. Palmer, *Slaves of the White God: Blacks in Mexico, 1570–1650* (Cambridge, Mass.: Harvard University Press, 1976), 72; Frederick P. Bowser, *The African Slave in Colonial Peru, 1524–1650* (Stanford: Stanford University Press, 1974), 88–93.

21. *Real cédula* (30 December 1595), AGI, Santo Domingo, leg. 116. The 1529 privileges are reproduced in RAH, *Colección de documentos inéditos* (segunda serie), 9:400.

22. I discuss below the origins and composition of these initial group of *señores de ingenio.*

23. Petición del cabildo de la Habana (Havana, 14 February 1604), ANC, Academia de la Historia, leg. 86, no. 334; Instrucciones a Alonso de Guibar (Havana, 1 September 1606), Archivo Histórico del Museo de la Ciudad de la Habana, ACAHT, 1605–9, f. 127.

24. ANC, PNH, ER, 1602, fs. 46, 623. The king requested additional information from the governor in a *Real cédula* (23 August 1603), in ANC, Academia de la Historia, leg. 85, no. 325.

25. Sugar mills' output in Puerto Rico and La Española oscillated between forty-five and sixty tons; from thirty or forty tons to as much as two hundred in Mexico; and from forty to one hundred tons in Brazil. The Xochimancas, a Jesuits' mill in Mexico, produced about 120 tons per year in the 1660s. These figures are all rough estimates, for production varied widely from mill to mill and from year to year, depending on climate and many other factors. As Schwartz claims, referring to Brazil, "the average productive capacity" of a sugar mill is "uncertain." See Rodríguez Morel, "Esclavitud y vida rural," 94; Morel, "Slavery and Sugar," 201; Chevalier, *Land and Society*, 77–78; Stuart B. Schwartz, *Sugar Plantations in the Formation of Brazilian Society: Bahia, 1550–1835* (New York: Cambridge University Press, 1985), 167–68; Frederic Mauro, *Le Portugal, le Bresil et l'Atlantique au XVII siècle (1570–1670)* (Paris: Fondation Calouste Gulbenkian, 1983), 239, 298–99; and Jean-Pierre Berthe, "Xochimancas: Les travaux et les jours dans una hacienda sucrière de Nouvelle-Espagne au XVIIe siècle," *Jahrbuch fur Geschichte von Staat, Wirtschaft und Gesellschaft Lateinamerikas* 3 (1966): 104.

26. "Relación de las cosas más necesarias e importantes que hay en el gobierno de Santiago de Cuba" (18 June 1617), BM, add. mss., 13992, fs. 529–32.

27. "Relación y cuenta que da don Luis de Coronado de los frutos y costos de las dos tercias partes del ingenio San Miguel (1676)," AGI, Escribanía de Cámara, leg. 84A, pieza 2, fs. 350–62.

28. "Certificación de Juan de Eguiluz" (Havana, 4 May 1611), AGI, Santo Domingo, leg. 116.

29. Marrero, *Cuba: Economía y sociedad*, 2:312.

30. Gabriel de Luján to the king (Havana, 31 April 1583), AGI, Santo Domingo, leg. 153.

31. Valdés to the king (Havana, 1 March 1604), AGI, Santo Domingo, leg. 100.

32. De la Fuente, "Población y crecimiento," 73–74.

33. "Relación de las cosas más necesarias" (18 June 1617), BM, add. mss., 13992, fs. 529–32.

34. A third production area emerged during the late seventeenth century around Santa Clara, in the central part of the island. The first *ingenio* seems to have been built there in 1697, but sugarcane fields appear in local documents since at least 1695; see Manuel D. González, *Memoria histórica de la villa de Santa Clara y su jurisdicción* (Villaclara: Imp. del Siglo, 1858), 471; and ANC, Protocolos Notariales de Santa Clara, 1691–96, f. 157v.

35. Juan Diez de la Calle, "Noticias sacras i reales de las Indias" (1646), BN, mss. 3023; Macías, *Cuba en la primera mitad del siglo XVII*, 64.

36. Marrero, *Cuba: Economía y sociedad*, 4:14; "Relación del obispado de Cuba, sus lugares y templos y de sus presidios y fuerzas y géneros (ca. 1650)," BN, mss. 3000. A 1687 manuscript describes Santiago de Cuba as a "very ordinary town" based on sugar production—"all that it produces is sugar"; see Gulielmus Hack, South Sea Cost, Peppys Island, and Bahama Banks (1687), BM, Sloane mss., 45.

37. These calculations are based on figures provided by Lutgardo García Fuentes, *El comercio español con América, 1650–1700* (Seville: Diputación Provincial, 1980), 216; and Hugette and Pierre Chaunu, *Seville et l'Atlantique (1504–1650)*, 8 vols. (Paris: SEVPEN, 1955–59).

38. Noël Deerr, *The History of Sugar*, 2 vols. (London: Chapman and Hall Ltd., 1949); David W. Galenson, *Traders, Planters and Slaves: Market Behavior in Early English America* (New York: Cambridge University Press, 1986), 6.

39. The "Portuguese period" of the slave trade is studied by Enriqueta Vila Vilar, *Hispanoamérica y el comercio de esclavos* (Seville: Escuela de Estudios Hispanoamericanos, 1977).

40. De la Fuente, "El mercado esclavista," 376–79.

41. After the epidemics of 1649, there were outbreaks of yellow fever in 1651, 1652, and 1654. I have estimated (see "Población y crecimiento," 67) that Havana's total population declined as much as 50 percent between 1648 and 1662. Several epidemic outbreaks took place in the mid-1670s as well, reinforcing the need for slaves. See "Petición del procurador general" (Havana, 29 December 1677), AGI, Santo Domingo, leg. 117, ramo 4.

42. "Acuerdo del Consejo" (2 October 1638), AGI, indiferente general, leg. 760; "Representación del procurador de la Habana" (30 December 1677), AGI, Santo Domingo, leg. 140, ramo 1; "Representación del procurador Francisco Carriego Valdespino" (Havana, 22 October 1684), AGI, Santo Domingo, leg. 117, ramo 4.

43. We lack a good series of sugar (or any other) prices for Cuba during the seventeenth century. I have calculated these averages from forty-two operations of sugar covering a total of 5,000 *arrobas*, scattered throughout the century in the Protocolos Notariales. Additionally, for the 1670s, I have used the figures included in the accounts of *ingenio* San Miguel (see note 27) and the register of the fleet commanded by Nicolás Fernández de Córdoba (1676), in "Certificación de los oficiales reales" (19 May 1676), AGI, Santo Domingo, leg. 140, ramo 1. It is noteworthy that the prices quoted in the last two documents coincide with those of the notarial records for the decade. A few prices are also quoted by Marrero, *Cuba: Economía y sociedad*, 4:257.

44. "Informaciones sumarias y demás diligencias sobre la averiguación de las talas hechas en el monte vedado de Cojímar" (1692), AGI, Santo Domingo, leg. 465, no. 2; "Consulta del Consejo de Indias" (25 March 1701), AGI, indiferente general, leg. 1; ACAHT, 1683–91, f. 648.

45. ANC, PNH, Escribanía Fornari, 1690, f. 135; 1691, f. 493; "Informaciones sumarias," AGI, Santo Domingo, leg. 465, no. 2.

46. "Representación del procurador Carriego" (Havana, 22 October 1684), AGI, Santo Domingo, leg. 117, ramo 4; Marrero, *Cuba: Economía y sociedad*, 4:30.

47. "Población de la Habana" (1691), AGI, Santo Domingo, leg. 111; Diego de Manzaneda to the king (Havana, 11 August 1691), ANC, Academia de la Historia, 91, no. 676.

48. Unless otherwise noted, the reconstruction of these family links and activities is based on databases built with Havana's town council records (ACAHT and ACAHO) and parish registries, located in Archivo del Sagrario de la Catedral de la Habana: Libro Barajas de Bautismos de Españoles, 1590–1600, Libro primero de bautismos de españoles, 1600–1623, and Libro barajas de matrimonios de españoles, 1584–1622. I have also used Francisco X. Santa Cruz y Mallen, *Historia de familias cubanas*, 6 vols. (Havana: Editorial Hércules, 1940–50). Information drawn from other sources is quoted specifically.

49. "Expediente de Alonso de Rojas" (1573), AGI, Santo Domingo, leg. 124. The reference

to "los cayos de Alonso de Rojas" appears in the manuscript of Benito Barrozo, "Derrotero de las Indias Occidentales y compendio de todas sus costas" (1689), BM, add. mss., 28496, f. 63.

50. ANC, PNH, ER, 1602, f. 745v. Together with other *vecinos*, Rojas commissioned Juan Gutiérrez del Rayo to petition the king and the Consejo de Indias for a subsidy to maintain their "horses and weapons," using the usual argument that life in Havana was very expensive, that the land was poor, and that they did not own "that much hacienda."

51. *Real cédula* (5 June 1581), AGI, Santo Domingo, leg. 99.

52. I have been unable to identify the link between Hernán Manrique, his equally notorious brother Gómez de Rojas Manrique, and the brothers Alonso de Rojas and Diego de Soto. Santa Cruz y Mallén mentions Manrique and his brother as "members of the family," but fails to establish their links to the other Rojas in the island. It is noteworthy that when Governor Gabriel de Montalvo, then in Bayamo (1575), named Diego de Soto his substitute in Havana, the latter appointed Gómez de Rojas Manrique for the position, a clear indication of strong bonds between the two. When problems between them arose later, Gómez de Rojas claimed that the problem was that Alonso de Rojas, brother of de Soto, strongly disliked him. See "Proceso contra Gómez de Rojas Manrique, capitán de la fortaleza de la villa de San Cristobal de la Habana" (1575), AGI, Justicia, leg. 41, no. 4; and Santa Cruz, *Historia de familias cubanas*, 1:316.

53. About Manrique's trip to Florida, see AGI, Contaduría, leg. 1174. See also Eugene Lyon, "Settlement and Survival," in *The New History of Florida*, ed. Michael Gannon (Gainesville: University Press of Florida, 1996). Concerning Manrique's activities in Drake's times, see Luján to the king (Havana, 4 May 1586), AGI, Santo Domingo, leg. 99, no. 138.

54. Francisco Morales Padrón, *Jamaica Española* (Seville: Escuela de Estudios Hispanoamericanos, 1952), 130; Wright, *Historia documentada de San Cristobal de la Habana en el siglo XVI*, 1:77.

55. For these activities of Manrique, see the following: "El capitán Juan de Parra con Hernán Manrique, por haber tratado y contratado con extranjeros" (1567), AGI, Justicia, leg. 979, no. 9, ramo 2; "Don Luis de Colón sobre que se le de cédula para que la Audiencia envíe preso a Hernán Manrique," AGI, Justicia, leg. 1001, no. 2; and Marrero, *Cuba: Economía y sociedad*, 2:261.

56. Marrero, *Cuba: Economía y sociedad*, 2:33; Juan de Tejeda to the king (Havana, 30 May 1593), AGI, Santo Domingo, leg. 99. The accounts of Manrique's administration of the sisa are in AGI, Contaduría, leg. 1011.

57. ANC, PNH, ER, 1595, f. 635; 1602, f. 104.

58. Expediente de José de Alarcón y de Pedroso, Caballero del Hábito de Santiago, AHN, ordenes militares (Santiago), leg. 193.

59. See the family dossier in ANC, Gobierno Superior Civil, leg. 1672, no. 83560.

60. Among those asserting that Juan Maldonado "el mozo" was the governor's son is Marrero, *Cuba: Economía y sociedad*, 2:316. Rather, he was son of Diego Maldonado y Salcedo, Caballero del Hábito de Santiago, and Juana de Salcedo. In 1604, Juan Maldonado "el mozo" declared to own a sugar mill "in company with Governor Don Juan Maldonado my uncle"; see ANC, PNH, ER, 1604, fs. 462 and 503.

61. Maria Millán de Bohorquez married twice, first with Diego de Lara, a Flemish mer-

chant resident in Havana, then with Guilisasti. The first daughter of the latter marriage was María, according to Santa Cruz y Mallen, *Historia de las Familias cubanas*, 4:207. The marriage record between Juan Maldonado and Maria Bohorquez does not include the names of her parents.

62. The 1582 military headcount is included in Diego Fernández de Quiñones to the king (Havana, 12 December 1582), ANC, Academia de la Historia, leg. 82, no. 110. Concerning the death of Casas, the merchant, see ANC, PNH, ER, 1592, f. 134.

63. See the family genealogies in AHN, Inquisición, leg. 1575, no. 772 and leg. 1304, no. 30.

64. Marrero, *Cuba: Economía y sociedad*, 2:319.

65. ANC, PNH, ER, 1591, f. 28; 1595, f. 1005. The 1591 reference is the first concrete reference I have seen about a sugar-producing unit in Havana.

66. These ordenanzas are reproduced by Marrero, *Cuba: Economía y sociedad*, 2:429–42.

67. This assertion is based on the inventories of 124 estancias sold in Havana's market between 1579 and 1610, as they appear in ANC, PNH, ER, 1579–1610.

68. Fernando Ortiz, *Contrapunteo cubano del tabaco y el azúcar* (Havana: Consejo Nacional de Cultura, 1963), 341.

69. ANC, PNH, ER, 1603, f. 362; 1601, f. 464v.

70. Maldonado to the king (Havana, 12 August 1598), AGI, Santo Domingo, leg. 116.

71. Schwartz, *Sugar Plantations*, 126. See also the introduction of Gil de Methodio Maranhão to Moacyr Soares Pereira, *A origen dos cilindros na moagem da cana: Investigação em Palermo* (Rio de Janeiro: Instituto de Açucar e do Alcool, 1955), 10–11.

72. ANC, PNH, ER, 1603, f. 362; 1598, f. 70v.

73. Duncan, *Atlantic Islands*, 10; Chevalier, *Land and Society*, 79.

74. Schwartz, *Sugar Plantations*, 127–28; Mauro, *Le Portugal, le Bresil et l'Atlantique*, 230. The great sugar historian Noël Deerr (*The History of Sugar*, 1:536) popularized the idea that these mills had been invented by Pietro Speciale, a prefectum from Sicily, in the fifteen century. His claim was later disproved in detail by Soares Pereira, *A origen dos cilindros*.

75. ANC, PNH, ER, 1606, f. 529v; 1609, f. 806.

76. "Relación de las cosas más necesarias" (18 June 1617), BM, add. mss., 13992, fs. 529–32. The copy of this letter in ANC, Academia de la Historia, leg. 104, no. 13, is dated (it seems by mistake) 1613.

77. "Causa seguida a Juan de Eguiluz" (11 February 1634), AGI, Santo Domingo, leg. 104; ANC, PNH, ER, 1652, f. 853.

78. In 1692 it was estimated that the production of 1,000 forms of sugar required from 1,500 to 3,000 cartloads of firewood. A free worker could cut up to twelve or fourteen cartloads per day and was paid around half a real per cartload, plus food. See "Informaciones sumarias," AGI, Santo Domingo, leg. 465, no. 2, fs. 52v, 63, and 89; "Memoria de lo que se ha gastado en el ingenio Río Piedras desde 22-XI-1655 hasta 23-XI-1656," AGI, Escribanía de Cámara, leg. 79A, pieza 1, f. 687.

79. Valdés to the king (Havana, 3 January 1604), AGI, Santo Domingo, leg. 100, ramo 2. Valdés was accused of placing obstacles in the production of firewood because he was personally interested in the business; in fact, he did provide wood for shipyards in the city.

See Gerónimo de Quero to the king (Havana, 29 December 1606), AGI, Santo Domingo, leg. 100, ramo 2; ANC, PNH, ER, 1604, f. 523.

80. ANC, PNH, ER, 1608, f. 548; 1603, f. 434v.

81. ANC, PNH, ER, 1603, f. 362. Another mill that incorporated a water-driven sawmill was the San Diego, of Juan Maldonado "el mozo" and his uncle, Governor Maldonado Barnuevo. ANC, PNH, ER, 1602, f. 663.

82. The discussion about Cuba is taken from my "Los ingenios de azúcar," 45. About the West Indies, see Richard Ligon, "Histoire de l'isle des Barbades," in *Recueil des divers voyages faites en Afrique et en l'Amerique* (Paris: Louis Billaine, 1674), 156. For the plantation period in Cuba, see Bergad, *Cuban Rural Society*, 151–57, who estimates that in Cárdenas (1860–78) cultivated acreage represented from 45 to 56 percent of the mill's total land.

83. This is largely based on the detailed description of the *ingenio* Nuestra Señora del Rosario (1603), but similar references to other crops appear as well in less detailed inventories. See ANC, PNH, ER, 1603, f. 362; 1608, f. 548; 1615, f. 131.

84. For an analysis of the system in Brazil, see Schwartz, *Sugar Plantations*, 295–312. This system is similar to the *colonato* that became popular in Cuba in the late nineteenth century; see Bergad, *Cuban Rural Society*, 277–84.

85. ANC, PNH, ER, 1604, f. 82v; 1610, f. 181v; 1603, f. 362.

86. ANC, PNH, ER, 1600, f. 99; 1602, f. 246.

87. ANC, PNH, ER, 1603, f. 433 and 495v.

88. Juan Maldonado, for instance, paid physician Francisco Salvador eighty *ducados* (880 *reales*) per year for "curing . . . the blacks of my mill" (ANC, PNH, ER, 1604, f. 40v.)

89. Typically, a white rural worker was paid five or six *ducados* per month, that is, from 500 to 700 *reales* per year. Six *ducados* was the amount paid monthly for the rent of a slave to work in a mill during the harvest as well. For examples, see ANC, PNH, ER, 1608, f. 56v; 1609, f. 143. Some of the salaries of the garrison are reproduced by Marrero, *Cuba: Economía y Sociedad*, 2:300–301.

90. For two early examples of these conditions, see ANC, PNH, ER, 1596, f. 596; 1599, f. 162.

91. El cabildo de la Habana to the king (Havana, 27 November 1597), AGI, Santo Domingo, leg. 116. The contracts with Rodríguez Quintero appear in ANC, PNH, ER, 1597, 225v, 232v, and 237.

92. As an example, in 1603 sugar mill owner Antonio de Ribera instructed the administrator of his mill to "contract with an officer who makes forms, because there are many in this city," to build a pottery in his mill capable of producing one or two thousand forms per year. ANC, PNH, ER, 1603, f. 362. For other examples of pottery works established around Havana in this period, see ANC, PNH, ER, 1599, f. 321.

93. A good example is the contract between sugar mill owner Ambrosio Gatica and Francisco García. The latter rented an *estancia* with a pottery works from Gatica for five years, for which he paid annually 500 pesos (4,000 *reales*), plus 300 forms to make sugar and the training of Gaspar *embuila*, a slave of Gatica, as a tile master. ANC, PNH, Escribanía Ortega, 1653, f. 622v. A pottery works that had expanded based on its location near several

mills was that of Luis Matías de la Cerda, a priest, which had eight slaves, four of them "tile masters." This unit was rented in 1652 for 1,200 pesos (9,600 *reales*) per year. ANC, PNH, ER, 1652, s/f (contract dated 5 October). For other examples, see ANC, PNH, ER, 1630, s/f (contract 15 January) and Escribanía Fornari, 1639, vol. 3, s/f (contract 3 October).

94. Locally produced kettles appear in the inventory of the mill Nuestra Señora del Rosario in 1603. ANC, PNH, ER, 1603, f. 362. The 1630 concession is mentioned by Marrero, *Cuba: Economía y sociedad*, 4:12. Sugar mill owners had requested a yearly supply of copper since at least 1611, according to a *real cédula* (24 February 1611), reproduced in Wright, *Santiago de Cuba*, app. 10, 118.

95. About the difficulties to collect the 1602 loan, see Governor Gaspar Ruiz de Pereda to the king (Havana, 22 August 1608), ANC, Academia de la Historia, leg. 86, no. 360; La ciudad de la Habana to the king (Havana, 7 June 1609), ANC, Academia de la Historia, leg. 86, no. 380; and Ruiz de Pereda to the king (Havana, 14 August 1611), ANC, Academia de la Historia, leg. 87, no. 397.

96. ANC, PNH, ER, 1601, s/f; 1606, f. 468v; 1607, f. 17.

97. ANC, PNH, ER, 1607, s/f; "Relación de los dueños de ingenios de azúcar" (Havana, 1610), AGI, Santo Domingo, leg. 100.

98. ANC, PNH, ER, 1605, f. 856v. About Méndez de Noroña and his social background, see "Demanda de naturaleza de Enrique Méndez y Diego de Noroña (1608)," AGI, Escribanía de Cámara, leg. 74a. In this naturalization request, Méndez de Noroña claimed to be "*señor* and owner of a sugar mill" and the owner of houses, cattle lands, and many goods.

99. ANC, PNH, ER, 1608, f. 548; 1615, f. 131.

100. The agreement between Manrique de Rojas and Casas was disputed frequently, even after Manrique's death in 1604. ANC, PNH, ER, 1600, f. 1036v; 1604, f. 294v.; 1605, f. 458; 1606, f. 277v. The company between Pérez and González Cordero is noted in ANC, PNH, ER, 1603, f. 62.

101. ANC, PNH, ER, 1603, f. 362. A description of the turtle business and its connection to the *ingenios* is provided in the "Relación del obispado de Cuba (ca. 1650)," BN, mss. 3000. For a concrete example of a contract of supply of turtle meat to a mill, see ANC, PNH, "Escribanía Fornari," 1681, f. 78.

102. ANC, PNH, ER, 1603, f. 362.

103. I have dealt with this issue in "Los matrimonios de esclavos en La Habana: 1585–1645," *Iberoamerikanisches Archiv* 16, no. 4 (1990).

104. ACAHT, 1599–1604, f. 648. The original version of these *ordenanzas* is located in ACAHO, 1603–9, f. 3v. There are copies in AGI, Santo Domingo, leg. 116, and ANC, Academia de la Historia, leg. 31, no. 289.

105. Modifications to the punishments originally prescribed by the *ordenanzas* can be found in ACAHT, 1599–1604, f. 504v., and 1609–15, f. 51v.

106. This section summarizes findings previously published in my "Los ingenios de azúcar." These results are based on a sample of forty inventories of mills from Havana's notarial records from 1640 to 1699.

107. In 1676 a Luis Ochoa y Aranda asked the town council to register "a new invention to mill cane" that he asserted would result in great savings of slaves and animals. The *cabildo*

agreed to grant him exclusive use for nine years if he received royal approval. I have been unable to locate any information on this issue in the AGI. The petition appears in ACAHO, 1672–83, f. 212v. Similar innovations elsewhere seem to have failed as well; see Schwartz, *Sugar Plantations*, 128–29.

108. ANC, PNH, ER, 1652, f. 853; "Escribanía Fornari," 1675, f. 196.

109. "Cuentas del ingenio San Miguel (1676)," AGI, Escribanía de Cámara, leg. 84A, pieza 2.

110. "Informaciones sumarias," AGI, Santo Domingo, leg. 465, no. 2, fs. 49, f. 74v.

111. "Relación jurada que da don Luis de Coronado," AGI, Escribanía de Cámara, leg. 84A, pieza 2, fs. 350–62. Hiring extra laborers was a common practice in other producing areas as well; see Berthe, "Xochimancas," 97; and Mauro, *Le Portugal, le Bresil et L'Atlantique*, 240.

112. Moreno Fraginals, *El ingenio*, 2:85.

113. Compare the results of Manuel Moreno Fraginals, Herbert S. Klein, and Stanley S. Engerman, "The Level and Structure of Slave Prices on Cuban Plantations in the Mid-Nineteenth Century: Some Comparative Perspectives," *American Historical Review* 88 (1983): 1215, with those presented in de la Fuente, "Los ingenios de azúcar," 55.

114. Berthe, "Xochimancas," 103.

115. Moreno Fraginals, *El ingenio*, 1:62. I have taken this quote from the English edition, *The Sugar Mill: The Socioeconomic Complex of Sugar in Cuba, 1760–1860* (New York: Monthly Review Press, 1976), 25.

116. Roland T. Ely, *Cuando reinaba su majestad el azúcar* (Buenos Aires: Editorial Sudamericana, 1963).

CHAPTER SIX

A Commonwealth within Itself

The Early Brazilian Sugar Industry, 1550–1670

Stuart B. Schwartz

Brazil did not generate much interest for the Portuguese so long as that distant shore was seen only as a place to obtain dyewood or tropical curiosities. By the 1530s, however, the introduction of sugar-cane and the beginnings of a sugar industry had begun to transform Brazil, especially its northeastern coast, into a colony of settlement. The sugar estates, because of their organization, and because of their socially and "racially" segmented populations, eventually determined much of the structure of the colony and of its society. Cuthbert Pudsey, an Englishman who visited Brazil in the early seventeenth century, captured the social character of the sugar mills, the political authority of their owners, and the way in which the mills themselves served as the poles of colonization: "Now they invent mills to grind the sugar reed, their slaves to plant and preen their reed that need be planted once in seven years. Founders to cast their kettles, masons to make furnaces, carpenters to make chests, another part is busy to erect churches. Every mill [has] a chapel, a schoolhouse, a priest, a barber, a smith, a shoemaker, a carpenter, a joiner, a potter, a tailor, and all other artificers necessary. That every mill is as a Commonwealth within [it]self and the lord of the mill Justicer and Judge within himself."[1] Pudsey, like other earlier observers, saw the sugar mills as determinants of the colony's character and its social trajectory, and believed that the health of the sugar industry set the parameters of the colony's success.

This essay examines the basic contours of the Brazilian sugar economy in the period from ca. 1550–1670, when it became the Atlantic world's primary producer of sugar. It begins in broad focus by placing Brazil within the context of the Atlantic trading system, and then narrows that focus to examine local conditions as well as the specific challenges of land, labor, and capital that confronted the early Brazilian industry and gave it its particular character and contours.

Second, it shows how this industry expanded rapidly until ca. 1620, and explains why that expansion slowed down even before the rise of new competitors in the Caribbean after 1650.

Brazilian Sugar and the Atlantic Trading System

The Brazilian coast presented excellent conditions for the production of sugar. Sugar could be grown in a variety of soils, but large areas of dark clay soil, the famous *massapé*, were accessible along the rivers near the coast. Sugarcane is a perennial, but its yield of juice diminishes with each cutting. It was said that cane planted in *massapé* could be cut for seven to ten years without replanting, and some mill owners (*senhores de engenho*) even bragged of cane cut for thirty or even sixty years, but such conditions were rare. Eventually, by the late seventeenth century much cane was planted in the sandier upland soils away from the coast, but *massapé* was always the preferred land for sugarcane before 1650. The Recôncavo of Bahia and the *várzea* (riverside lowlands) of Pernambuco had both the appropriate soils with large areas of *massapé* and the advantage of rivers such as the Capibaribe, Ipojuca, and Berberibe in Pernambuco and the Subaé, Cotegipe, and Sergimerim in Bahia that supplied water to power the mills and provided for easy transport to the port. Access to water transport was particularly important because in the rainy months the *massapé* became an impassable quagmire. The coast of northeastern Brazil also had appropriate rainfall, receiving between one and two milimeters a year for sugarcane cultivation and the region was not subject to freezing. Thus, while good conditions for sugar production had existed on Madeira or São Tomé, Brazil offered an unequalled combination of location, climate, soils, water, forests to supply firewood, and other supplies. The Brazilian colony needed only to resolve the problems of capital and labor in order to become a major producer.

While there is some evidence that sugar was being produced in Brazil by the 1510s, and that Brazilian sugar was reaching the market in Antwerp in those years, it was during the period of the lord proprietors or *donatários* after 1534 that the sugar industry began to flourish. By the 1540s, Portuguese colonists and government officials had constructed *engenhos* along the coast. Technicians and specialists, some of them probably slaves, were brought from Madeira and the Canary Islands to build and operate the mills. Capital was first found in Europe from both aristocratic and merchant investors. A Portuguese noble, the Duke of Aveiro, invested in the captaincy of Porto Seguro, the Lisbon-based Italian merchant, Lucas Giraldes set up a mill in the captaincy of Ilhéus, and an Aachen merchant residing in Antwerp, Erasmo Schetz, financed a large

Engenhos *of Recife and Olinda. This early seventeenth-century map shows Olinda, the capital of Pernambuco, and the port area of Recife with shipping in the roadstead. Along the lowlands or* varzea *of the rivers, the* engenhos *and cane fields are depicted. From Diogo do Campos Moreno,* Livro que dá rezão ao todo Estado do Brasil *(1612). Courtesy of the Biblioteca Pública do Porto.*

engenho in the southern captaincy of São Vicente with the help of his agent and relative Jan van Hilst (João Veniste).[2] In Pernambuco, the donatary or Lord Proprietor, Duarte Coelho took an aggressive role in initiating the industry, bringing artisans and specialists from the Atlantic islands, asking in 1542 for royal permission to import Africans as slaves, and seeking investors in Portugal. The first *engenho*, Nossa Senhora da Ajuda, was constructed by his brother-in-law, Jerónimo de Albuquerque, but other mills were built by Duarte Coelho himself, by men such as Cristóvão Lins, an agent of the Fuggers, and one by the New Christian (convert from Judaism) Diogo Fernandes in partnership with other "companions from Viana, poor folk."[3] In most of the captaincies, especially in Ilhéus, Espírito Santo, and Bahia, however, attacks by the indigenous peoples and internal conflicts between the donataries and the colonists disrupted the growth of the industry. Sugar only took a firm hold in the Recôncavo, the excellent lands around the Bay of All Saints in the captaincy of Bahia, after the arrival of Tomé de Sousa as governor general in 1549. His efforts and the subsequent military campaigns of his successor, the third governor, Mem de Sá

(1557–72), resulted in destruction of the native peoples and the granting of many land grants (*sesmarias*), some of which served as the basis for the building of mills. The Brazilian sugar industry, concentrated in the captaincies of Bahia and Pernambuco, flourished after 1570. From that date to the middle of the seventeenth century Brazilian sugars dominated the European market.

Since newly created *engenhos* were exempted from the tithe (*dízimo*) for ten years by a series of laws designed to stimulate the industry, and because many *senhores de engenho* found ways to continue to avoid taxation thereafter, it is difficult to establish the growth of the sugar economy or to estimate production based on the receipts collected from the tithe or other official sources. The best we can do is to infer growth based on the accounts of various observers who reported, between 1570 and 1630, on the number of mills and the total production. While these observations are not consistent and are sometimes contradictory, they certainly provide a rough outline of the industry's progress.

By 1570, there were sixty *engenhos* in operation along the coast with the largest numbers concentrated in Pernambuco (twenty-three) and Bahia (eighteen). Together these two captaincies accounted for over two-thirds of all the mills in the colony. During the next twenty years, the predominance of those two captaincies increased so that by 1585 when the colony had 120 *engenhos*, Pernambuco (sixty-six) and Bahia (thirty-six) accounted for 85 percent of the total. These captaincies predominated throughout the colonial period, but other captaincies—Ilhéus. Espírito Santo, São Vicente—also produced sugar for export. Considerable income was generated in these years of expansion. A royal official, Domingos Abreu e Brito, who visited Pernambuco in 1591 reported sixty-three *engenhos* that produced an average of 6,000 *arrobas* of sugar each for a total of 378,000 *arrobas*. At an average price of 800 *réis* per *arroba*, this amounted to a total value of the crop of over 30:240$000.[4] A report on Brazil from the first decade of the seventeenth century stated, "The most excellent fruit and drug of sugar grows all over this province in such abundance that it can supply not only the kingdom [Portugal] but all the provinces of Europe, and it is understood that it yields to His Majesty's treasure about 500,000 *cruzados* and to private individuals about an equal amount."[5] This would indicate for the colony as a whole a sugar production value of 400:000$. This estimate may be too high, but by the end of the first decade of the century, Brazilian income was about 50 percent above the colony's cost to the Crown.

From the sixty *engenhos* in the colony, reported by Pero de Magalhães de Gandavo in 1570, there was a considerable growth to 120 *engenhos* in 1583 and then to 192 in 1612, reported by the military investigator, Diogo de Campos Moreno. By 1629 the colony had 346 *engenhos*. The annual rate of growth had been highest between 1570 and 1585 when Pernambuco (8.4 percent) and Bahia (5.4 percent) led the way.

This expansion seems to have been driven by favorable prices and a growing demand in Europe in the late sixteenth century and opening decades of the seventeenth century. Local prices for white sugar at the mill in Bahia rose from about 500 réis per arroba in 1570 to almost 1$600 by 1613. Good harvests and peace in the Atlantic because of the truce between Spain and the United Provinces after 1609 led to a general climate of prosperity and expansion. Joseph Israel da Costa, a man with experience in northeastern Brazil, reported to the Dutch West India Company that by 1623 the captaincies of Pernambuco, Paraíba, and Itamaracá had 137 engenhos that were "active and operational (moentes e correntes)" and produced almost 660,000 arrobas, an average of 4,800 (seventy tons) per engenho.[6] The expansion came to an abrupt, if temporary, halt during the general Atlantic depression from 1619 to 1623 when sugar prices fell precipitously, declining so much that no one in Bahia came forward to assume the tithe contract because of the "low reputation and poor repute of the sugars."[7] Although conditions in the Atlantic market had improved by 1623, the outbreak of hostilities between Holland and Spain led to new problems for Portugal and its colonies that after 1581 were also ruled by the Spanish Hapsburgs. The Dutch attacked Salvador, the royal capital of Brazil, in 1624 and held it for a year. During the fighting, considerable damage was done to Bahian sugar estates and much sugar was lost to the invading and the liberating armies. During the 1620s, the Portuguese merchant fleet in the Brazil trade became a major target of Dutch naval action and hundreds of ships were sunk or taken as prizes. The subsequent Dutch seizure of Pernambuco in 1630 and the extension of their control over most of the northeast disrupted the sugar industry in the area at least temporarily and also removed a large proportion of Brazilian sugar production from Portuguese control.

There were regional variations in the pattern of development. The industry in Rio de Janeiro grew at a somewhat different rhythm than the sugar economies of the northeast. It expanded rapidly between 1610–12 and 1629, the number of engenhos growing from fourteen to sixty, a rate of growth of almost 8 percent a year. This expansion seems to have resulted from a technological change, the adoption of the vertical three-roller mill, which made it easier and less expensive to construct new engenhos.[8]

The Brazilian sugar industry adapted the technology of the Mediterranean and Atlantic sugar industries to local conditions. Despite the industrial organization of sugar production, most of the processes involved were done by hand: agricultural labor, heating and clarification, and purging; complex machinery was used only to crush the cane and the extract juice.[9] The only major technological innovation, which occurred in the period under discussion, was the introduction of the vertical roller mill that crushed cane. The first mills either had used a large milling stone in an edge runner arrangement, or more commonly, horizontally arranged rollers driven by water or animal power. This system was unable to extract the juice effectively from the cane and a second stage of processing was necessary using large screw presses called *gangorras*. In 1591, for example, *engenho* Sergipe in Bahia was using this system, a horizontal mill (*moenda*) and two presses. These were still at use at that mill in 1612.[10] This method was expensive, inefficient, and because of the impurities that the crushing process produced, it made the subsequent clarification process more difficult. A major change was introduced in the first decade of the seventeenth century with the introduction of the vertical three-roller mill. The *engenho de tres paus*, sometimes called *de palitos*, or *de entrosas*, comprised of three vertical rollers, allowed smaller producers and marginal areas to enter the industry at lower cost. It eliminated the need for secondary presses, made the crushing of the cane more efficient, and apparently made the establishment of a mill less costly. This innovation was supposedly introduced by a priest who had been in Peru sometime in the period from 1610 to 1614, but later a Portuguese technician, Gaspar Lopes Coelho, applied for compensation in 1620 claiming that he was inventor of the process and offering to build more of the new units in the area of Maranhão that the Portuguese were just conquering.[11] This may be an instance of an idea introduced from abroad and then adapted and employed by a local technician. In any case, the innovation spread relatively fast. Even in São Vicente, inventories made as early as 1615 listed *engenhos de tres palicos*.[12] This technological change affected sugar production in all the captaincies, but Rio most of all. It made the building of *engenhos* less costly, obviating to some extent the need for tax exemptions to promote construction, and opening the possibility of mill ownership to a wider and less affluent range of colonists. This was an innovation that contributed to the rapid growth of the industry and was perhaps the only major technological change until the end of the century.

By the end of the seventeenth century (ca. 1689), Jesuit priest Andreoni, who wrote under the pseudonym Antonil, reported 528 *engenhos* in Brazil that produced about 1,295,000 *arrobas* or 18,500 tons. At that time, the 146 *engenhos* in Bahia had an average production of about fifty-one tons while the 246 *engenhos* of Pernambuco averaged only twenty-six tons. The scale of production in Rio de Janeiro was even smaller. The captaincy had only 136 *engenhos*, averaging thirty-eight tons a year.

Annual productivity varied widely, but by 1610 Brazil produced 10,000 tons and by the 1620s it produced from 1–1.5 million *arrobas* or about 15,000 to 22,000 tons a year, although it rarely reached the higher figure. Matias de Albuquerque, a governor of Pernambuco, perhaps exaggerated when he estimated in 1627 that Brazil sent about 75,000 crates (*caixas*) a year to Portugal, which at eighteen *arrobas* per crate equaled 1,350,000 *arrobas* or about 20,000 tons. Antonil's later estimate for 1710 thus falls into the range that had already been established in the 1620s. This capacity did not change markedly until the mid-eighteenth century.

The Brazilian sugar economy was particularly vulnerable to the political and economic vicissitudes of the Atlantic world. The depression of the early 1620s brought on by the beginning of the Thirty Years War in 1618, the reopening of hostilities with the Dutch after 1621, currency manipulations by various European governments, and overstocking by European markets seriously affected the Brazilian sugar economy for a decade. One observer estimated that between 1626 and 1627 alone, 20 percent ($^{60}/_{300}$) of the ships in the Brazil trade had been taken by the Dutch with a loss of over 270,000 *arrobas* or almost 4,000 tons. By 1630, the disruption caused by Dutch attacks and by unstable prices had lowered profits for Bahian planters from 30 to 50 percent of the levels in 1612 and the tithe (*dízimo*) of that captaincy had also fallen in value by 30 percent. Frei Vicente do Salvador, Brazil's first historian, asked in 1627 what good it was to make sugar if the earnings did not equal the costs.[13] This is a refrain that sugar producers have repeated since the seventeenth century.

The problem, of course, was never simply that of productive capacity, but of the price of sugar as well. Whatever the levels of productivity, the success of the industry and of individual planters depended on the price of sugar. Price series based on European values are often deceptive because European prices were often three times higher than the price at the mill in Brazil. Thus it is difficult to establish the profitability of the industry and its ability to generate capital. While evidence is sketchy, there are enough observations to allow us to establish a general outline.

The secular trend of Brazilian sugar prices was upward from 1550 to about 1620. After the latter date, a general economic crisis (1619–21), the reopening of hostilities between the Dutch and the Spanish Hapsburgs who ruled Portugal and its empire (1580–1640), and a general European market contraction all contributed to a fall in the price of Brazilian sugar. Locally, these events were driven home by the Dutch attack and capture of Salvador, Bahia (1624–25), the destruction of a number of mills and disruption of the harvests in Bahia of 1624–26, and by the Dutch seizure of many ships carrying Brazilian sugar. This situation drove up the price of sugar in Europe, but lowered it in Brazil where planters could find no one willing to carry their produce. A Brazilian planter

TABLE 6.1. Estimates of the Bahian Sugar Crops, 1612 and 1629

Year	Number of Engenhos	Estimated Average Production per Mill (Arrobas)	Sugar		Price per Arroba (Réis)	Value (Réis)
			Type	Arrobas		
1612	55	4,700	White	144,760	1,287	186,306$120
			Muscavado	72,380	771	55,804$980
			Panela	41,360	480	19,852$800
			Total	258,500		Total 261,963$900
						Average per engenho 4,762$980
1612	55	3,700	White	113,960	1,287	146,666$520
			Muscavado	56,980	771	43,931$800
			Panela	32,560	480	15,628$800
			Total	203,500		Total 206,226$900
						Average per engenho 3,749$580
1629–30	84	3,700	White	174,048	714	124,270$270
			Muscavado	87,024	373	32,459$952
			Panela	49,728	170	8,453$760
			Total	310,800		Total 165,183$982
						Average per engenho 1,966$476

Source: Stuart B. Schwartz, Sugar Plantations in the Formation of Brazilian Society: Bahia, 1550–1835 (New York: Cambridge University Press, 1985), 177.

with 3,000 *arrobas* of sugar would have suffered a loss of 45 percent between the value of his crop in 1611 and its value in 1623. We can see the effect of the Atlantic market on Brazilian producers in table 6.1 that compares the value of the Bahian sugar crop in 1612 when prices were high, and in 1630 when they had fallen. Using an estimate of the ratio of the sugar grades made by contemporaries (white 56 percent; *muscovado* 28 percent; *melles* 16 percent) and the prices current for each grade, it is clear that the value of the crop had fallen about 20 percent, but that the average earnings per mill had been halved. Even if we deflate the average productivity per mill to compensate for the introduction of the three-roller mill after 1612, average income per mill would still be over a third lower.

This situation continued through the 1620s and only after 1634 did sugar prices begin to recover, stimulated to some extent by the disruption of production caused by the Dutch invasion of Pernambuco. Even though prices began to slip again after 1640, they stayed well above 1$000 per *arroba* until the last decades of the century. By the 1640s, however, the rise of competing sugar economies, first on Barbados, and then in the Dutch and French Caribbean, and the introduction of exclusionist policies, such as the English Navigation Acts of 1651, changed the relationship of Brazilian sugar to its traditional markets. Whereas Brazilian sugars had supplied about 80 percent of the London market in 1630, by 1670, that figure had dropped by half. Moreover, in Brazil itself, fighting with the Dutch in the 1630s and 1640s destroyed many mills and cane fields and disrupted colonial shipping. To pay for the war, the Portuguese government increasingly taxed sugar, adding an additional burden on the planter's finances.

Sugar in Dutch Brazil

Perhaps the most obvious example of the impact of European politics on the Brazilian sugar economy was the twenty-five year Dutch occupation of northeastern Brazil. The Dutch capture of Pernambuco and the captaincies of the northeastern coast (1630–54) disrupted the sugar industry in that area and brought considerable pressure on the Portuguese controlled sugar economy in the rest of Brazil. During the period of invasion, the burning of *engenhos* and cane fields by both invaders and resisters put sixty of the 166 *engenhos* of the region out of operation by 1637. The Dutch West India Company (WIC) eventually confiscated many of the mills abandoned by those Portuguese who joined the resistance or who fled to Bahia. These properties were then sold to Dutch or Portuguese investors as the Company sought to vertically integrate the industry by controlling the production as well as the commercialization of sugar. Although the Dutch controlled Pernambuco and its adjacent captaincies until 1654, the nearly ten-year Luso-Brazilian revolt against their rule, beginning in 1645, severely disrupted agricultural production.

The WIC had targeted the Brazilian northeast because of the attraction of the sugar economy. The Dutch and other northern Europeans had traditionally carried a large proportion of Brazilian sugars to European markets, and they particularly resented their exclusion from this trade imposed by the Spanish Hapsburgs after 1605. With the resumption of Dutch-Spanish hostilities in 1621 and the formation of the WIC in that year, Brazil became an attractive military and economic target.

Once in control of Pernambuco, the WIC sought to reinvigorate and stimulate the sugar economy. They were partially successful, especially during the enlightened and astute administration of Governor Johan Maurits of Nassau (1637–44), who offered the Portuguese residents religious toleration and a voice in local affairs in order to keep them employed in the sugar industry. The WIC extended credit to the planters, both the Portuguese who stayed and the Dutch who acquired mills, as a way of stimulating the industry. The policy was only moderately successful. Even during periods of peace, the captaincy rarely produced half of its estimated capacity of 15,000–20,000 *caixas* a year, and its share of total Brazilian production fell to only about 20 percent, and sometimes to as low as 10 percent.[14] Between 1631 and 1651, Dutch Brazil exported about 25,000 tons of sugar, or an average of about 1,200 tons a year. Private merchants exported about two-thirds of this sugar, and the remainder was exported by the WIC itself.[15] This was far below the region's capacity, and although in the years of relative peace (1637–44) exports considerably exceeded the mean, the overall performance of the industry was seriously compromised by the political and military situation.

A number of Dutch accounts provide a picture of the status of the sugar economy during their rule.[16] A report from ca. 1637 on the Dutch controlled captaincies of Pernambuco, Itamacará, Paraíba, and Rio Grande identified 217 *engenhos*, but many of these had been confiscated when their owners had fled with the Luso-Brazilian forces, and others were inoperative due to depredations of the campaigning armies, or abandonment by their owners or managers. For example, in the district of Olinda there were sixty-seven mills, but twenty of them were *fogo morto*, and five had been confiscated and resold.[17] Overall, the anonymous observer calculated that there were 150 *engenhos* in the four conquered captaincies, but only ninety-nine were functioning. A rough calculation is that by 1639, of the 150 mills in the region, about one-third were inoperative (*fogo morto*) and some sixty-eight (46 percent) had been confiscated and resold by the WIC.[18]

While many Portuguese planters and cane farmers remained on their properties under Dutch rule, the policy of confiscation and resale of abandoned mills, and the profits they hoped to make in the sugar industry, led a number of Dutch and some Jews to enter the industry. In 1637 and 1638, fifty-one mills were sold off to Dutch merchants and administrators, to be paid for in installments. The report of Adrien Van der Dussen of 1637 listed mills like that of *engenho* Marapatigipe in Ipojuca, owned by Miguel van Meerenburch and Martius de Conten, which was supplied by four cane farmers, three of them Portuguese and the fourth, Abraham van Molligen, a Hollander. The area of Itamacará to the north of Pernambuco witnessed an especially heavy Dutch penetration of the industry. Of the twenty-two mills listed, ten had Dutch or other foreign owners, and of the

Brazilian sugar mill and planter's residence, painting by Frans Post (1612–80). Post, a Dutch landscape artist, was brought to Brazil during the Dutch occupation of Pernambuco. This painting shows in detail the organization and activities of a Brazilian mill, including the drying platform and the crating of sugar. Courtesy of the Museum Boijmans van Beuningen, Rotterdam.

seventy cane farmers that supplied these estates, about a third were Dutch or other foreigners ($^{22}/_{70}$). There were other examples of this penetration. Seven of the eight mills in the parish of Goiana belonged to non-Portuguese.[19] But despite these acquisitions, it was commonly said that the Dutch really never learned how to manage the *engenhos* themselves and remained dependent on Portuguese expertise. As the Portuguese planter and confidant of Governor Maurits of Nassau, Gaspar Dias Ferreira, stated in 1645, "God created the various nations among men, and endowed each one with a different disposition and ability for various occupations. . . . As for the Dutch nation, he gave them no aptitude for Brazil. If this observation seems unjust, show me the Hollander who up to the present day in Pernambuco who was a workman in making sugar or who wished to learn it, or any other position in a sugar mill. . . . There are but few Flemings who devote themselves to the sugar industry or to the maintenance of the mills in Brazil, and only rarely do they own them, and thus both the Negroes and the sugars have to pass through the hands of the Portuguese."[20]

FIGURE 6.1. Sugar Exports and Slave Imports, Dutch Brazil, 1630–1651

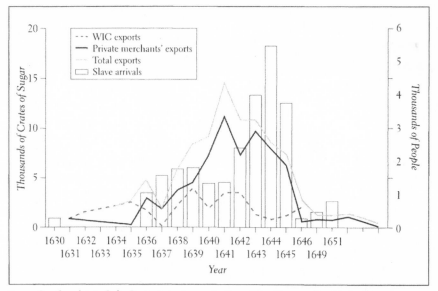

Source: Based on data in Pedro Puntoni, *A misera sorte: Escravidão africana no Brasil holandês e as guerras do tráfico no Atlântico sul (1621–48)* (São Paulo: HUCITEC, 2000).

As a *modus vivendi* developed between the remaining Portuguese planters and cane farmers and the Dutch, the WIC sought to stimulate recovery of the industry through a policy of loans and credit arrangements. These allowed planters to acquire the necessary equipment and slaves that the company began to import from the Guinea coast and Angola. Best estimates of the total number of slaves imported is about 26,000 over a twenty-year period from 1631 to 1651, with the trade particularly strong in the decade from 1635 until the outbreak of the revolt against the Dutch in 1645. As can be seen in figure 6.1, this period of intense slave importation coincided with the high point of sugar exports from Dutch Brazil, which crested between 1639 and 1644 and then fell off precipitously as the fighting in the countryside resumed.

The political situation worsened with the withdrawal of Count Maurits of Nassau in 1644 and with new pressures from the WIC to enforce collection from the planters indebted to it. Some of the Portuguese planters most heavily indebted to the WIC, particularly João Fernandes Vieira and André Vidal, were among the principal leaders of the revolt. The War of Divine Liberation (1645–54), in which Portuguese residents of the colony, aided secretly at first by the home government, rose against the Dutch, caused further destruction of the sugar industry as mills were abandoned, destroyed, or confiscated, and slaves took advantage of the situation to flee to Palmares or other runaway maroon communities. In addition, the war at sea disrupted Portuguese sugar commerce. The Dutch captured some 220 ships in the Brazil trade in 1647–48 alone.

After 1645 the Dutch lost control of the countryside and were progressively forced to abandon the captaincies beyond Pernambuco. "Sugar" was not only the password of the rebels, but also the objective of the contending sides. Moreover, the war was not only fought over sugar, but "financed by it as well."[21] By 1648, over 80 percent of taxes in Pernambuco were derived from sugar production and commerce. Even after the war, taxes on sugar were used to pay for the rebuilding of Recife, and there was also a long series of legal battles between those who had abandoned their *engenhos* and wanted them back and those who had purchased them from the Dutch. There was little capital left for expansion of the sugar industry or other economic sectors. Whatever profits had been possible in the sugar industry of Pernambuco were negated by these conditions. The sugar economy of Pernambuco never fully recovered from the Dutch interlude and its effects. Bahia surpassed it and then remained the principal producer in Brazil until the nineteenth century.

To some extent the Dutch hiatus in northeastern Brazil was not only a cause but also a result of the economic conjuncture of the 1630s. The price of sugar began to rise again after 1634. The improved conditions of Atlantic commerce and the rise in the price of sugar and other colonial commodities gave Brazilian merchants and *senhores de engenho* a renewed sense of security, but these same conditions also created a new and more serious challenge. Rising sugar prices of the 1630s and early 1640s had attracted the interest of the small Caribbean island colonies of the English, French, and Dutch. Shifting from tobacco and other crops, colonists on Barbados actually sought advice and expertise in Pernambuco, and by 1643 sugar from Barbados was on sale in Europe. After the WIC abandoned Pernambuco in 1654, its interest and capital were shifted to the Caribbean as well. With their own growing colonial sources of supply, France and England began to limit Brazilian sugar imports. The English Navigation Acts of 1651, 1660, 1661, and 1673 and Colbert's policies in France aimed at stimulating a French colonial sugar sector effectively drove Brazilian sugar from these markets. In the 1630s, 80 percent of the sugar sold in London had come from Brazil, and by 1690 that ratio had fallen to only 10 percent. The loss of these markets could not be recovered in Portugal itself. Its population was simply too small to absorb what Brazil could produce.

Still another negative effect of Caribbean competition was a rise in labor costs and an expansion of the slave trade. The Dutch had already made attempts to secure their own sources of slaves for their Brazilian enterprise with attacks on

The war for sugar. During the Portuguese rebellion against the Dutch (1645–54), the sugar plantations of Pernambuco became the focus of military operations. Both sides armed slaves for defense. Here a Portuguese force is repelled by armed slaves. From Matheus van den Broeck, Journael, ofte historiaelse beschrijvinge (Amsterdam, 1651). Courtesy of the John Carter Brown Library at Brown University.

the Portuguese African ports. They seized El Mina in 1638 and held thereafter and in 1641 captured Luanda, from which they were expelled in 1648. The new sugar economies now also needed labor, and increased European demands and activities on the African coast drove up the price of slaves in Brazil.

In the second half of the seventeenth century, the Brazilian sugar economy was challenged by competition that increased the supply of sugar in the Atlantic market and created new demands for slave workers. The result in Brazil was lower sugar prices and higher slave costs. Between 1659 and 1688, the price of sugar in Lisbon fell by over 40 percent. Brazil's problem was not production. Even after the Dutch War in 1654, it still had the capacity to produce 18,000–20,000 tons, more than any competitor. Moreover, it still had comparative advantages, but international political and economic conditions and their effects on Portugal's fiscal policies combined to create a situation of crisis. Then too, nature did not help. Periodic problems such as droughts and excessive rains, the irregularities of the fleet system, and various "calamities" created problems in the 1660s and 1670s. More importantly, the War of the Restoration for independence from Spain (1641–68) and Portugal's foreign policy commitments to its allies were financed to a large extent by increasing taxation on sugar at the very moment that industry was facing lower earnings and higher costs. Various forced "voluntary" contributions, such as the dowry for Catherine of Bragança as part of her marriage negotiation with Charles II of England, and other similar imposts and assessments, weighed heavily on the sugar economy and caused constant complaints in the municipal councils of Brazil about the "miserable status" of the colony, but the Portuguese Crown had little choice but to tax this major source of revenue to pay for its commitments.

By the 1680s the economy had reached a low point. Portugal like the rest of western Europe was in a general recession that led to a devaluation of Portuguese currency in 1688 and to an increased search for new sources of revenue. Hides and tobacco exports became regular items on the fleets arriving from Brazil and the search for mines increased, but when João Peixoto Viegas penned his famous memorial in 1687, the sugar economy seemed to be beyond recovery. He complained that Brazil had contributed more to the Portuguese empire than any province of Portugal itself, but foreign competition, royal policies, and general economic conditions had caused its ruin. His doomsday forecast was premature. War in Europe (1689–97) and (1701–13) once again disrupted Atlantic commerce and raised the prices for colonial products. The struggles of England and France were usually profitable for Brazil. White sugar that had sold in Bahia for 800 *réis* in 1689 was selling for 1$440 in 1695. Although prices stabilized after 1700, conditions for the Brazilian sugar economy remained good until the 1720s, although the competing demand for slave laborers in the Caribbean began to create an upward pressure on slave prices by 1670. Meanwhile the discovery of gold in Minas Gerais between 1693 and 1695 also began to alter the whole nature of the Luso-Brazilian economy. Brazil, after all, was no Caribbean island and its potential for economic diversity and diversification was great. Sugar remained regionally important in the coastal northeast and it continued to comprise a large proportion of Brazil's export value into the eighteenth century, long after Brazil had lost its predominant share of the European market for sugar.

The Brazilian Sugar Trade

The foregoing description makes clear the importance of Brazil's integration into the European market system. Despite important work by Mauro, Kellenbenz, and Stols, there has been no general study of the commercial aspects of the sugar trade from Brazil in its early stages.[22] Although sparse, some documentation for such a study does exist. The records of the merchant Miguel Dias Santiago, who shipped sugar from Bahia (1596–98) and from Pernambuco (1599–1601) during the period of the sugar industry's rapid expansion, provide one of the few detailed sources on the cargoes shipped, the taxes paid, and the costs of carrying the sugar.[23] Also from Bahia, for the period 1608–18, the records kept by the administrators of *engenho* Sergipe provide a close accounting with similar information.[24] In addition to these Bahian sources, there is also now available the *Livro das saidas dos navios e urcas, 1595–1605*, which records the customs activity of Pernambuco. Together these sources present an excellent basis for some generalizations about sugar shipping from Brazil during the period of the industry's rapid growth.[25]

These shipping records indicate that at the end of the century, the predominant ship type carrying Brazilian sugar was the rather large, round-bottomed hulks or what the Portuguese called the *urca*, a ship favored by the Hanseatic and Baltic merchants. All of the ships registered by the Pernambucan customs house were *urcas* and 52 of the 101 ships on which Miguel Dias sent his cargoes between 1596 and 1602 were also of this type.

By the decade 1608–18, that situation had changed: the predominant type of ship was no longer these *urcas* more popular with the northern Europeans, but rather the smaller and speedier caravels favored in Portugal and southern Spain. Of forty-two vessels sailing from Bahia with sugar in this period, thirty-five of them were caravels. To some extent this change was due to the shift toward Portuguese flag vessels and the fact that non-Portuguese carriers were being squeezed out of the trade. The caravels had their problems. They were small and nimble sailors, but under attack their only recourse was to run. That is why the Jesuit António Vieira called them "schools of cowardice." Another observer complained that they often arrived so heavily laden from Brazil that their decks were almost awash and their tiny crews were unable to handle the heavy sugar crates.[26]

Until the 1590s many vessels from northern Europe carried the sugar under Portuguese license, principally to northern European ports. Antwerp, where the Schetz enterprise was based, was a major receiving port. It had developed its refining industry with the sugars of Madeira, Canaries, and São Tomé, and by the 1560s was regularly receiving Brazilian sugars.[27] A number of Flemish agents, some married to Portuguese women, lived at various Brazilian ports where they were actively engaged in shipping sugar cargoes and dyewood. The predominance of Antwerp lasted until the political crisis of 1578–85, and although the trade resumed after that date, Antwerp increasingly lost its place to Amsterdam in the Brazil sugar trade.

An important aspect of this transition and of the Brazilian sugar economy was the role of the Sephardic Jews and the so-called "New Christians," that is, those Spanish and Portuguese Jews and their descendants who had converted or were forced to convert. Starting in 1595, members of this community established themselves in Amsterdam, and although prior to 1648 they played only a secondary role in the Dutch economy as a whole, they quickly predominated in the colonial trades, particularly those of Portugal.[28] Linked to relatives and coreligionists in Lisbon and to various ports in the South Atlantic as well as the Indian Ocean, Portuguese New Christians in Brazil also became deeply involved in the production of sugar as mill owners, cane farmers, technicians and skilled workers, as well as merchants. Miguel Dias, for example, had business ties with his merchant brothers in Lisbon, cousin of Diogo Fernandes, who was the manager of *engenho* Santiago in Camarajibe, Pernambuco, which was owned by Bento Dias Santiago, who was probably also a relative of Miguel. The Inquisitorial investigations carried out in Brazil in 1591–95 and 1618 are replete with denunciations of New Christians from the sugar-growing rural areas as well as urban merchants.[29]

The records of the New Christian merchant Miguel Dias Santiago for the years 1595–1601 are some of the best available sources about the patterns of the early trade. During this period, he shipped over 200 tons of sugar to Europe. Along with an accounting of the number and weight of the crates shipped, Dias usually included the place of origin of the ship captains, which usually served as a guide to their port of embarkation.[30]

Patterns of the destinations emerge from these records. Portugal was represented by Lisbon, Sezimbra, Matozinhos, and Vila do Conde, with nine ships sailing from these ports. What is truly impressive, however, is the variety of foreign ports sending ships to load Brazilian sugar. Not only Holland but also the Baltic ports of Riga, Bremen, Copenhagen, and Malmö (Melma) were sending vessels to Bahia, as was the Venetian Adriatic port of Ragusa. This diversity was not atypical. Symbolic of the far-reaching attraction of Brazilian sugar was the case of a ship from Danzig, owned by subjects of the king of Poland that in 1623 delivered goods in Lisbon and then sought and received permission to sail to Bahia to load sugar.[31] The distribution in Pernambuco also reflects the importance of the Baltic and northern European ports. Of the thirty-one ships that carried sugar from Recife, over 60 percent (nineteen) originated in Hamburg, with others sailing from Antwerp, Bergen, and Lubeck. This trade was theoretically done under Portuguese license and control.

In Portugal itself, although Lisbon was the principal destination of Brazilian sugars, other ports such as Porto and Viana do Castelo also developed a regular trade with the colony. Brazilian sugar, in fact, had opened Portuguese trade by breaking the state-controlled commercial system that had grown in the century around the spice trade from the Indian Ocean. These smaller Portuguese ports now became active players in the trade. Viana do Castelo had an active merchant community, and by the first decade of the seventeenth century there were about seventy ships from that port dedicated to the Brazil trade. Most of these were medium size ships of 80 to 150 tons capable of carrying 300 to 450 crates of sugar. This trade was vital to the port's existence and about 85 percent of its customs revenue derived from Brazilian sugar in this period.[32]

Faced with the uncertainties of maritime commerce and the vulnerability of the sugar ships, especially during the recurrent hostilities in the Atlantic, various techniques were developed as insurance measures. Instruments of exchange, such as letters of credit and bills of exchange authorizing a trade of goods rather than exchange for currency, were commonly used.[33] As a kind of insurance, cargoes were often divided between numbers of vessels. The administrators of *engenho* Sergipe usually loaded between eight and twelve crates with 100–150 *arrobas* on any single ship. In 1611, for example, a total of 136 crates with 1,871 *arrobas* were shipped to Portugal on thirteen ships.

The risks were high, but the characteristic of the Brazilian sugar trade was its private nature. Merchants and planters preferred the dangers of this trade to the heavy hand of government intervention. Although taxes were imposed in the 1590s to pay for the costs of providing some protection to shipping, and by 1605 merchants with Lisbon-bound cargoes were required to purchase insurance, royal efforts to force the use of larger ships or suggestions in 1586 and 1615 to establish a convoy system were firmly resisted by the merchants in the sugar trade. It was only the stunning losses of Portuguese shipping between 1647 and 1648 that finally cleared the way for the establishment of the fleet system organized by the Brazil Company, which in return for its provision of protection of the two annual fleets was given monopoly control over basic food imports to Brazil. As was to be expected, the price of imports rose in the colony, planters complained that sugar prices were set too low, and merchants from the smaller Portuguese ports complained of the new centralization of trade on Lisbon, the primary destination of the fleets. With the sailing of the first fleet in 1650, the age of private sugar trade and of the caravel's predominance came to an end.[34]

Finally, it should be noted that the role of the sugar merchants was probably crucial in the financing of the industry's early stages if later patterns are a guide. We are particularly handicapped in establishing this fact since the notarial records from early Brazil are essentially lacking, but by the second half of the seventeenth century merchants provided about 25 percent of the money at loan and may have provided an even higher percentage before the institutional lenders such as religious orders, convents, and the charitable brotherhood of the Misericórdia had sufficient funds to do so. Merchants extended credit and carried standing accounts for sugar growers allowing them to buy slaves, tools, and equipment as an advance against their production. This availability of credit was an essential element in the early growth of the industry.

The Art of Making Sugar in Brazil

From this general survey of the Brazilian sugar economy let us narrow our focus to the specific realities of making sugar in that colony. The complex and difficult process of sugar making influenced in many ways the social organization and hierarchies of the colony as well as the particular solutions to the challenges of sugar production. Sugar making was an art, the result of a series of integrated processes: cultivation, milling, cooking, purging, and crating. Each had its particular labor requirements and each was essential to the ultimate success of the engenho. These sugar mills were called engenhos (ingenious), it was said by antonomasia, because they were a "spacious theater of human ingenuity," and "marvelous machines that require art and great expense."[35] With some regional variations, the engenhos of Brazil followed a similar method of operation with very few major changes until the late eighteenth century. We can, therefore, use the pattern in Bahia as an example of the process, recognizing that there were slight regional differences within the general mode of operation.

In a spirit of festival, the harvest or safra commenced when the mills began to turn in late July or early August after the mill itself and the workers were blessed and the protection of the saints was invoked.[36] During the safra, the cane was cut during daylight hours, but the mills began to operate at 4 P.M. and continued to about 10 A.M. the following morning, thus making the working day eighteen to twenty hours long. The work went on in shifts. For the slaves, the rhythm of labor soon became exhausting. Their "service is an incredible thing," said Israel da Costa. "Sleepy as an engenho slave" was a common expression, and "industrial accidents" were frequent. The evidence of this can be seen in the inventories of many engenhos that listed slave women with one arm. These were milling women (moedeiras) who had become tired or inattentive while feeding the cane in the mill and had lost their limbs as a result. Cuthbert Pudsey, the English observer, wrote, "If by occasion a Negar be laimed as they make no more account of them then beasts, then they put him to feed the mill or to rasp cassava roots on the wheel; they use their slaves very strictly in making them work immeasurably, and the worse they use them the more useful they find them, such is their dispositions, as by experience they find kind usage perverts their manners."[37]

In Bahia, the safra lasted until the heavy winter rains in May made the *massapé* impossible to traverse and began to lower the sucrose content in the cane. The *engenhos* operated over a period of 270–300 days a year, although with stoppages for religious observance, repairs, and shortages of cane or firewood, that figure could be reduced by about one-third. The church required the *engenhos* to stop for Sundays and holy days, but many *senhores de engenho* tried to avoid these religious obligations, which were responsible for about three-quarters of the lost days. In 1592 João Remirão testified before the Inquisition in Bahia: "in his *engenho* on all Sundays and saint's days, the mill operated after sundown . . . which is the general use and custom in this captaincy among all the mill owners and managers without exception."[38] The *senhores de engenho* argued that because the cane had to be milled within twenty-four to forty-eight hours after being cut, and because the juice extracted then had to be processed immediately, the mills could not stop without damaging the work of the days preceding and following those of religious observance. The Jesuits, in particular, and the church, in general, condemned such self-serving arguments, but the repetition of complaints indicates that many *senhores de engenho* ignored the church's directives.[39]

The extended length of the *safra* gave Brazil a considerable advantage over its competitors in the Caribbean, where the harvest season lasted an average of only 120–180 days. It also made sugar production in Brazil particularly well suited to slavery, since between the milling cycle and the planting period there was virtually no "dead time" and slaves could be employed in some aspect of sugar making almost continually.

The key to the success of the harvest cycle actually lay in the preparation of the cane fields. Sugarcane took fourteen to eighteen months to mature after first planting and then usually nine to ten months thereafter to produce the second growth or *rattoons*. A *senhor de engenho* or general overseer had to be able to regulate the planting and cutting of the cane so that each field belonging to the *engenho* and those cultivated by *lavradores de cana* (dependent cane farmers) could be cut at the appropriate moment and also so that there was never too much or too little cane at the mill. Cane not cut at the right moment produced less sugar, and once cut the juice of the cane would dry or go sour rapidly if not processed. Thus the problem of regulating and managing the operation of field and factory demanded skill and experience. A good sugar master (*mestre de açúcar*) who could control and predict how the various activities would mesh and who by art and intelligence had mastered the ratios and volumes of the various parts of the process was essential for success. This job was usually well paid, but there are references to mills where this position was filled by slaves, as the mill owners sought to lower their costs.

In the fields, slaves planted the cane by hand. Plows were rarely employed in sugar cultivation in Brazil probably because the *massapé* soils of Bahia and Pernambuco made their use difficult. Once the cane was planted, groups of slaves did the disagreeable job of weeding the cane at least three times. Then during the *safra*, groups of twenty to forty slaves cut the cane. They often worked in pairs, a man to cut the canes and a woman to bind them into sheaves. Each pair had a quota expressed in "hands" (*maos*). In the time of Antonil's report (ca. 1689) the quota at *engenho* Sergipe in Bahia was twelve canes in each sheaf, ten sheaves in each "finger," five fingers in each "hand" (*mao*), and seven "hands" or 4,200 canes as the daily quota. The cut cane was then taken to the *engenho* in ox carts or in small boats.

The mill or *engenho* (whence the name of the whole estate was derived; the word "plantation" was never used) was powered either by water wheels or animal traction. Those that used water power, because of the costs of building a waterwheel, holding tanks, and an aqueduct or *levada*, were more expensive to construct but had a greater productive capacity. Ambrósio Fernandes Brandão, author of the *Diálogos das grandezas do Brasil* (1618), estimated the cost of setting up an *engenho* at 10,000 *cruzados* (4:000$) without counting the construction of buildings or operating expenses for the first year. A so-called *engenho real* could produce 10,000 *arrobas* a year or even more, although few did so. Animal-powered mills sometimes called *trapiches* or *engenhocas* were usually turned by teams of oxen. They averaged 3,000–4,000 *arrobas* a year but were cheaper to build initially.[40] It was estimated in 1639 that in Pernambuco a *trapiche* could process about thirty cartloads of cane and produce half a ton (twenty-five to thirty-seven *arrobas*) per day while an *engenho real* could mill forty-five cartloads and produce a maximum of one ton per day (fifty to seventy-five *arrobas*).[41] As already noted, the introduction of *engenhos* of three vertical rollers in the early seventeenth century was quickly adapted throughout Brazil. The new construction was adapted to both water and animal power.

The juice squeezed from the cane was then passed through a series of kettles and teaches in the boiling house, where through a process of clarification and evaporation the liquid was purified and the impurities skimmed off. The iron and copper kettles, in a 1663 set of instructions to a *feitor-mor*, were referred to as "the most important things at the *engenho*" and were major items of expense and in constant need of repair.[42] There was no local copper available, so its importation was vital to the industry's health.[43] This may explain to some extent the flurry of Portugal's diplomatic efforts to establish friendly trade with Sweden, a major copper source, after 1641. The process of clarification depended on the heat of large furnaces beneath the kettles, known to the planters as "great open mouths" because they swallowed unlimited amounts of firewood. On Bahian *engenhos* the cost of firewood was usually about 20 percent of operating expenses. Until the introduction of the more fibrous *cana caiena* in the late eighteenth century, Brazilian *engenhos*, which processed the *cana crioula*, rarely made use of *bagaço* (the hulks of the pressed cane) as a fuel and depended instead on the seemingly unlimited forest resources of the colony for fuel. The result was a destruction of large tracts of the Atlantic forest.[44]

The work in the boiling house demanded considerable knowledge and skill. Under the direction of the *banqueiro*, the workers at each of the *caldeiras* moved the clarifying liquid through the kettles with large ladles until the purified and thickened fluid could be poured into large clay forms that were then placed in a separate building, the purging house (*casa de purgar*) where they were arranged in long rows. The crystallizing sugar in the forms was periodically covered with moistened clay. The water in the clay then percolated through the forms of crystallizing sugar, further draining impurities and producing a form in which white sugar predominated. The residue was reprocessed to make lower grades of sugar and the molasses drained from the forms was distilled to make *cachaça*. Padre Antonil, with an eye to both theology and profit, pointed out that dirty mud turned the sugar white just as the filth of sins mixed with tears of repentance could cleanse our souls.[45] Brazil's concentration on the production of this white, "clayed" sugar gave the colony a comparative advantage over its Caribbean competitors who tended to produce brownish *muscavado* sugars.

After four to six weeks the forms were emptied on a large platform (*balcão*) and under the sun, the white sugar separated from the brown *muscavado* and the lower grades under the direction of slave women, the so-called "mothers of the platform *mães de balcão*." Depending of the quality of the sugar and the skill of the purgers, the ratio of white to brown was usually 2:1 or 3:1. Still, making sugar, said João Peixoto Viegas in his famous memorial, was like the act of procreation: one had to wait until the end of the process to see the result.[46]

Brazil specialized in producing the white sugar that was more highly valued than *muscavado*, but which also tended to eliminate the need for further refining. Thus its metropole, Portugal, unlike Holland and England, did not develop a refining industry until the eighteenth century.[47] The Brazilian *engenhos* also produced lesser grades of sugar and from the molasses they made alcohol, or, as it was called regionally, *cachaça* or *geribita*. During difficult times, Brazilian *senhores de engenho* argued that they only met their expenses in the making of sugar and depended on the sale of *cachaça* for profit. Some regions like Rio de

Janeiro came to specialize in the production of *geribita*, which was used in the African slave trade, but in the seventeenth century, the production of white sugar predominated in the colony.

Finally, under direction of the crater (*caixeiro*), the tithe was subtracted, and when necessary division was made between the *engenho* and *lavradores de cana*. The separated sugar was then packed in large wooden crates weighing in the seventeenth century about 200–300 kilograms (14–20 *arrobas*). These were then registered by the crater, marked with the weight, quality, and sign of ownership, and then transported by ox cart or boat to the main port.

From the foregoing description it is clear that the Brazilian sugar industry paralleled the other Atlantic sugar economies in its basic elements and structure. A Brazilian *engenho* needed a large labor force, some of it possessing considerable experience or skills. On the average, *engenhos* in Bahia and Pernambuco had 60–70 slaves as part of the workforce but also drew on the labor of the slaves of the dependent cane farmers so that the total effective number of workers per mill was about 100–120. Each mill also required adequate supplies of the raw material, sugarcane, large amounts of fuel, usually in the form of firewood, as well as food to feed the labor force, and a variety of materials and equipment. In this the Brazilian sugar industry reproduced the patterns established by its Atlantic and Caribbean predecessors.

Three Keys to the Brazilian Sugar Economy

Three key elements determined the nature of the Brazilian sugar economy and its success and gave it its peculiar contour and character. These elements, namely, the structure of ownership, the supply of labor, and access to credit, are all related to a lack of capital in the early stages of the industry that contributed to patterns of organization and practice that persisted in Brazil for centuries.

The first of these elements lay in the structure of production and ownership. Brazilian sugar mills were owned by the state, by institutions, or private individuals. In the earliest days of the industry a few mills had been built with royal funds as a means to encourage settlement and economic growth. As late as 1587, there was still a royal *engenho* in Bahia, in Pirajá close to the city, but it was leased to a private individual.[48] By the next century, however, the Crown had withdrawn from direct participation and preferred to stimulate the industry by granting lands and tax exemptions to private investors.

Institutional owners held some sugar mills, the most important of these being the Religious Orders, particularly the Jesuits, Carmelites, and the Benedictines. The Jesuits, present in Brazil after 1549, were originally supported by royal

subsidies and private bequests.[49] Although at first reluctant to engage in plantation agriculture, especially that using slave labor, because of the possible dangers to their vows of poverty and Christian charity that such activities implied,
the Jesuits found by the beginning of the seventeenth century that agriculture
and stock raising could provide an economic basis for their missionary and
educational activities. In Bahia they began to develop two small mills in the first
decade of the seventeenth century, but a major breakthrough took place when
the Jesuit College of Bahia and that of Santo Antão in Lisbon acquired by
bequest *engenho* Sergipe in Bahia and *engenho* Santana in Ilhéus, both of which
had belonged to Mem de Sá, a former governor of Brazil. Although the ownership of these estates was the cause of long litigation that pitted the two Jesuit
colleges against each other as well as against other claimants, these mills, especially *engenho* Sergipe, "Queen of the Recôncavo," were important assets. Later
in the seventeenth century both the Jesuit college of Olinda and that of Rio de
Janeiro also acquired sugar estates.[50]

Other religious orders also became involved in the sugar economy. The
Franciscans, Carmelites, and Benedictines in Bahia all cultivated sugarcane at
various times and the Benedictines and Carmelites eventually had their own
mills.[51] The Benedictines, only established in Brazil after 1581, acquired cane
fields in the Bahian Recôncavo when a cane farmer, Gonçalo Anes, took vows in
that Order. Adding to that property by purchase, they eventually erected a mill,
São Bento dos Lages, sometime prior to 1650. By the mid-seventeenth century,
over 60 percent of the income of the Bahian Benedictines was derived from
sugar. In Pernambuco, the Benedictines of Olinda owned *engenho* Musurepe,
which was functioning from the second decade of the seventeenth century,
while the Benedictines of Rio de Janeiro depended on *engenho* Guaguaçu.
Although the records of these institutionally owned estates have sometimes
survived and thus provide the best documentation available on the colonial
sugar economy, these mills were the exceptions in terms of ownership. Still,
the Benedictines apparently knew how to manage their estates effectively (see
table 6.2).

The vast majority of sugar mills were privately owned. Partnerships were not
unknown and a few of the earliest mills were joint ventures in which a number
of investors pooled their resources, but individual ownership was the most common form. Eventually, the ownership of more than one mill also became common, a situation caused to some extent by technological bottlenecks created by
the limited capacity of the mills and the problems of transporting cane long
distances. Thus the tendency to increase capacity by creating a new unit was
common, resulting in individuals and families owning more than one mill.

TABLE 6.2. Profitability of *Engenho* São Bento dos Lajes, 1652–1714
(Values in *Milréis*)

Slaves	Expenses[a]	Earnings	Profit	Annual	Average
1652–56	87	16,018	44,239	28,221	7,055
1657–60	113	7,152	20,020	12,868	4,289
1662–67	115	6,632	14,076	7,444	1,861
1700–1703	117	4,140	14,356	10,216	3,405
1711–14	—	3,881	15,326	11,445	3,851

Source: Estados of the Mosteiro de São Bento da Bahia, Arquivo da Universidade do Minho. Congregação de São Bento, 136.

[a] Expenses have been increased by 20 percent to include the costs of slave replacement that were not reported by the Benedictines in their records.

Although sugar mills provided the economic foundation for a number of aristocratic planter families who remained the social elite for centuries, more usual was a history of rapid turnover and volatility of ownership. One of the distinguishing features of the sugar economy was this insecurity and turnover, a sign of the difficulties of plantership. For those individuals and families that were successful, the local avenues of power and prestige were fully in their hands. Prior to 1650 the municipal councils of Olinda, Salvador, and Rio de Janeiro as well as prestigious lay brotherhoods like those of the Misericórdia were dominated by the *senhores de engenho*. They came to see themselves as an aristocracy worthy of respect and deference despite the fact that the origins of most were not noble, and in fact many were the descendants of New Christians.[52] In Bahia, for example, they constituted over 20 percent of the mill owners for those mills recorded between 1587 and 1592.

Those men (and some women) who had neither the capital nor credit to set up a mill turned instead to the growing of sugarcane. From its beginnings of the Brazilian sugar industry had been characterized by the existence of cane farmers (*lavradores de cana*) who supplied cane to the *engenhos*. Even the original instructions for government carried by the first royal governor, Tomé de Sousa, in 1549 had recognized their existence and had sought to establish rules for their relationship with the *senhores de engenho*.[53]

It would appear that the Portuguese experience in the Atlantic islands, especially Madeira, had been particularly important in establishing the utility of cane farmers. Small-scale producers seem to have been part of the sugar industry there from the period of expansion after Diogo de Teive built the first *engenho* in 1452. In this regard, the *Livro do almoxarifado das partes de Funchal* (1494) is particularly important. It lists 221 *lavradores de cana* in the captaincy of Funchal, but records only sixteen *engenhos*. Other sources indicate that Madeira had perhaps eighty *engenhos* at the time. Clearly, these figures suggest that many people who cultivated cane did not own an *engenho*. These *lavradores de cana* included a few *fidalgos*, but most were of artisan background or held some administrative position on the island. They included a few foreigners—Flemish and Genoese—but the vast majority were Portuguese. Thus, as Alberto Vieira demonstrates in this volume, the islands had numerous small and medium producers, men and women, many of whom were linked by blood or marriage to each other or to the owners of the *engenhos*. Traces of the existence of such cultivators also exist in the Canary Islands, where as early as 1508 the municipal council of Tenerife sought to regulate the relations between the owners of the sugar mills and cane growers. In Santo Domingo and Puerto Rico there is also some evidence of the early existence of cane farmers, but this class does not seem to have survived long in those colonies. In Brazil, however, they became a regular and essential aspect of the sugar economy whose existence had profound implications on the structure of the economy and the operation of slavery. Before 1650, *lavradores de cana* cultivated the largest portion of the sugarcane produced in Brazil.[54] This was evidence of the diffusion of investment and risk that characterized the early Brazilian sugar industry.

The explanation for the existence and importance of cane farmers in Brazil is puzzling. Certainly, the tradition of small producers established in Madeira created a precedent, as did long-standing Portuguese practice of rural contracts or *arroteias*, but the key in Brazil may have been the relative shortage of capital for *engenho* construction in the initial stages of colonization and the desire of the Crown to stimulate settlement by presenting opportunities to potential colonists. In a way, the cane farmers are evidence of the shortage of capital in the formative stage of the colony. The Crown's attention to their existence, and its demands that those who received land grants to build the first *engenhos* provide protection and benefits to dependent cane farmers, underscored their importance to the project of colonization and to the establishment of the sugar industry. As early as 1548, the correspondence between the manager of *engenho* São Jorge in São Vicente and the absentee owner noted the presence of cane farmers, but he also presented arguments why milling their cane was costly and perhaps unnecessary.[55] This was a tension that remained in the Brazilian sugar economy into the nineteenth century, but prior to 1650, the *lavradores de cana* were the most distinguishing feature of that economy.

While the term *lavrador* was used for any kind of farmer in Brazil, the *lavradores de cana* were, in fact, an agricultural elite, ranking just below the status of the *senhores de engenho* and often sharing many of the same social origins, features, and aspirations; but also because of the nature of their dependency, often in conflict with the mill owners. The nature of their relationship and their status depended on their tenure and access to land (see figure 6.2). *Lavradores de cana* who held their land by grant (*sesmaria*) or purchase were in effect small-scale landowners and were in the best position for bargaining with mill owners. Those with this so-called "free cane" usually divided the sugar produced from his or her cane, one-half to the mill and one-half to the *lavrador,* and they could negotiate other advantages such as the loan of oxen, help with the transport of cane, or preference in the schedule of the mill. A *lavrador de canas* with many *tarefas* of free cane was often treated with "much coddling" by the *senhores de engenho* who needed his cane. The majority of the *lavradores de cana* did not have this advantage. They produced "captive cane" and held a *partido de cana* in which they rented land, and then were required to bring it to the owners' *engenho* and pay ⅓ or ¼ of their half of the sugar produced as a land rent. A *partido de terço* put great pressure on the *lavrador* but was preferred by the *engenhos.* In 1601, for example, the Conde de Linhares ordered the manager on his Bahian *engenho* to rent cane lands to *lavradores* on the ⅓ arrangement.[56] Usually, only *lavradores* with considerable resources in slaves and capital could accept the *partido de terço.* Figure 6.2 presents the arrangements that character-ized the relationship between mill owners and *lavradores.*

Lavrador contracts varied over time depending on local custom and on the current state of the sugar economy. In the period of expansion in the late sixteenth and early seventeenth centuries, many people were willing to accept the burdens of the *terço* or *quarto* contracts, but as the industry encountered difficulties the situation changed. By the close of the seventeenth century con-tracts of ⅕ were common in Pernambuco and in Bahia contracts of 1/10 or even 1/20 were used.[57] The length of tenure of a *partido* varied, although terms of nine or eighteen years were common. Often *lavradores de cana* were required to provide firewood (*lenha*) for the processing of their cane and worst of all, at the end of the tenure, all improvements made to the land became the property of the *engenho.* Moreover, sales of land with obligations to provide cane to a particular *engenho* were often the cause of legal battles when, after a series of subsequent sales or transfers, the *engenho* sought to insist upon the original obligations. These arrangements and disadvantages contributed to the instabil-ity of the *lavradores de cana* as a class. Over a period of eighteen *safras* (1622–50) at *engenho* Sergipe in Bahia, 128 individuals appeared as *lavradores* but only 41 percent (53) appeared in more than one *safra* and only 19 percent (24) appeared in more than five.[58] When *lavradores* were forced to surrender the lands that

FIGURE 6.2. Mill Owner and Cane Farmer Arrangements

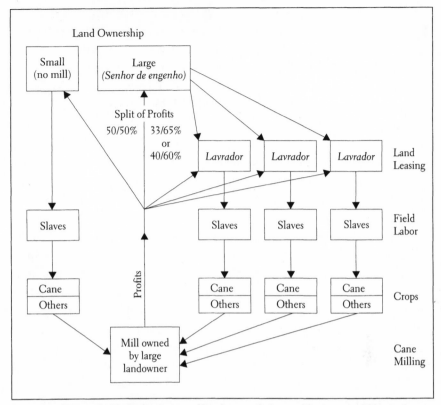

Source: David Watts, *The West Indies: Patterns of Development, Culture and Environmental Change since 1492* (Cambridge: Cambridge University Press, 1987), 187.

they had worked and improved for years they often objected and resisted either physically or in the courts. Finally, there were also other forms of the rental of cane lands, usually in multiples of nine years, with various forms of obligation or payment attached to them.

The relationship between *senhores de engenho* and *lavradores de cana* was complex because of their need for each other as well as the conflict inherent in their relationship. An *engenho* might have as many as thirty *lavradores* supplying cane in a single harvest but the average number of *lavradores de cana* per *engenho* in the Brazilian northeast was probably three to four. In Pernambuco there were 250 *lavradores* supplying cane to about 166 mills in 1639. This situation allowed many people relatively easy entry into the sugar economy, often with the hope of social mobility. Startup costs for a *lavrador de canas* were about one-third of those for a mill owner. From the viewpoint of the *senhores de engenho* the existence of *lavradores de cana* was a way of sharing the risks and financial burdens of sugar production. In Bahia, about one-third of the slaves employed in sugar production were owned by the *lavradores de cana* rather than by the *engenhos*. *Senhores* wanted and needed *lavradores* but feared that when *lavradores* acquired their own lands they would be able to negotiate better arrangements for milling their cane or they would build their own *engenhos* and thus create competition for cane and firewood. One strategy was then to sell land to *lavradores*, but with restrictions forcing the buyer to provide his cane to the *engenho* of the seller in perpetuity or to pay other penalties if cane was brought elsewhere. *Lavradores* responded with their own strategies, often illegally bringing "captive cane" to other *engenhos*, especially during poor years when the demand for cane was high and many people could not meet their obligations.

This situation finally created a crisis in Bahia in the 1660s when Bernardino Vieira Ravasco, brother of the famous Jesuit Padre António Vieira, a *senhor de engenho* and secretary of the state of Brazil, led a movement within the municipal council of Salvador to limit the construction of new *engenhos*. He argued that competition had placed the *senhores* at the whim of the *lavradores* who, now driven by "vanity or deceived by greed," sought to build new *engenhos*, thereby ruining themselves and others. This suggestion met serious competition from many *senhores de engenho* who argued that if the hopes that *lavradores* had of becoming *senhores de engenho* were eliminated, then none would be willing to serve as *lavradores de cana*. Eventually in Bahia in 1681 and 1684 the Crown issued laws that limited construction of new *engenhos* to within 1,500 *braças* (about two miles) from existing ones. The effect of this was to stimulate the opening of new sugar areas further from the coast. Similar laws were issued for other captaincies. While *senhores de engenho* disliked the potential competition of new *engenhos* and the comparative advantage that *lavradores de cana* had when many *senhores* competed for their cane, they also realized that without the hope of social mobility that few people would accept the burdens of growing sugarcane. The *lavradores de cana* were a permanent aspect of the Brazilian sugar economy and in its early stages a measure of its economic status.

The *lavradores de cana* thus worked under a variety of arrangements and varied considerably in terms of their wealth and social status. It is difficult to estimate the average size of a *partido de cana*. At *engenho* Sergipe, the majority seemed to hold under six hectares of land, about half of it devoted to cane, but land sales indicate that larger parcels between thirty and two hundred hectares were also purchased. We have no good quantitative source on the holdings of *lavradores* in the seventeenth century, but a *partido* of a few hectares and per-haps five or six slaves was probably common.[59]

These variations in wealth and capital indicate the range of *lavrador* social positions. Counted among *lavradores de cana* were some of the most prestigious individuals in the colony, many of whom were related by blood or marriage to the *senhores de engenho*. The religious orders at times cultivated cane and provided it to *engenhos*. Individual clerics also were *lavradores de cana* and some, like one priest in Pernambuco in 1639 with sixty *tarefas* of cane, operated on a large scale. There were also a good number of women, often widows, who participated in the sugar economy. Noticeable, however, until the late eigh-teenth century was the fact that *lavradores de cana* were almost invariably white. Free blacks and mulattos simply did not have the credit or capital to take on the burdens of this agriculture. Their absence underlines the relatively high social status of the *lavradores de cana* as potential planters. It was a status that few of them actually achieved, but the promise was always an attraction.

Overall, the *lavradores de cana* and *senhores de engenho* were united by interest and by their dependence on the international market. Together they formed the "nerves of the body politik," in the words of Wenceslão Pereira da Silva in 1738. Antonil admonished *senhores* to treat their *lavradores* well, and one administrator at *engenho* Sergipe reported in 1623 that he had to treat the *lavra-dores* carefully because "in this land everything is respect and courtesies."[60] But many *senhores* abused their power. In last analysis, each side needed the other. The *lavradores de cana* were in many ways proto-planters, owning oxen, slaves, and sometimes land. They were often drawn from the same social strata as the great planters, and shared many of the same attitudes. They cooperated in conflicts with the merchants and in seeking a moratorium on their debts, a general concession that was achieved in Bahia in 1663 in a law that prohibited foreclosure of an *engenho* for debts less than its total value; the law was extended to Bahian *lavradores de cana* in 1720 and to other captaincies thereafter.

The second characteristic of the early Brazilian sugar industry was its rela-tively long dependence on an indigenous labor force and its gradual shift to Africans. For the first seventy years or so, the industry depended on indigenous workers. This, too, suggests the lack of capital or credit to finance the more expensive importation of African workers as slaves. African and Afro-Brazilian slaves eventually predominated in the sugar economy, but the process by which that happened took place over a long period of over half a century.[61]

The transition from Indians to Africans as laborers was a key element in the expansion of the Brazilian sugar economy at the end of the century. With the demands of sugar agriculture growing by the 1560s, Indian labor could no longer be obtained by barter as when the principal Portuguese activity was the gathering of brazilwood. Indians refused to work for wages or sometimes demanded goods like firearms that the Portuguese were reluctant to supply. Moreover, for many groups such as the Tupinambá, agriculture was considered the work of women, and men refused to do it. Portuguese attempts to acquire native workers by ransoming war captives and then holding them as temporary slaves was increasingly opposed by the Jesuits, who claimed that Indians in Jesuit-run villages could provide labor to the engenhos more efficiently and with fewer abuses. By 1600, they claimed to have 50,000 Indians under their control and available to both the Crown and the colonists. The Jesuits claimed to be able to provide 400–500 workers a month to the settlers for a salary of 400 réis per worker that usually went unpaid.[62] Colonists and Jesuits disputed control of Indian labor after the 1550s and the policies of both proved disruptive of Indian life. Meanwhile, the Crown increasingly legislated against the enslavement of Indians with laws in 1570, 1595, and 1609. During this period, nevertheless, Indians, both enslaved and free, were the primary labor force in the sugar economy, and they remained so until the first decades of the seventeenth century.

Demography was also a major factor in the transition. Diseases, first smallpox and then measles, decimated the Indian populations between 1559 and 1563. Thousands died, whole villages were abandoned, and many groups fled to the interior, spreading the disease. The Portuguese responded by sending new columns into the backlands to bring in more workers, and by moving groups from one captaincy to another, but such policies were costly, as were the military operations needed to confront the Indian resistance they provoked. The susceptibility of Indians to disease made sugar planters reluctant to invest in acquiring more Indians or in training them in the technical aspects of making sugar.

The transition of the labor force from Indians to Africans took place slowly over a period of about half a century. In the early years, the relatively low cost of Indian workers had facilitated the industry's rapid growth, but as overall expenditures associated with Indians rose, Africans were increasingly seen as cost-effective replacements. African slaves were first acquired as early as the 1540s. Many seem to have been skilled workers, and some undoubtedly had already labored on engenhos in Madeira or São Tomé. At Engenho São Jorge in 1548, for example, there were only seven or eight Africans, serving as sugar master, purger, and kettlemen. The records of engenho Sergipe show that only 7 percent of its workforce was African in 1574, but this rose to over 37 percent by 1591; by 1638 the workforce was totally African or Afro-Brazilian.[63] In Pernambuco, two-thirds of sugar laborers were Indian in 1580, but the demand for Africans was increasing, despite their being more expensive to obtain. The Portuguese, already believing that the Indians were more susceptible to disease and likely to run away, considered the Africans as stronger and better workers; at the same time the cost of acquiring Indians was on the rise. In 1572, at *engenho* Sergipe in Bahia, an African worker was valued at 25$000, while an Indian with similar skills averaged only 9$000, but in the long-run Africans were proving to be a more profitable investment. The skilled African workers at *engenho* São Jorge, for example, had eliminated the need to hire salaried employees.

The transition from a native American labor force to one composed primarily of Africans and their descendants was paralleled by a second transition from mostly free, white, skilled workers to sugar-making specialists and artisans who were either slaves or free people of color.[64] In the early stages of the sugar industry in Brazil, often as many as twenty whites had worked for an annual salary or provided services for wages. Sugar masters, kettle men, overseers, blacksmiths, carpenters, shipwrights, and stonemasons all were needed. Workers were paid differentially according to not only their skill but to their ethnicity as well; whites were always paid more generously than blacks or mulattos, and Indians paid least of all for the same tasks. Salaries for such workers constituted one of the largest expenses for an estate. Given the lack of specie, some mills only settled accounts with their salaried employees every two or three years, but over time, a general tendency emerged to replace these white artisans with either slaves or former slaves who had gained their freedom, and for whom such occupations provided a means of social mobility. Access to these positioned served as an incentive to *engenho* slaves. Planters favored mulattos and native-born blacks (*crioulos*) with these positions. From the planters' point of view, the replacement of free white workers with slaves or by freedmen who could be paid less than whites was another way of facing the costs of plantership. We have little direct evidence of this transition. The letter of 1548 from the administrator of *engenho* São Jorge in São Vicente to the absentee owner noted with pride that the *engenho* was now saving 30$000 a year by using an African as sugar master, since that was the amount usually paid to a master brought from Madeira.[65] The records of Engenho Sergipe in Bahia are also suggestive of the process. Whereas early accounts had usually identified employees by their color or ethnicity when there was a multi-ethnic workforce made up of whites, native Americans, and Africans and Afro-Brazilians, by 1670 racial distinctions had all but disappeared in the account books because the vast majority of workers were now either black or mulatto; thus such distinctions no longer served much of a purpose. This shift to an Afro-Brazilian skilled labor component was a result of the intensification of the Atlantic slave trade and the demographic shifts it generated, which created an opportunity for planters to reduce their operating expenses by turning to a growing Brazilian population of mixed origin. A century or so later, planters would complain that the ignorance of these skilled workers was the cause of the sugar industry's inability to compete, but in the seventeenth century, planters saw the use of these workers as a necessary, and quite positive, response to the costs of plantership.

Finally, the access to capital and credit and the level of profitability were key factors in the success of the sugar economy. How did this industry grow in its early stages and what was the effect of this growth, as Brazil became the world's leading supplier of sugar? In 1618, the New Christian Ambrósio Fernandes Brandão argued that many Portuguese who had made a fortune in India returned to Portugal to spend it and live the good life, but rarely did someone who had become wealthy in Brazil return to the home country. The reason was that the wealth of Brazil was in realty and thus not so moveable. Despite occasional observations about the opulent life style of the great planters, many of them lived rather simply, sinking their fortunes into the building of their estates. Planters were always complaining of their indebtedness and the costs of plantership, but clearly considerable wealth was created at least in the first seventy years of the industry's growth.

Calculating that wealth, however, remains a difficult task. The issue is complicated by a lack of documentation for the period in question and by accounting practices that mixed capital-stock expenditures with current expenses.[66] Planters simply calculated annual income against expenses to know how they were doing. This often gave them a false impression of their economic status. Then too, Brazil and its metropolis, Portugal, were chronically lacking money in circulation, especially in the period prior to 1580. This led to practices such as the exchange of goods for services or for other commodities, payments over time, and a dependence on complex credit arrangements. The manager of *engenho* São Jorge put it clearly in 1548: "For here there is no circulation of money and one must by force give things on credit for a year and before being paid wait for two years. In this way, anyone who has an engenho here pays all his workers in goods." This situation changed somewhat between 1580 and 1620 when the Portuguese in Brazil got access to Peruvian silver by contraband through Buenos Aires, which in 1605 the Crown estimated to be as much as 500,000 *cruzados* in coin and bar a year.[67] But this door was closed after 1621 and the previous conditions of shortage returned. Thereafter, the planters generally saw the lack of circulating specie as a major cause of their continual indebtedness, and it complicated any attempt by them to calculate the creation and distribution of assets, as it complicates any modern attempt to make these calculations. Thus for the first half century of the sugar industry we must depend on occasional observations, a few contemporaneous estimates, and comparison by inference with latter patterns.[68]

In the early years of the industry, while some capital came from noble or foreign merchant investors or the state, many of the mills were set up depending on credit extended by merchants in the sugar trade. In this period, land was often acquired by grant and labor acquired by the capture of Indians, which kept down the original fixed capital costs and provided an impetus for the formation of capital. Still, buildings and machinery had to be constructed, kettles and sugar forms had to be bought or made, livestock acquired, boats and ox-carts to move the cane built, and cane lands prepared or contracted. One source of capital for the sugar industry seems to have been governmental offices. The recent studies of João Fragoso for the development of the sugar economy of Rio de Janeiro reveal that the majority of Rio's sugar planter families established before 1620 had held administrative offices that had apparently been used to open doors for the accumulation of wealth or the gaining of other advantages that had then made plantership possible.[69] Succeeding generations owned sugar mills and commonly held posts in the municipal council of Rio de Janeiro, thereby continuing the union of office and fortune. Royal office, the holding of tax contracts, and municipal office all generated capital that was invested in the sugar industry. Similar patterns seem to have existed in Bahia and Pernambuco.

Those wishing to enter the business of sugar making usually found specie to be scarce, and thus credit was essential to begin operations for both planters and cane farmers—the latter sometimes depending on the former for access to it. If later patterns can be used as a model, many plantations were set up with an outlay of only about one-third of the necessary capital, the rest being supplied by credit. This allowed people of relatively modest means to aspire to the status of senhor de engenho, and it meant that their returns were considerably higher than those implied by the ratio of capital to annual income.

Credit was obtained from a variety of sources. The charitable brotherhoods (misericórdias), convents, and other religious institutions were the major sources, loaning money on easy terms of about 6.25 percent to low-risk or high-profile borrowers. These loans were often very long term. Less advantaged borrowers contracted loans at much higher effective rates from merchants who found ways to avoid the limitations on usury. Many senhores established an engenho, depending primarily on credit, but this often led to later conflicts with merchants over foreclosure for debt. The lack of notarial records for this period, from Pernambuco or from Bahia, are a serious impediment to determine the nature of credit arrangements. What is noticeable is that the Amsterdam notarial registers, where many transactions involving New Christian investors linked to the Brazil trade and to the Portuguese imperial economy appear, reveal virtually no evidence of direct investment in the production of sugar.[70] It would appear that credit was being extended for the most part by local merchants and correspondents in the colony rather than from the European sources.

During the rapid growth of the industry after 1570, a number of observers spoke of the wealth and opulence of the sugar planters, their taste for lavish hospitality, high living, and the symbols of a noble life style. In Antonil's often-cited expression, to be a *senhor de engenho* in Brazil was equivalent to having a title among the nobility of Portugal. But prestige was not the same as wealth. Despite a taste for luxury, planter returns on capital do not seem to have been as extraordinarily high as have been projected by some modern estimates that have overestimated output and underestimated costs.[71] Labor was an essential element of these expenses, both as a fixed cost in the form of the purchase, replacement, feeding, and care of slaves, perhaps about 25 percent of yearly expenditures, but also in the form of salaries paid to sugar-making specialists, artisans, and occasional workers, or about 20 to 30 percent of annual costs. As we have seen, this was one area where sugar planters sought to cut expenses.

In the early seventeenth century, an *engenho* could be set up for 8,000–10,000 *cruzados* (3,600$). By the end of the century the average value of a Bahian *engenho* was about 15,000, not counting the slaves, and perhaps 18,000–20,000 *cruzados* with them. Capital was distributed among various assets (buildings, equipment, and livestock), with land consistently the most valuable one, usually constituting half the *engenho*'s total value. The slave force was usually around 20 percent of the capital value. During this period, a return of 2,000$ to 3,000$ on an *engenho* worth 20,000$, or a return of 10 to 15 percent, was considered very good and not always achieved. The Jesuit *engenho* Sergipe do Conde in Bahia, one of the few *engenhos* for which we have account books, ran at a deficit for many years in the seventeenth century, a fact that has long confused historians. But its problems seem to have been the result of mismanagement and accounting procedures that took no notice of capital investments and counted them as operating expenses, and it may have been a special case.[72] In the same period, the *engenho* São Bento dos Lages of the Benedictine Fathers in Bahia produced excellent earnings in the mid-seventeenth century, as did the Benedictine *engenhos* in Pernambuco.

Throughout the seventeenth century period an annual return on capital of between 5 and 10 percent for the industry as a whole was probably common, although in periods of expansion higher rates were possible. *Lavradores de cana* faced even more difficult odds. We have Father Estevão Pereira's estimate of 1635 that the *partidos* or dependent cane fields of the *lavradores de cana* supplying *engenho* Sergipe were valued at 16:000$. It was calculated that these lands produced 500 *arrobas* of whites and 250 *arrobas* of *muscavado*, which at 800 and 360 *réis*, respectively, would generate a return of 490$000, or about a 3 percent return on capital. Surely others did better than this and some *lavradores* eventually built their own mills, but the opportunities for quick fortunes seem limited.

Still, cash flow may not be the best way to evaluate the business of making sugar. Much of the industry's early gains may have been in the form of capital creation as the value of assets grew more quickly than income, suggesting a high rate of savings. We should remember that many of the first *engenhos* in the previous century acquired lands by *sesmaria* and Indian workers by capture at relatively little monetary cost, so that capital value grew rapidly. The clearing of land, the building of chapels, houses, and buildings, the construction of aqueducts and waterwheels all increased capital value and represented the building of personal wealth. This in turn created assets that allowed for an expansion of credit. Here the importance of familial and other personal ties so common in Early Modern commerce also played a role, explaining the active participation of the New Christians in all aspects of the industry that linked merchants to planters, managers, and artisans.

For the industry as a whole, the period between 1560 and 1620 probably witnessed the greatest gains in wealth, with a considerable slowdown thereafter as sugar prices declined and costs increased as a result. The foundational generation of planters had acquired much of their land by grant and their labor by capture or as unpaid or modestly paid workers from the Jesuits. This process had kept their expenditures down and thus their gains increased. By 1620 or so, the best lands close to the littoral had been occupied so that new expansion had to be made on lands further from the coast where transportation costs would be higher. *Sesmarias* became less common and new lands were increasingly acquired by purchase. Royal measures to eliminate Indian slavery and Jesuit opposition to it made the acquisition of indigenous laborers more difficult and more expensive and only the introduction of the three-roller mill kept the process of expansion moving forward, although now at a slower pace. With the crisis of 1623 and the subsequent fall of sugar prices in the Atlantic market, and then with the Dutch invasion of 1630 and the disruption it caused, including the rising level of slave resistance and escape, the Brazilian sugar industry moved into a new stage of stability and slow expansion in which the exigencies of war and politics played a role more important than the benefit and blessings of climate and rainfall. By the time that the new Caribbean competitors in

Barbados, Suriname, Jamaica, and Martinique were challenging Brazil's predominant position, the sugar industry was already experiencing considerable difficulties created by the internal organization of the industry and the social organization and strains it had created. Sugar remained the single most valuable agricultural commodity of Brazil until the middle of the nineteenth century, and sugar planting remained a difficult and sometimes profitable business through the eighteenth century. But the heady days of the late sixteenth and early seventeenth century never returned in quite the same way, although the hope and memory of them lingered in the minds of those who could aspire to the title *senhor do engenho* and to the wealth, power, and authority it had come to represent.

NOTES

Abbreviations

AHU   Arquivo Histórico Ultramarino (Lisbon)
ANTT   Arquivo Nacional da Torre do Tombo (Lisbon)
CSJ   Cartório dos Jesuítas

1. Cuthbert Pudsey, *Journal of a Residence in Brazil*, ed. Nelson Papavero and Dante Martins Teixeira, in the series, *Dutch Brazil*, 3 vols. (Petrópolis: Petrobras, 2000), 3:25. Pudsey's impression was not singular. Frei Vicente do Salvador, Brazil's first historian, noted that in Brazil things that were inverted for the whole colony did not form a republic, but rather each house seemed to be one. See discussion in Fernando A. Novais, "Condiçoes da privacidade na colônia," in *História da vida privada no Brasil*, ed. Laura de Mello, 4 vols. (São Paulo: Companhia das Letras, 1997–98), 1:13–40.

2. Hermann Kelenbenz, "Relaçoes econômicas entre Antuérpia e o Brasil no século XVII," *Revista de História* 33 (1968): 295; Eddy Stols, "Um dos primeiros documentos sobre o engenho dos Schetz em Sáo Paulo," *Revista de História* 33 (1968).

3. José Antônio Gonsalves de Mello and Cleonir Xavier de Albuquerque, eds., *Cartas de Duarte Coelho a El Rei* (Recife: Imprensa Universitaria, 1967), 71.

4. Domingos Abreu e Brito, *Um inquerito a vida administrativa e económica de Angola e do Brasil (1591)*, ed. Afredo de Albuquerque Felner (Coimbra: Imprenta da Universidade, 1931), 58–59.

5. "Provincia do Brasil," ANTT, Convento da Graça de Lisboa, tomo VI F. This document is analyzed in Artur Teodoro de Matos, "O império colonial português no início do século XVIII," *Arquipélago* 1, no. 1 (1995).

6. Memorial of Joseph Israel da Costa, Algemein Rijksarchief, Loketkas 6, Staten Generaal West Indische Compagnie.

7. AHU, Bahia, *papéis avulsos* caixa 1, 1st ser. uncat.

8. On the still unresolved question of the invention of the three-roller vertical mill, see John Daniels and Christian Daniels, "The Origin of the Sugarcane Roller Mill," *Technology and Culture* 29, no. 3 (1988).

9. G. B. Hagelberg, "Sugar and History: A Global View," in *Slaves With or Without Sugar*, ed. Alberto Vieira (Funchal: Centro de Estudos Atlânticos, 1996).

10. ANTT, CSJ, maço 13, doc. 4. In the harvest of 1611–12 at *engenho* Sergipe, the following entry was made in the account book: "to an artisan who helped Sebastião Pereira to make a gangorra for twelve days at 320rs"; see ANTT, CSJ, maço 14, doc. 4.

11. Coelho's original petition was discussed by the Colonial Council (Conselho da Fazenda) in July 1620. See Archivo General de Simancas, secretarias provinciales Portugal 1473, fs. 38–39v. His request was refused and he petitioned again in February 1622. Consulta, Conselho da Fazenda, AHU, cod. 34, fs. 29v–30. The idea of making newly conquered Maranhão into a sugar-producing region led the noble, António Barrieros, nephew of the Bishop of Brazil, to build "one of the old style *engenhos* or two of those they make nowadays," to help populate the region in a few years. The discussion of his request in council indicated recognition of the willingness of colonists to build the new style mills. See Consulta, Conselho da Fazenda, AHU, cod. 32, fs. 58v.–60.

12. See Suely Robles Reis de Queiroz, "Algumas notas sobre a lavoura do açúcar em São Paulo no período colonial," *Anais do Museu Paulista* 21 (1967).

13. See the discussion in Stuart B. Schwartz, *Sugar Plantations in the Formation of Brazilian Society: Bahia, 1550–1835* (Cambridge: Cambridge University Press, 1985), 170–73.

14. For a discussion of the effects of the fighting on sea and land on the sugar industry in Pernambuco, see Evaldo Cabral de Mello, *Olinda restaurada: Guerra e açúcar no nordeste, 1630–1654*, 2d ed. (Rio de Janeiro: Topbooks, 1998), esp. 89–141.

15. See Pedro Puntoni, *A míseria sorte: A escravidão africana no Brasil holandés e as guerras do tráfico no Atlântico sul, 1621–1648* (São Paulo: HUCITEC, 1999), 81–82. See also Hermann Wätjen, *O domínio colonial holandés no Brasil* (São Paulo: Editora Nacional, 1938).

16. See José António Gonsalves de Mello, ed., *Fontes para a história do Brasil holandês: A economia açucareira* (Recife: MEC/SPHAN, 1981).

17. "Breve discurso sobre o estado das quatro capitanias conquistadas," *Fontes Históricos do Brasil Holandés*, 77–129.

18. See the estimate provided by Puntoni, *A míseria sorte*, 78.

19. Van der Dussen, "Relatório sobre o estado das capitanias conquistadas no Brasil," *Fontes Históricos do Brasil Holandés* 1, 137–232.

20. "Cartas e pareceres de Gaspar Dias Ferreira," *RIAGHP* 31 (1886) and *RIAGHP* 32 (1887), cited in C. R. Boxer, *The Dutch in Brazil* (Oxford: Oxford University Press, 1957), 143. The same point is made by José António Gonsalves de Mello, *Tempo dos flamengos: Influência da ocupaçao holandesa na vida e na cultura do norte de Brasil*, 2d ed. (Recife: Secretaria da Cultura, 1978), 134–35. This statement of Dutch ineptitude was often repeated by the Portuguese but contested by some of the Dutch who argued that admittedly with instruction from specialists from Brazil, the English and Dutch eventually learned sugar making well enough to become major competitors to the Portuguese.

21. Evaldo Cabral de Mello, *Olinda restaurada.*

22. The essential books on the early sugar trade are Frédéric Mauro, *Le Portugal et l'Atlantique au XVIIe siècle* (Paris: SEVPEN, 1960); Hermann Kellenbenz, *Unternehmerkräfte im hamburger Portugal-und Spanienhandel, 1590–1625* (Hamburg: Hamburgischen Bücherei, 1954); and Eddy Stols, *De Spaanse Brabanders of de handelsbetrekkingen der Zuidelijke Nederlanden met de Iberische Wereld, 1589–1648,* 2 vols. (Brussels: Paleis der Academiën, 1971). To these should now be added Leonor Freire Costa, *O transporte no atlântico e a companhia geral do comércio do Brasil (1580–1663),* 2 vols. (Lisbon: Comissão Nacional dos Descobrimentos Portugueses, 2002).

23. Public Record Office (London), SP 9/104. These records have been studied more intensively by José Antônio Gonsalves de Mello in "Um mercador cristão novo e seu livro de contas: Miguel Dias Santiago," in *Gente da nação: cristãos novos e judeus em Pernambuco, 1542–1645* (Recife: Fundação Joaquim Nabuco and Ed. Massangana, 1989), 35–49.

24. ANTT, CSJ, maço 11, no. 5. This is a register of the lading of the *engenho's* sugar shipping; it includes the name of the ship, the name and place of origin or residence of the captain, the cargo expressed as the number of crates and the weight of the sugar, and the costs in fees and taxes.

25. J. A. Gonsalves de Mello, "Os livros de saídas das urcas do porto do Recife," *Revista do Instituto Arqueológico, Histórico e Geográfico Pernambucano* 58 (1993). See also Manuel António Fernandes Moreira, *Os mercadores de Viana e o comércio do açucar brasileiro no século XVII* (Viana do Castelo: Preifeitura Municipal, 1990), for a view from a small Portuguese port in the mid-seventeenth century.

26. Archivo General de Simancas, Guerra antigua 690.

27. See John Everaert, "Les barons flamands du sucre à Madère," in *Flandre et Portugal: Au confluent de deux cultures,* ed. J. Everaert and E. Stols (Antwerp: Fonds Mercator, 1991). On the continuity of the sugar market from Bruges to Antwerp, see W. Brulez, "Brugge en Antwerpen in de 15e en 16e eeuw: een tegenstelling?," *Tijdschrift voor geschiedenis* 83 (1970).

28. This point is forcefully made by Jonathan Israel, "The Economic Contribution of Dutch Sephardi Jewry to Holland's Golden Age, 1595–1713," in *Empires and Entrepots: The Dutch, the Spanish Monarchy and the Jews, 1585–1713* (London: Hambledon Press, 1990).

29. Still the best study of the Brazilian New Christians is Anita Novinsky, *Cristãos novos na Bahia* (São Paulo: Ed. Perspectiva, 1972). See also José Gonçalves Salvador, *Os cristãos novos: povoamento e conquista do solo brasileiro* (São Paulo: Livraria Pioneira, 1976).

30. A typical entry read as follows: "The lading made by me, Miguel Diaz in Bahia of All the Saints on the galleon St. Andre which is captained by Bartholomew Balde, a *vezino* of Ragusa. On behalf of Senhor Manuel Gomes da Costa and consigned to him or to whoever bears his certified message."

31. AHU, Bahia papéis avulsos caixa. See also *Calendar of State papers, Venice,* no. 859 (24 February 1599), on Venetian shipping to Brazil.

32. Fernandes Moreira, *Os mercadores de Viana,* 20–27.

33. See Anna Amélia Vieira Nascimiento, *"Letras de risco," e "carregações" no comércio colonial da Bahia, 1660–1730,* Estudos Bahianos 78 (Bahia: Centro de Estudos Bahianos, 1977).

34. The caravel was actually banned from the Brazil trade in 1648. See AHU, cod. 14, f. 146v. I provide a fuller description of the end of private trade and the role of the Brazil Company in *Sugar Plantations*, 180–81. On the Brazil Company we now have the extensive study of Freire Costa, *O transporte no Atlântico* and also her more general and perceptive essay on Portuguese commerce. See Leonor Freire Costa, *Império e grupos mercantis: Entre o Oriente e o Atlântico (século XVII)* (Lisbon: Horizonte, 2002).

35. Domingos de Loreto Couto, "Desagravos do Brasil e glorias de Pernambuco," *Anais da Biblioteca Nacional do Rio de Janeiro* 24 (1902): 171.

36. During the Dutch occupation of Pernambuco, the custom of the blessing of the mill and the workers by a priest at the beginning of the harvest was so entrenched that even Dutch mill owners allowed it, to the consternation of the governing board of the Christian Reformed Church, which objected to the presence of such "superstition," and to the fact that the *safra* usually began on a Sunday. See the "Atas da Classe," in Frans Leonard Schalviwijk, "A Igreja Reformada no Brasil holandês," *Revista do Instituto Arqueológico, Histórico, e Geográfico Pernambucano* 58 (1993): 168, 172, 178.

37. Cuthbert Pudsey, *Journal of a Residence in Brazil*, 31.

38. ANTT, Inquisição, no. 10,776. The original states: "que no dito seu engenho sempre em todos os domingos e sanctos moendo seu engenho despois do sol posto . . . que usão e costuma geralmente nesta capitania a todos os senhores e feitores de engenho sem excepçao." A Jesuit mission to the Bahian Recôncavo in 1619 reported that the custom was to continue labor of the day prior to a Saint's day until mid-day of the Saint's day. The Jesuits tried to stop the practice that "broke the commandments just to make an extra four *tarefas* of cane [a month]." See St. Louis University, Vatican Film Library, roll 159, bras. 8 (1619). The account book of 1611–12 for *engenho* Sergipe in Bahia records a payment of over 7,000 *réis* for "a sentence in the ecclesiastical court for the *engenho* to operate on holy days," indicating either a fine for past practice or an exemption purchased. See ANTT, CSJ, maço 14, doc. 4.

39. In Inquisition investigations made in Bahia and Pernambuco in the 1590s, a number of planters were accused of not making their slaves work on Saturdays, a supposed sign of Jewish observance by the planters. See Elias Lipiner, *Os judaizantes das capitanias de cima* (São Paulo: Brasiliense, 1969), 71.

40. Soares de Sousa, *Tratado*, notes a number of *engenhos* in Bahia that operated with two oxen driven mills (*moendas*), apparently an adjustment that permitted greater milling capacity without the necessity of building a waterwheel. Soares also mentions the existence in Bahia of an *engenho* with two waterwheels, but this is the only reference I have seen to such an arrangement. There is a drawing of a two-*moenda trapiche* by the Dutch artist Frans Post who observed it in Pernambuco. For a more detailed discussion, see Ruy Gama, *Engenho e technologia* (São Paulo: Duas Cidades, 1979), 149–50.

41. Van Dussen, *Relatório*, 93–96.

42. José António Gonçalves de Mello, "Um regimento de feitor mor de engenho de 1663," *Boletim do Instituto Joaquim Nabuco* 2 (1953): 80–87.

43. Hapsburg Spain prohibited the import of copper from nations at war with Spain, and in 1638 advised Portugal to do the same, ordering the Desembargo do Paço to discuss how

copper could be obtained for Brazil's *engenhos*. See ANTT, Desembargo do Paço, Liv. 18, f. 14, Miguel de Vasconcelos to Desembargo do Paço (10 February 1638).

44. Shawn Miller, *Fruitless Trees: Portugese Conservation and Brazil's Colonial Timber* (Stanford: Stanford University Press, 2000).

45. Andre João Antonil, *Cultura e opulência do Brasil por suas drogas e minas*, ed. and trans. Andrée Mansuy (Paris: Institut des Hautes Études des Ameriques Latines, 1968), liv. 2, cap. 12.

46. João Peixoto Viegas, "Parecer e tratado feito sobre os excessivos impostos que cahirão sobre as lavouras do Brasil," *Anais da Biblioteca Nacional do Rio de Janeiro* 20 (1898).

47. Sugar refining in Portugal was prohibited in 1551 by a royal order that expressed concern for deforestation that might result. See Mauro, *Portugal*.

48. Gabriel Soares de Sousa, *Tratado descritivo do Brasil em 1587*, 4th ed. (São Paulo: Editora Nacional 1971), 146.

49. Dauril Alden, "Sugar Planters by Necessity, Not Choice: The Role of the Jesuits in the Cane Sugar Industry of Colonial Brazil, 1601–1759," in *The Church and Society in Latin America*, ed. Jeffrey A. Cole (New Orleans: Center for Latin American Studies, Tulane University, 1984), provides a detailed analysis of Jesuit sugar estates. See also Schwartz, *Sugar Plantations*, 96–97, which differs in some details on productivity.

50. The complex legal history of these estates is summarized in Schwartz, *Sugar Plantations*, 488–97.

51. Stuart B. Schwartz, "The Plantations of St. Benedict: The Benedictine Sugar Mills in Colonial Brazil," *The Americas* 39, no. 1 (1982).

52. On this point, see Schwartz, *Sugar Plantations*, 264–67; on Pernambuco, see Evaldo Cabral de Mello, *A Fronda dos Mazombos. Nobres contra Mascates. Pernambuco 1666–1715* (São Paulo: Companhia das Letras, 1995), 128–30; for Rio de Janeiro, see João Fragoso, "A nobreza da república: Notas sobre a formação da primeira elite senhorial do Rio de Janeiro (séculos XVI e XVII)," *Topoi: Revista de História* 1 (2000): 45–122.

53. I have written extensively on the *lavradores de cana* in Schwartz, *Sugar Plantations*, esp. 295–312. See also my earlier article, "Free Labor in a Slave Economy: The Lavradores de Cana of Colonial Bahia," in *Colonial Roots of Modern Brazil*, ed. Dauril Alden (Berkeley: University of California Press, 1973). An excellent and important study that takes up the theoretical implications of the *lavradores de cana* is Vera Lúcia Amaral Ferlini, *Terra, trabalho e poder* (São Paulo: Brasiliense, 1988).

54. Ferlini, *Terra*, 171. Ferlini believes that prior to 1650 *lavradores de cana* provided almost all the cane, but I believe this to be an error based on the atypical situation of *engenho* Sergipe, whose records served as a basis for her study and on a misreading of the 1639 inventory of Van der Dussen, which tended not to record the cane supplied by the *engenho* itself.

55. See the text in Eddy Stols, "Um dos primeiros documentos sobre o engenho dos Schetz em São Vicente," *Revista de História* 76 (1968).

56. Countess of Linhares (23 March 1601), ANTT, CSJ, maço 13.

57. Antonil, *Cultura*, liv. 1, cap. 3.

58. Buescu, *História econômica*, 110–12.

59. A calculation for Bahia in 1817 indicates the average cane farm size at between five and ten hectares. While some *lavradores de cana* actually had no slaves and a few had over forty, the average for the 478 *lavradores* listed in that year was about ten or eleven slaves, although almost 60 percent of them held fewer than ten slaves and 25 percent held fewer than five. See Schwartz, *Sugar Plantations*, 451–55.

60. ANTT, CSJ, maço 70, no. 87

61. I have detailed this process in Stuart B. Schwartz, "Indian Labor and New World Plantations: European Demands and Indian Responses in Northeastern Brazil," *American Historical Review* 83, no. 3 (1978); and in a comparative context, with Russell Menard, in Russell R. Menard and Stuart B. Schwartz, "Why African Slavery? Labor Force Transitions in Brazil, Mexico, and the Carolina Lowcountry," in *Slavery in the Americas*, ed. W. Binder (Würzburg: Königshausen and Neumann, 1993). See also John Monteiro, *Negros da terra* (São Paulo: Companhia das Letras, 1994).

62. Belchior Cordeiro, "Coisas notaveis," 77.

63. Schwartz, *Sugar Planations*, 65–72. The pace differed in other captaincies. On the Benedictine *engenho* Guaguaçu in Rio de Janeiro, in 1652, there were twenty-five Indians among the eighty-three adult slaves and by 1657 there were only fourteen of the eighty-six adults. See Schwartz, "Plantations of St. Benedict," 12.

64. I have treated this question in some detail in *Sugar Plantations*, 313–37.

65. Stols, "Um dos primeiros," 418.

66. For example, Father Estevão Pereira, in his famous accounting of *engenho* Sergipe, has the following entry that demonstrates the confusion of capital stock replacement and current expenses: "Every year more or less it is necessary to replace at least five slaves for those that die and they cost at the minimum 35$000 each . . . 175,000$000." Other than the manioc flour that the *engenho* receives as rent, which is all given to the slaves, another, at least 200, *alqueires* are needed, which usually cost at least 200 *réis* each $40,000. See "Dase rezão da fazenda que o Colégio de Santo Antão tem no Brasil," in Antonil, *Cultura*, 513–26.

67. King to Viceroy, Dom Pedro de Castilho (26 November 1605), Biblioteca da Ajuda (Lisbon), 51-VII-8, f. 220–220v. On the illegal and contraband trade through Buenos Aires, see Luís Ferand de Almeida, *A diplomacia Portuguesa* (Coimbra: Universidade de Coimbra, 1957), 78–80, 303–6.

68. An excellent review of the problem of credit is Jacob Price, "Credit in the Slave Trade and Plantations Economies," in *Slavery and the Rise of the Atlantic System*, ed. Barbara Solow (Cambridge: Cambridge University Press, 1991).

69. Fragoso, "A nobreza da republica."

70. See E. M. Koen et al., "Notarial Records in Amsterdam Relating to the Portuguese Jews in That Town up to 1639," *Studia Rosenthaliana* (1967–79). See James Boyajian, "New Christians and Jews in the Sugar Trade, 1550–1750: Two Centuries of Development in the Atlantic Economy," 471–84; and Ernst Pijning, "New Cristians as Sugar Cultivators and Traders in the Portuguese Atlantic, 1450–1800," 485–500, both in *The Jews and the Expansion of Europe to the West, 1450–1800*, ed. Paolo Bernardini and Norman Fiering (New York: Berghan Books, 2001).

71. Celso Furtado, for example, estimated production at 2,000,000 *arrobas* for 1600 and

believed that returns of 80 percent on invested capital could be obtained in a good year with very low expenditures for salaries and other expenses. See Celso Furtado, *The Economic Growth of Brazil* (Berkeley: University of California Press, 1965).

72. Schwartz, *Sugar Plantations*, 229–34, devotes considerable attention to the problem of *engenho* Sergipe's accounts. This was also the subject of an earlier study by Frédéric Mauro, "Contabilidade teórica e contabilidade prática na América portuguesa no século XVII," *Nova história e novo mundo* (São Paulo: Ed. Perspectiva, 1969). Mauro and others made much use of the famous "Dase rezão," written in 1635 by the Jesuit Estevão Pereira, who managed *engenho* Sergipe. This document was written as a self-defense after he had been accused of mismanagement by his order; while valuable, it must be used with caution. Dauril Alden, *The Making of an Enterprise*, 419, tries to come to Pereira's defense, but Pereira's own Jesuit colleagues held him responsible for losses during his administration of the estate. Although apparently never formally charged, his career in the order does not seem to have progressed thereafter.

CHAPTER SEVEN

The Atlantic Slave Trade to 1650

Herbert Klein

The forced migration of Africans in the Atlantic slave trade is traditionally associated with the rise of sugar production in the Old and
New Worlds. But, in fact, the slave trade evolved independently of
the expansion of the sugar economy. For the first 160 years, the
Atlantic boat trade in African slaves was correlated with a host of
different factors, from the use of Africans in domestic slavery in
Europe and Spanish America, to the evolution of sugar and other
products for the European market in the Atlantic islands and America. It is only
after 1600 that the movement of Africans across the Atlantic became so inti-
mately tied to the expansion of American sugar production. Moreover, until
1700, Africa earned more from the exportation of gold, ivory, and pepper than it
did from slaves.

Though of limited importance, slavery still existed in Europe in 1492. Like
almost all complex societies in world history until that time, the nations of
Europe had known slaves, and slavery in earlier centuries had been a fundamental labor institution. From the sixth century B.C. until the eighth century A.D.,
under the Greek city-states and the Roman Empire slave labor had been almost
as important as peasant labor in the production of goods for local and long
distance markets. Under the Islamic states of the Mediterranean world from the
eighth century onward, slavery also had been important, though less tied to
production and more associated with the state and private household economies. But in fifteenth-century Christian Europe, as in most such societies, slavery was primarily domestic slavery, which meant that the labor power of the
household was extended through the use of these workers.

Equally, slavery existed in the African continent from recorded times. But like
Medieval Christian Europe, it was a relatively minor institution in the period
before the opening up of the Atlantic slave trade. It could be found as a domestic
institution in most of the region's more complex societies, and a few exceptional

states may have developed more industrial forms of slave production. But African slaves were to be found outside the region as well. With no all-embracing religious or political unity, the numerous states of Africa were free to buy and sell slaves and to even export them to North African areas. Caravan routes across the Sahara had existed from recorded times, and slaves formed a part of Africa's export trade to the Mediterranean from pre-Roman to the modern times. But a new dimension to that trade occurred with the expansion of Islam in the eighth century. As the Islamic world spread into India and the Eastern Mediterranean, Islamic merchants came to play an ever more important part in the African slave trade. The frontier zones of the sub-Saharan savannas, the Red Sea region, and the East Coast ports on the Indian Ocean, in turn, became major centers for the expansion of Moslem influence. From the ninth to the fifteenth century, a rather steady international slave trade occurred, with the majority of forced migrants being women and children. Some six major and often interlocking caravan routes and another two major coastal regions may have accounted for as many as 5,000 to 10,000 slaves per annum in the period from 800 to 1600 A.D., accounting for anywhere from 3.5 to 10 million Africans who left their homelands.[1]

There also existed an internal slave trade. Given the use of slaves for domestic and social purposes within Africa, the stress in the internal slave trade was even more biased toward women. For both these long-term trades, the whole complex of enslavement practices, from full-scale warfare and raiding of enemies to judicial enslavement and taxation of dependent peoples, had come into use and would easily be adjusted to the needs of the Atlantic slave trade when this came into existence in the early fifteenth century. Although the number of persons who were forcibly transported was impressive, these pre-1500 northern and eastern African slave trades still fit in with a level of production and social and political organization in which slave trading remained an incidental part of statecraft and economic organization.

The arrival of the Portuguese explorers and traders on the sub-Saharan African coast in the early 1400s would ultimately represent a new development in the history of the slave trade in Africa in terms of the intensity of its development, the sources of its slaves, and the uses to which these slaves would be put. But initially there was little to distinguish the Portuguese traders from the Moslem traders of North Africa and the sub-Saharan regions. Portuguese interest was initially directed toward subverting the North African Saharan routes by opening up a route from the sea. Their prime interest was gold, with slaves, pepper, ivory, and other products as only secondary concerns. Even when they began shipping slaves in the early 1440s, they were mainly sent to Europe to serve as domestic servants. Africans had already arrived at these destinations via the overland Muslim controlled caravan routes, and thus the new trade was primarily an extension of the older patterns. The Portuguese even carried out extensive slave trading along the African coast, primarily to supply the internal African slave market in exchange for gold, which they then exported to Europe. Major imports in the early Portuguese trade to the Gold Coast were North African dyed cloths and copper ingots and bracelets, all items that local consumers and smiths had often purchased from Moslem sources.[2] It was the volume of these goods that was new to these the local markets, not the goods themselves. Thus the major impact of the coming of the Europeans to Africa was the addition of new trading routes rather than strange or exotic products. Whereas the Niger River, flowing mostly north toward the Sahara, had been the great connective link for the peoples of West Africa until then, now the Senegal, Gambia, and other local rivers running west and south toward the Atlantic coasts became major links to the outside world. So intense and widespread did this trading become over most of West Africa that the Portuguese language quickly became the basis of a trading patois that was spoken throughout the region.

Just as the beginnings of the Portuguese slave trade had complemented a traditional trading system, the first use of Atlantic slave trade Africans by Europeans was in traditional activities. For the first half of the century, the European slave ships that cruised the Atlantic shoreline of Africa carried their slaves to the Iberian peninsula. The ports of Lisbon and Seville were the centers for a thriving trade in African slaves, and from these centers slaves were distributed rather widely through the Western Mediterranean. Though Africans quickly became the dominant group within the polyglot slave communities in the major cities of the region, they never became the dominant labor force in the local economies. Africans were used no differently than the Moorish slaves who preceded and coexisted with them, and were to be found primarily in urban centers, and worked mostly in domestic service, and even the wealthiest European masters owned only a few slaves. Probably the largest such concentration of urban slaves was found in Lisbon, which by the 1630s had an estimated 15,000 slaves out of a total urban population of 100,000 persons and an established community of some 2,000 free colored.[3] But Seville also had a significant number, for in 1565 there were 5,327 slaves out of a population of 85,538 and most of these were Africans.[4]

But the major importer of African slaves would not be Europe itself. Just as Portugal was opening up the African coast to European penetration, its explorers and sailors were competing with the Spaniards in colonizing the Eastern Atlantic islands. By the 1450s, the Portuguese were developing the unpopulated

Azores, Madeira, the Cape Verde Islands, and São Tomé, while the Spaniards were conquering the previously inhabited Canary Islands by the last decade of the century. Sugar became the prime output on Madeira Island by the middle of the fifteenth century, and by the end of the century was Europe's largest producer.[5] The Cabo Verde islands also became a center for African slave activity as well as a minor entrepot in the developing Atlantic slave trade.[6] But it was the island of São Tomé, in the Gulf of Guinea, that most exclusively tied African slave labor to sugar. In terms of plantation size, the universality of slave labor, and production techniques, this was the Atlantic island closest to what would become the American norm. By the 1550s there were some sixty mills in operation on the island producing over 2,000 tons per annum and some 5,000 to 6,000 plantation slaves, all of whom were Africans. Eventually American competition, Dutch invasions, and a series of major African slave revolts destroyed the local sugar industry.[7]

Given the high costs of attracting free European labor across the Atlantic and the ultimate abandonment of American Indian slavery even by the Portuguese, it was African slave labor that would sustain the agricultural export industries created in America. In the central provinces of the Spanish American empire, with their dense settled Indian peasant populations—above all Peru and New Spain (Mexico)—the need for European or African laborers was relatively limited. But even here, European diseases decimated Indian populations, especially along the coasts, and the lack of a poor white migration created a need for skilled and unskilled urban workers. With an excellent supply of precious metals, and a positive trade balance with Europe, the Spaniards of America could afford to experiment with the importation of African slaves to fill in the regions abandoned by Amerindian laborers. They found African slaves useful for the very reasons that they were kinless and totally mobile laborers. Indians could be exploited systematically but they could not be moved from their lands on a permanent basis. Being the dominant cultural group, they were also relatively impervious to Spanish and European norms of behavior. The Africans, in contrast, came from multiple linguistic groups and had only the European languages in common and were therefore forced to adopt themselves to the European norms. African slaves, in lieu of a cheap pool of European laborers, thus added important strength to the small European urban society that dominated the American Indian peasant masses.

The northern Europeans who followed the Iberians to America within a few decades of the discovery had even fewer Indians to exploit than the Portuguese and were unable to develop an extensive Indian slave labor force, let alone the complex free Indian labor arrangements developed by the Spaniards. Nor did they have access to precious metals to pay for imported slave labor. Unlike the Iberians of the sixteenth century, however, they did have a cheaper and more willing pool of European laborers to exploit, especially in the crisis period of the seventeenth century. But even with this European labor available, peasants and the urban poor could not afford the passage to America. Paying for that passage through selling one's labor to American employers in indentured contracts thus became the major form of colonization in the first half-century of northern European settlement in America. The English and the French were the primary users of indentured labor, helped by a pool of workers faced by low wages within the European economy. But the end of the seventeenth-century crisis in Europe, and especially the rapid growth of the English economy in the last quarter of the century, brought a thriving labor market in Europe and a consequent increase in the costs of indentured laborers. With their European indentured laborers becoming too costly, and with no access to American Indian workers or slaves, it was inevitable that the English and the French would also turn to African slaves, especially as they discovered that sugar was one of the few crops that could profitably be exported to the European market on a mass scale.

That Africans were the cheapest available slaves at this time was due to the opening up of the West African coast by the Portuguese. Given the steady export of West African gold and ivory, and the development of Portugal's enormous Asiatic trading empire, the commercial relations between western Africa and Europe now became common and cheap. Western Africans brought by sea had already replaced all other ethnic and religious groups in the European slave markets by the sixteenth century. Although Iberians initially enslaved Canary Islanders, these were later freed as were the few Indians who were brought from America. Moslems who had been enslaved for centuries were no longer significant as they disappeared from the Iberian peninsula itself and became powerfully united under independent states of North Africa. The dominance of the Turks in the Eastern Mediterranean also closed off traditional Slavic and Balkan sources for slaves. Given the growing efficiency of the Atlantic slave traders, the dependability of African slave supply, and the stability of prices, it would be Africans who would come to be defined almost exclusively as the available slave labor of the sixteenth century.

Although the relative importance of African slaves was reduced within Spanish America in the sixteenth and seventeenth centuries, African migration to these regions was not insignificant and began with the first conquests. Cortés and his various armies held several hundred slaves when they conquered Mexico in the 1520s, while over a thousand African slaves appeared in the armies of Pizarro and Almagro in their conquest of Peru in the 1530s, and in their subsequent civil wars in the 1540s. Although Indians dominated rural life everywhere, Spaniards found their need for slaves constantly increasing. This was especially true in Peru, which was initially both richer and lost a progressively higher proportion of its coastal populations to European diseases in areas that were ideal for such European crops as sugar and grapes. Already by the mid-1550s there were some 3,000 African slaves in Peruvian viceroyalty, half in the city of Lima. This same balance between urban and rural residence, in fact, marked slaves along with Spaniards as the most urban elements in Spanish American society.[8]

The needs for slaves within the Peruvian viceroyalty increased dramatically in the second half of the sixteenth century as Potosí silver production came into full development, making Peru and its premier city of Lima the wealthiest zone of the New World. To meet this demand for Africans a major slave trade developed, especially after the unification of the Portuguese and Spanish Crowns from 1580 to 1640 gave the Portuguese access to Spanish American markets. Initially most of the Africans came from the Senegambia region between the Senegal and Niger Rivers, but after the development of Portuguese Luanda in the 1570s important contingents of slaves from the Congo and Angola began arriving.

Though a major component of urban population and the dominant workers in gold mining, some African slaves were also used in agriculture. To serve such new cities as Lima, Spaniards developed major truck farming in the outskirts of the city that were worked by small families of slaves. Even more ambitious agricultural activity occurred up and down the coast in specialized sugar estates, vineyards, and more mixed agricultural enterprises. In contrast to the West Indian and Brazilian experience, the slave plantations of Peru were likely to be mixed crop producers. On average the plantations of the irrigated coastal valleys, especially those to the South of Lima, had around forty slaves per unit. The major wine and sugar producing zones of the seventeenth century, such as Pisco, the Condor, and Ica Valleys, contained some 20,000 slaves. In the interior, there were also several tropical valleys where slave estates specializing in sugar could be found.[9]

But it was in all the cities of the Spanish continental empire that the slaves played their most active economic role. In the skilled trades, they predominated in metalworking, clothing, and construction and supplies, and were well represented in all the crafts except the most exclusive, such as silversmithing and printing. In semi-skilled labor they were heavily involved in coastal fishing, as porters and vendors, in food handling and processing, and were even found as armed watchmen in the local Lima police force. Every major construction site found skilled and unskilled slaves working alongside white masters and free blacks of all categories as well as Indian laborers. They also predominated in the semi-skilled and unskilled urban occupations, from employment in tanning works, slaughter houses, and even hat factories to being the dominant servant classes. All government and religious institutions, charities, hospitals, and monasteries had their contingent of half a dozen or more slaves who were the basic maintenance workers for these large establishments.[10]

As the city of Lima grew, so did its slave and free population and by 1640 there were an estimated 20,000 colored persons in the city and about 40,000 persons of African descent in the colony.[11] This growth was initially faster than the white and Indian participation in the city, and by the last decade of the sixteenth century Lima was half black and would stay that way for most of the seventeenth century. Equally, all the northern and central Andean coastal and interior cities had black populations so that by 1600 they accounted for half their total populations. As one moved further south into the more densely populated Indian areas their relative percentage dropped, though black slaves could be found in the thousands in Cuzco, and even the interior city of Potosí, which was dominated by Indian workers, was estimated to have some 6,000 blacks and mulattos, both slave and free, in 1611.[12]

African slaves in the viceroyalty of Mexico were also to be found from the first moments in the armies, farms, and houses of the Spanish conquerors. As in Peru, the first generation of slaves probably numbered close to the total number of whites. They were also drawn heavily into the local sugar and European commercial crop production in the warmer lowland regions, which were widely scattered in the central zone of the viceroyalty. Also, given the discovery of silver in the northern regions of the viceroyalty where there were few settled Indians, Africans were even initially used in silver mining. In a mine census of 1570, 45 percent of the laboring population comprised some 3,700 Africans slaves, double the number of Spaniards, and just a few hundred fewer than the Indians.[13] But the increasing availability of free Indian wage labor lessened the need for more expensive African slave labor, and they disappeared from the mines by the end of the century. Given the more extensive Indian population in Mexico, Africans were used less than in Peru and their relative importance declined over time. Though slaves performed many of the same urban tasks in Mexico City as they did in Lima, the former was essentially an Indian and mestizo town and slaves never achieved the same importance in the labor force.[14]

The relative significance of Mexican slavery was well reflected in the growth of its slave population, which peaked at some 35,000 slaves by 1646 when they represented less than 2 percent of the viceregal population.[15] In contrast, the number of slaves within Peru had reached close to 100,000 by mid-century, where they accounted for between 10 and 15 percent of the population. By 1650, Spanish America, meaning primarily Peru and Mexico, had imported some 250,000 to 300,000 slaves, a record that they would not repeat in the next century of colonial growth.

The major demand for African slaves, after 1600, came from Portuguese America and the marginal lands that the Spaniards had previously neglected, above the lesser islands of the Caribbean. With no stable Indian peasant populations to exploit, and few alternative exports in the form of precious metals, successful colonization in these zones required the export of products that Europe could consume, which would eventually lead to sugar production and the massive use of African slave labor. The first to develop the plantation slave model were the Portuguese, who took possession of the eastern coastline of South America in the early sixteenth century. First relying on Indian slave labor to produce sugar, by the mid 1580s Pernambuco alone reported 2,000 African slaves, composing one-third of the captaincy's sugar labor force.[16] With each succeeding decade the percentage of Africans in the slave population increased. By 1630 some 170,000 Africans had arrived in the colony and sugar was now predominantly a black slave crop. The early decades of the seventeenth century would prove to be the peak years of Brazil's dominance on the European sugar market, and it was this very sugar production monopoly that excited the envy of other European powers and led to the rise of alternative production centers.

It was the Dutch who opened up much of this world to slave trading. Initially they were Brazil's major commercial link with northern Europe before turning hostile after the beginning of their wars for independence in the 1590s. As early as 1602 they established the East Indies company to seize control of Portugal's Asian trade, followed in 1621 by the West Indies company organized to take Portugal's African and American possessions. The Dutch West Indies Company finally captured and held Recife and its interior province of Pernambuco in 1630. They next captured Portugal's African possessions: first the fortress of Elmina on the Gold Coast in 1638, and then Luanda and most of the Angolan region in 1641.

These conquests profoundly affected the subsequent history of sugar production and African slavery in America. For Brazil, the Dutch occupation resulted in Bahia replacing Pernambuco as the leading slave and sugar province, in the reemergence of Indian slavery to replace the African slaves lost to the Dutch, and in the ensuing Indian slave trade opening up the interior regions of Brazil to exploitation and settlement. For the rest of America, Dutch Brazil would became a major source for the tools, techniques, credit, and slaves that would carry the sugar revolution and its slave labor system into the West Indies, especially after the fall of Dutch Pernambuco in 1654.

The opening of the Lesser Antillean islands and the northeastern coast of South America to northern European colonization represented the first systematic challenge to Iberian control of the New World. By 1650, the English had 44,000 whites on their West Indian islands (compared to 53,000 in the settlements of New England),[17] while the French islands of Martinique and Guadeloupe held 15,000 white settlers by the end of the 1650s. At mid-century, tobacco and indigo were the primary exports from these islands, and though slaves were present from the beginning, they were still outnumbered by the whites. The Dutch assistance finally made sugar a far more viable proposition, especially as the opening up of Virginia tobacco production led to a crisis in European tobacco prices.

The transformation that sugar created in the West Indies was truly impressive. The experience of Barbados, the first of the big production islands, was typical. On the eve of the introduction of sugar in 1645, there were only 5,680 African slaves. By 1680 there were over 38,000. At this point Barbados was both the most populous and the wealthiest of England's American colonies and averaged some 265 slaves per square mile, compared to less than 2 slaves per square mile in the island of Jamaica, which had been captured from the Spanish in 1655.[18] The slave ships were bringing in over 1,300 Africans per annum to Barbados and by the end of the century this tiny island contained over 50,000 slaves.[19] That model was quickly replicated by the French. As of 1670 Martinique, Guadeloupe, St. Christopher, and the recently settled western half of the island of Santo Domingo were all producing sugar. By the early 1680s these islands, including Santo Domingo, contained over 18,000 slaves.[20]

The growth of West Indian slavery led to these powers entering directly into the transportation of African slaves in their own ships. But initially it was the Portuguese who dominated this trade and were quick to develop these new trade routes. In the 1450s they obtained exclusive Christian rights from the Pope for dealing with Africa south of Cape Bojador. In 1466 they settled the island of Santiago in the Cape Verde islands; in 1482 they built the fort of São Jorge da Mina (Elmina in present-day Ghana); by 1483 they were in contact with the kingdom of the Kongo just south of the Congo River in Central Africa; in 1493 they had definitely settled the island of São Tomé in the Gulf of Guinea; and by 1505 they had constructed the fort of Sofala on the Mozambique coast of East Africa.

At first the Portuguese treated their African contacts as they had the North Africans they encountered. They raided and attempted to forcibly take slaves and plunder along the coast they visited. Thus when they landed at Rio de Oro, just south of Cape Bojador, in 1441 they seized several Berbers along with one of their black slaves. In 1443 a caravel returned to the same Idzāgen Berbers to exchange two aristocratic members of the group for gold and ten black slaves. In 1444 and 1445 merchants and nobles of the Algarve outfitted two major expeditions against the Idzāgen; the first brought back 235 Berber and black slaves who were sold in Lagos. Thus began the Atlantic slave trade.

Not only did the second and later expedition encounter serious hostility from the now-prepared Berbers, but attempts to seize slaves directly from the black states on the Windward coast ended in military defeat for the Portuguese. The result was that the Portuguese moved from a raiding style to peaceful trade, which was welcomed by Berber and African alike. In 1445 came the first peaceful trade with the Idzāgen Berbers at Rio de Oro in which European goods were exchanged for African slaves. Trade with the Idzāgen led to the settlement of a trading post (called a "factory") at Arguim Island off the Mauritanian coast; and after 1448 direct trade began for slaves and gold with the sub-Saharan West African states.[21]

As long as the Portuguese concentrated their efforts in the regions of Mauritania, Senegambia, and the Gold Coast, they essentially integrated themselves into the existing network of North African Muslim traders. The Moslems had brought these coasts into their own trade networks and the Portuguese tapped into them through navigable rivers that went into the interior, especially the Senegal and Gambia Rivers, or through the establishment of coastal or offshore trading posts: Arguim Island, the Cape Verde islands off the Senegambia coast, and the Guinean Gulf islands of São Tomé and Príncipe. Even their establishment of São Jorge da Mina (Elmina) on the Gold Coast, in 1481, fit into these developments. Although Portuguese slave trading started slowly at about 800 slaves taken per annum in the 1450s and 1460s, it grew close to 1,500 in the next two decades and to over 2,000 per annum in the 1480s and 1490s, about a third of whom were sold to Africans themselves in exchange for gold.[22]

But a major structural change occurred after 1500, with a combination of the effective settlement of the island depot and plantation center of São Tomé in the Gulf of Guinea and the beginning of intense trade relations with the kingdom of the Kongo after 1512, which brought West-Central Africa into the Atlantic slave trade in a major way for the first time. The Kongolese were located by the Congo River and were unconnected to the Moslem trade before the arrival of the Portuguese. The kingdom also sought close relations with the Portuguese and tried to work out government control of the trade. Portuguese sent priests and advisers to the court of the Kongolese king, and his representatives were placed on São Tomé. These changes occurred just as the Spanish conquest of the Caribbean islands and the Portuguese settlement of the Brazilian subcontinent was getting under way and thus opened the American market for African slaves.

All these changes found immediate response in the tremendous growth of the Portuguese slave trade. After 1500 the volume of the trade passed 2,600 slaves per annum, and after the 1530s these slaves were shipped directly to America from the entrepot island of São Tomé just off the African coast. This latter development marked a major shift in sources for African slaves for America. The acculturated and christianized blacks from the Iberian peninsular had been the first Africans forced to cross the Atlantic. Now it was non-Christian and non-Romance language speakers taken directly from Africa, the so-called *bozales*, who made up the overwhelming majority of slaves coming to America.

Another major change came about in the 1560s as a result of internal African developments. Hostile African invasions of the kingdom of the Kongo led to direct Portuguese military support for the regime and finally in 1576 to their establishment of a full-time settlement at the southern edge of the kingdom at the port of Luanda. With the development of Luanda came a decline in São Tomé as an entrepot, for now slaves were shipped directly to America from the mainland coast and from a region which was to provide America with the most slaves of any area of Africa over the next three centuries. By 1600, the Atlantic slave trade was finally to pass the northern and eastern African export trades in total volume,[23] though it was not until after 1700 that slaves finally surpassed in value all other exports from Africa.[24]

The Portuguese did not hesitate to conquer and evangelize. As early as the 1480s their agents had reached the Christian kingdom of Ethiopia in Eastern Africa, and by 1490 they sent an army to support a Christian pretender to the Jolof throne in Senegambia, which failed in its effort. In 1491 Christian missionaries were sent to the kingdom of the Kongo, south of the Zaire River, and there had more successes, even enthroning Affonso, a powerful leader, and a christianized African as head of the state in 1506. But within a generation the missionaries were expelled and the Kongo reverted to traditional religious beliefs. In 1514 they sent missionaries to the Oba of Benin, and this attempt ended in failure. Finally in their desire to control the Shona gold fields of East Africa, in 1569 they mounted an unsuccessful thousand-man expedition that included missionaries to expel the moslemized Swahili traders and christianize the local miners.

The only results of all these Portuguese efforts at penetration and conversion was the unintended one of the creation of a mixed Afro-Portuguese free merchant class that claimed Portuguese identity and adopted Catholicism, but rejected the sovereignty of the Portuguese state. Some of these communities not only occupied key settlements along the coast, but often penetrated deep into the interior. They colonized the Benguela highlands in Angola and even created mini-states with African followers and slave armies in the interior of Mozambique on their "estates" or *prazos*. This model, in turn, was followed to a lesser extent by the creation of local Afro-English and Afro-French merchant groups along the West African coast in the seventeenth and eighteenth centuries. In each case, these were racially mixed elites who intermarried with members of the local African establishments and were deeply involved with the regional African states and societies and who no longer obeyed the commands of the European states that had fostered them.

The Portuguese failures of colonization and christianizing brought the end to any thought of actual conquest and colonization in Africa as well. This forced acceptance of African autonomy was due to the difficulties of maintaining troops, missionaries, and bureaucrats in the African environment because of extraordinarily high European death rates, as well as the military balance that existed between Europeans and Africans. Respect for religious and political autonomy therefore became the norm in African-European dealings, and the Portuguese and those Europeans who followed them were even forced to deal with the many Moslem groups of the Western Sudan region in peace.

This does not mean that the coming of the Europeans to the coasts of Africa had little impact on internal African society and economy. But this impact varied depending on the nature of the local society and its previous contacts with the non-African world. In the region from Senegambia to the Cameroons, and in East Africa, it was less a revolutionary event than in the more isolated Central African regions of the Congo and Angola. The former regions had long-term international contacts with the Mediterranean and Middle Eastern worlds that were not severed by the new European boat trades. Even at their most intensive contact with the Gold Coast, Europeans took only a minority of the local output by sea, with a majority of the gold still crossing the Sahara. Nor was the political impact of the European arrival in this zone very profound. The slave trade initially was a minor movement of peoples with little influence over warfare or raiding. Thus the great Songhai empire in the Upper Niger region—the largest empire in West Africa—was weakened by the coming of the Europeans, but it was actually destroyed by a Moroccan invasion from across the desert. In 1591 a revived Moroccan state not only drove the Europeans out of most of its coastal cities and even killed one Portuguese king in doing so, but they headed south across the desert and seized Timbuktu from the Songhai. Equally, Swahili and

Arabic traders had dominated the East African trade routes long before the arrival of the Portuguese, never fully controlling local gold exports.

As long as the slave trade remained small, which was the case through most of its earlier period, it had a relatively limited impact even on the internal African slave markets. Estimates of all the slave trades to 1600 suggest that the Atlantic slave trade took only a quarter of all slaves leaving Africa and was still considerably smaller than the trans-Saharan slave trade. It was only in the course of the seventeenth century that the Atlantic route forged ahead as the dominant slave trade, accounting for close to two-thirds of all Africans leaving the continent. At the end of the fifteenth century the Atlantic slave trade involved the shipment of no more than 800 to 2,000 slaves per annum, all of whom were being sent to Portugal or its Atlantic island possessions such as Madeira and São Tomé. Portuguese extraction of slaves was estimated to have risen to some 4,500 slaves per annum in the first decades of the sixteenth century as slave shipments to Spanish America had begun. This movement was still not that different in volume from the slave trade going across the Sahara or out of the Red Sea ports at this time. It was also of a far different dimension than the nearly 80,000 Africans per annum shipped to America at the height of the Atlantic slave trade in the 1780s.

Despite its still small volume compared to later developments, the Atlantic slave trade by the middle of the seventeenth century was one of the most complex economic enterprises known to the preindustrial world. It was the largest transoceanic migration in history up to that time; it promoted the transportation of people and goods among three different continents; it involved an annual fleet of several hundred ships; and it absorbed a very large amount of European capital. The trade was closely associated with the development of commercial export agriculture in America, and Asian trading with Europe. It involved complex and long-term credit arrangements in Europe and Africa and was carried on by a very large number of competing merchants in an unusually free market.

Given the high entry costs to trading, and the initial lack of detailed knowledge of the various African and American markets, the earliest period of the slave trade was one in which the state played a major role. Although slaves were shipped off the African coast by private European traders from the 1440s onward, the organization of an intensive slave trade took some time to develop. Although the Portuguese were rich enough to allow private contractors to develop some part of the early trade, both they and all the Europeans who followed used heavy state control in the form of taxation, subsidization, or monopoly contracts to get the trade going and control its flow of forced workers to America. In almost every case, the state was needed to subsidize the trade in order to get it organized. The Spaniards even declared it a royal monopoly and eventually developed a complex exclusive arrangement called the *asiento* for selling the right to deliver slaves, a system that lasted until the end of the eighteenth century. Though the Spanish contract holders subcontracted to private or foreign monopoly company firms, the trade was still heavily controlled by the state. Even the Portuguese finally resorted to state monopoly companies in the eighteenth century to get the trade going to colonies that were underdeveloped and lacked the capital to finance the trade.[25]

It was the relative ability of the American importing colonies to pay for their slaves that determined whether a slave trade could develop. In the case of Spain, it was the silver and gold mined by the Indians which would pay for the forced migration of African slaves. The trade was a very controlled one, but only for state taxing purposes, as private individuals from all over Europe were given exclusive contracts to carry slaves to the American colonies (the so-called *asiento*) in return for paying the Crown a fixed fee and taxes on each slave delivered. In the Portuguese case, their early dominant position in African trade gave them a decided advantage in the slave trade by lowering their costs of entry. In turn, the very rapid development of a sugar plantation economy based initially on American Indian slave labor in Brazil permitted them to generate the capital needed to import African slaves. But all other trades required some use of monopoly companies to get the system to provide slaves to American colonies that did not have the capital or credit to pay for the imported slaves.

From the fifteenth century until the early sixteenth century the Portuguese dominated the trade in gold, ivory, and slaves from Africa. By the early seventeenth century, between their own American needs and their supplying the Spanish American colonies, they were probably shipping some 3,000 to 4,000 slaves per annum. But this monopoly was challenged as early as the late sixteenth century by the French and the British. French and British free traders intermittently visited the African coast from the middle decades of the sixteenth century, but they and the Dutch did not become a major presence with forts and permanent trading links until the seventeenth century.

The costs of entry into the trade was so high, however, that only some kind of government support and a corresponding monopoly arrangement seemed capable of opening up a continuous and successful trade for these late-arriving Europeans. Using the model of the successful Dutch and English East Indies Companies, every slave trading nation but Spain and Portugal between 1620 and 1700 experimented with joint-stock monopoly trading companies. All succeeded in opening up systematic trade for the first time, but all would eventually fail and be replaced by free traders from their respective nations.

While French interlopers had involved themselves in the Atlantic slave trade from the early sixteenth century, serious French participation began only with the development of the monopoly trading companies in the second half of the seventeenth century. After many partial and incomplete attempts, the French finally organized their first monopoly company in 1664, and by 1672 the French government offered a bounty for every slave transported to the French West Indies. By the 1690s most of France's African trade was in the hands of private entrepreneurs, though it was not until the 1720s that free traders finally succeeded in definitively breaking the Company's control over trading.[26]

The Dutch West India Company was initially the most successful of these early monopoly companies, the one most involved in delivering slaves to colonies of the other European powers, and the one that shipped the most slaves to America. From its founding in 1621 it operated both as a commercial company and as a military organization with quasi statelike powers. It seized major territories from the Portuguese, produced sugar in Brazil, and became a major slave trader in the Gold Coast and Angola. It even made war on the Spaniards and succeeded in capturing one of the American silver fleets. But by the 1670s it was reduced to a few American possessions and to its Gold Coast forts, with the Portuguese having retaken most of their lost territories. Free trade in slaves was finally permitted in the 1730s. From the 1620s to the 1730s the company moved 286,000 slaves from Africa. It even held the Spanish *asiento* at one time, and from the seventeenth century until 1729 some 97,000 of the slaves that it shipped to America were delivered to the Spanish colonies.[27]

The English Royal African Company grew out of a series of earlier English monopoly companies and was put together in 1672. Like all such African adventures it was required to invest heavily in fixed costs, such as forts and armaments. The English probably started trading with Africa in the 1550s and tried some slave trading to the Spanish colonies in the 1560s, but they only established their first fort in Africa in the 1630s. From 1672 until 1713 the Royal African Company transported over 350,000 slaves to the English colonies. The company gave up its monopoly in 1698.[28]

The ultimate failure of all these monopoly companies was due to their high fixed costs in forts and ships and/or their obligations to deliver a fixed number of slaves into a given region no matter what the demand or the costs, obligations that were often too expensive to maintain. They usually tied up too much capital for too long a period and found it increasingly difficult to raise new funds. In the free trade era these companies were replaced in all trades by temporary associations of merchants who joined together to finance individual voyages. Thus merchants in the sending port committed their capital to relatively short periods and/or spread it over many different slaving voyages. Moreover, they delivered slaves only in the quantities demanded in the New World and to zones that were capable of paying for them with cash or exportable products that could be sold for a profit in Europe.

Although some formal joint stock companies were established, it was more common to form a trading company as a partnership of from two to five merchants. If it was two partners, both usually worked actively in the enterprise, but if it was more than this number there was usually an active partner who organized the expedition and a group of more or less passive partners. Interestingly, most of these associations engaged in other trades as well as that of slaves, indicating the diversification of risk of the entire transaction. The contract that the partners signed, or that founded the joint stock company, were usually for seven years, which was the time needed to completely close the books on a slaving expedition.

But given the high costs of entry into the trade, many of the partnerships or joint stock companies offered stock or shares in individual voyages that they financed. Thus while one of these slave-trading companies might undertake several voyages, each voyage attracted a different set of investors. The owner and outfitter of the ship (called an *armateur* in French) sold parts of the expedition and/or the ship to outside investors. In so doing he thus formed a mini-company that handled just that one expedition. Often these investors were other outfitters, and it was common for the principal company or association to itself invest as temporary shareholders in ships outfitted by other companies. The remaining shares in an expedition came from the Captain, who was allowed to invest on his own account.[29]

The actual purchase of the ship, collection of the cargo, and the arrangement of final papers and insurance typically took some four to six months to arrange. After the owners, the second most important person was the captain. Whether he bought shares or not, most captains were given 2 to 5 percent of the sale of all slaves he delivered in the Americas. Successful captains could obtain a respectable fortune in just two to three voyages. It was the captain who had the most responsibility, being both in charge of sailing the ship and of doing all the trading in Africa as well. But mortality of captains and crew was high, and though many carried out repeat voyages, there were great risks involved. Among the 186 Dutch captains employed by the Dutch West Indies Company in the seventeenth and early eighteenth centuries, the average was for 1.4 voyages per captain, but this was highly concentrated since two-thirds of the captains made only one voyage. Moreover, crew and slave mortality were high, and something like 11 percent of the captains died in these early voyages.[30]

A large complement of subofficers and skilled persons were needed in the crew, with the three leading skilled persons being the ship's doctor, the carpenter, and the cooper or barrel maker. The first cared for and evaluated the health of the slaves. The carpenter designed the holds for the slaves when they were collected, and the cooper was in charge of the crucial water casks. The average French and Dutch slaver in the seventeenth and eighteenth centuries took from thirty to forty sailors for their crew, the majority of whom were poorly paid common seamen.[31] Between the needs of coastal trading in Africa and the potential for violence and the need for tight security on the African coast and in the Atlantic crossing, all slave traders carried double the number of crew that a normal merchant ship of their tonnage would carry. Moreover, these slave traders by the late seventeenth century were moving toward a norm of shipping a third to half the tonnage of a typical West Indian merchant vessel.

In all slave trades where the data are available on crews, tons, and slaves, there is the same high correlation between the numbers of slaves carried and the number of sailors manning the ships. Even where tonnage cannot be made comparable, there is the same difference seen between slavers and regular merchant ships. Thus for some twelve ships engaged in the slave trade to the Spanish Indies in 1637, the average of 7.7 slaves per sailor for these seventeenth-century Spanish American ships was quite similar to a sample of 525 French slavers from the first half of the eighteenth century, which carried 7.5 slaves per crewmen. Moreover, as was to be expected, in all the slave trades the number of slaves per crew kept increasing over time, reflecting an increasing efficiency of the slave ships, reaching the 9.5 range for almost 1,500 slavers in the second half of the eighteenth century.[32]

The biggest outfitting expense of these slaving voyages was always the cargo, which averaged between 55 and 65 percent of total costs. This made the slavers unique in almost all the major commodity trades. In France, which has the best data on costs, the cargo accounted for two-thirds of the total costs, and the ship and its crew a mere one-third. There was some variation depending on whether or not the ship was newly built. This explains why the average value of the outfitted slaver per ton was six times the average value per ton of the much larger direct-trade ships.[33]

The African consumer market was unusual in that all the Europeans had to import foreign goods to make up their cargoes. Top on the list were East Indian textiles that were made up of cotton cloths of white, or solid blue and/or printed design. Also from Asia came cowry shells produced in the Maldives archipelago, just off the southern coast of India. Important as well were armaments, which were sometimes produced at home, but often purchased abroad, and Swedish produced bar iron used by African blacksmiths to make local agricultural instruments. Knives, axes, swords, jewelry, gunpowder, and various national- and colonial-produced rums, brandies, and other liquors were also consumed along with Brazilian-grown tobacco. No one nation could produce all these goods, and over time purchases shifted from nation to nation. Early on the French tended to buy English arms, the English preferred cheaper Dutch-produced arms, and everyone bought their cloths from the Dutch, French, and British traders to Asia. Even when Europeans used African products to purchase slaves, these in turn were bought with European or Asian or even American manufactured goods. All these goods were purchased by traders for hard currencies.

In a major study of African trade in the seventeenth century, it has been estimated that textiles made up 50 percent of the total value of imports into Africa. Next in importance after textiles came alcohol at 12 percent; manufactured goods 12 percent; guns and gunpowder at 8 percent; tobacco at 2 percent and bar iron 5 percent. So important was the East Indian textile component of the trade that it explains the rise of the chief African slave trading ports in the French and English trades. While La Rochelle and Le Havre had been major slaving ports in the seventeenth century, by the beginning of the eighteenth century Nantes rose to be the primary port, much as Liverpool would be in the eighteenth century.

The trip out from Europe to Africa took anywhere from three to four months. Many ships stopped at other European ports for more cargo on the outward leg, or temporarily stopped to provision in the southern European ports or the Canary Islands. Moreover, the length of the trip also depended upon which area of Africa was to be the prime trading zone. Reaching Gorée, a major trading zone in the Senegambia region, for example, left another trip as long again to reach Angola.

The region selected for trade by each European national depended upon local and international developments. Rough spheres of influence were slowly established, with the English, Dutch, and the Portuguese most dominant as residents on the African coast with their permanent forts or factories. But no African area was totally closed to any European trader, and there was an extensive published contemporary literature and general European knowledge on the possibilities of local trade everywhere in Western Africa. The local forts maintained by some European powers were not military centers, but were commercial stations that facilitated local commerce with the Africans and had little inland activity. Many of these forts would allow foreign traders access to their resources. Even the Portuguese, who were the Europeans most likely to concentrate on a limited set of regions in southern Africa, would conduct trade in other quite open and competitive regions.

Although forts were well established by the middle of the seventeenth century along the Gold Coast and selected other regions, there was no major "bulking" or warehousing of slaves. The cost of slave maintenance to the Europeans was prohibitive, and would have made final costs quite high. There was little agricultural activity around the fort and it was virtually impossible for them to maintain slaves in storage once the majority of the slave ships had left for America. On the other hand, hinterland traders could easily absorb slaves into their own agricultural or industrial production as they waited for the return of the slave ships. It is estimated that feeding a slave on the coast for a year would have increased his or her price by 50 percent.

In the overwhelming majority of cases it was the Africans who controlled the slaves until the moment of sale to the captain. Only occasionally in the era of free trade did a local European purchase slaves on his own account for resale to the slaver captains. This had been more common in the earlier age of the monopoly companies, but even then had accounted for only a small volume of sales. It was the norm everywhere for slaves to be purchased in relatively small lots directly from the African sellers.

African slave traders came down to the coast or to river banks in a relatively steady and predictable stream to well-known trading places. The cost of moving the slaves in caravans to the coast was relatively cheap—only the costs for food for the slaves and the salaries of the guards, and the costs of purchase for any slaves lost in transit because of death or escape (a loss for which we have no systematic data for any African interior trade route). The slaves also could be used to move goods at no cost, with each male slave head carrying up to twenty-five kilograms of goods and women up to fifteen kilograms.

Given alternative local uses of slaves, inland traders arriving by caravan could respond to low European prices by holding these slaves off the market and using them as workers in agriculture or industry for any time period needed until prices rose again. Equally they could be sold to local consumers at any time on the trip, and from the few eyewitness accounts, this seems to have been a common experience. Eventually many of these slaves would then be resold into the Atlantic trade if demand were strong.

This trading system meant that all European traders tended to spend months on the coast or traveling upriver gathering their slaves a few at a time. Even the ports of Luanda and Benguela, the only African centers that maintained a large resident white population, still required a stay of several months for a ship's voyage to Brazil to complete their complement of slaves. It was typical in most trading areas for the captain to leave the ship in one spot and take small craft to trade inland, leaving another officer in charge of the ship. He was usually accompanied by the ship's doctor, who examined each slave for disease prior to purchase. During the several months needed to purchase the several hundred slaves each ship carried, the already purchased slaves were held ashore as long as possible to prevent the outbreak of disease on the ship, but even so death rates were relatively high for this "coasting" period. Good data are available on the "coasting" experience from the Dutch, French, and English trades. The seventeenth-century Dutch West Indies Company ships averaged 100 days on the coast picking up slaves, which was comparable to thirty-four British vessels in this period that averaged 95 days. In the era of the free traders, these rates often doubled.[34]

In provisioning for the voyage, all trades used common African foods and condiments along with dried foods and biscuits brought from Europe. They also brought with them lime juice for combatting scurvy. The Europeans all tried to supply standard foods that local Africans could consume, though this varied from region to region. Most used European- or American-produced wheat flour or rice to make a basic gruel that would then be seasoned with local condiments and supplemented with fresh fish and meats as well as dried versions of these foods. In the Sahel region, African millet was preferred to rice, while those from the delta of the Niger preferred yams. All the condiments used came from Africa, including the palm oil and the peppers, and all trades provided biscuits for both crew and slaves.

Even in the earlier seventeenth- and eighteenth-century trade, when much more European dried foods were used, Dutch slave captains purchased fresh vegetables and small livestock on the African coast, along with the ever-necessary supply of fresh water. As early as 1684 the Portuguese enacted provisioning acts for the slave trade, which mostly dealt with space utilization and water rations.[35] The French estimated that they needed one cask or barrel of water for every person aboard ship, and all trades gave drinking water three times per day, even when meals might only occur twice a day. The ship carrying the 600 slaves for a two-month voyage would thus need one water cask per slave (weighing between sixty and sixty-five kilograms per cask), which meant that some forty tons of water casks were loaded for the slaves alone. On some coasts water was not readily available and often had to be obtained in regions far from where the slaves were obtained. Finally the maintenance of the casks and the guarantee of their quality was an important part of the responsibility of the captain and the carpenters.

Almost all slave traders housed and organized daily life of the slaves in the same manner. The decks were divided usually into three separate living quarters, one for males, one for boys, and one for women and children. Sick slaves were usually isolated in their own compartment. Depending on the number taken, and the number ill, these compartments could be expanded or reduced. Slaves were usually shackled together at night to prevent rebellion and movement, but were then brought up to the deck during the day. On deck they were forced to exercise, often accompanied by African musical instruments. In the Dutch trade, for example, all captains purchased African drums so as to force the slaves to dance as a form of exercise. Usually the Africans stayed the entire day on deck and had their meals there if the weather permitted. At this time the crew went into their quarters and cleaned them out, often using vinegar and other cleansing agents. While all females were given simple cloths to wear, in some trades the males were left naked if the weather permitted. All slaves were washed everyday with sea water.

As is obvious from these details, it was the aim of all traders to keep the slaves and their quarters as clean as possible since there was a general awareness of the correlation between cleanliness and disease. Beyond this all slave trades carried a ship's doctor to care for the slaves and crew and their illnesses. Nevertheless the details given of the medical cabinets of these "doctors" show little of value for fighting the standard diseases that struck both crew and the slaves. Mortality and morbidity were high among the slaves and little beyond maintaining clean food and water supplies and guarantying the sick provided any effective remedy for these diseases.

After arriving in America, the slaver had to clear local customs and health registrations before the slaves were sold to local planters. Among the French it was the custom for the captain to sell his slaves directly, using a local agent who took a commission on all sales. The sale usually began immediately upon docking or within a week of landing. Slaves were either sold directly from the ship or brought to a special market on land. Usually the slaves were sold one at a time, with occasional sales of several to one buyer. Once agreeing on the price, usually only about a 20 percent down-payment was made, the rest to be paid in eighteen to twenty-four months. This second payment was often made in colonial goods, not in cash, which was always scarce in all American colonies. It was the agent who determined the credit worthiness of the local purchaser and it was he who was required to collect the final payments. He was also required to obtain a return cargo for the slave ship if this were possible. More often than not the ship was sent off with only a limited cargo, and finally returned to Europe some fifteen to eighteen months after having left Europe.[36]

Despite the myth of the so-called "triangle trade," the leg of the slaving voyage between America and Europe was the least important part of the slaving voyage; similarly, slave ships were not a significant element in the transportation of slave-produced American goods to the European market. In fact, most of the West Indies goods were shipped to Europe in boats specifically designed for that purpose, were both larger than the typical slaver and were exclusively engaged in this bilateral trade. Since many in the slavers' crews were supernumeraries after the slaves were sold in America, and the ships cost was a relatively minor part of the original expenditures of outfitting the slaving voyage, it often happened that slavers ended their voyage in the New World and the captain and a few crewmen returned to Europe on their own. Even when they did return to Europe they waited only a short time to return, making no effort to wait for the availability of American goods; more often than not such ships returned in ballast.

It usually took three full years to complete a voyage with the merchant needing to sell the constantly arriving colonial goods (by which most of the slaves were paid for) to local importers. The bulk of the credit sales were completed by the end of six years, though outstanding bills sometimes were never paid. It was the last three of the six years when the profit of the trade was made. From the work of the European economic historians, it is now evident that slave trade profits were not extraordinary by European standards. The average 10 percent rate obtained was considered a very good profit rate at the time, but not out of the range of other contemporary investments.

By whatever definitions used, the sale of African slaves was done in relatively open-market conditions. Although early in each European trade there are cases of ignorant slave captains seizing local Africans who appeared before them on the coast, these practices stopped quickly. European buyers were totally dependent on African sellers for the delivery of slaves. European traders never seriously penetrated beyond the coast before the late nineteenth century. The coastline itself was often lightly populated and had few slaves. Slaves in numbers sufficient to fill the holds of the slave ships only arrived to the coast via African merchants willing to bring them from the interior. The complexity of this exchange was such that it explains why slaves were purchased in such small numbers on the coast and why Europeans took months to gather a full complement of them for shipment to America. Given this balance of resources and power relations, the Europeans quickly discovered that anything but peaceful trade was impossible. Those who did not adapt were rapidly removed from the trade, sometimes by force.

Europeans also had to deal with the special demands of established states if they wished to purchase slaves. Almost all traders paid local taxes for their purchases. They also dealt with the African traders as autonomous and powerful foreigners who controlled their own goods and markets and quickly adapted to local trading practices. All locally built forts paid ground rents and local taxes as well. It was estimated that in the Kingdom of Whydah on the Slave Coast, in the late seventeenth century, European slave traders had to pay the equivalent of thirty-seven to thirty-eight slaves (this cost valued at £375) per ship in order to trade for slaves in the kingdom.[37]

Given that both an internal and international slave trade existed prior to their arrival, the Europeans found it convenient to adjust to well-established local African markets and trading arrangements already in place. In many cases, the Europeans only deepened preexisting markets and trade networks. Africans were also quick to respond to European needs beyond the slaves themselves. Coastal Africans developed specialized production to feed and clothe the slaves arriving to the ports and to supply provisions for the European trading posts and their arriving ships. New trade routes were opened as European demand increased beyond local coastal supplies, and with it more long-distance trading became the norm everywhere. It was said that white and black Moslem traders from the Saharan region finally began trading at Whydah at the beginning of the eighteenth century, as the complex Saharan routes were now linked to the European Atlantic ones in the so-called Benin gap—the open savanna lands in the Lower Guinea Coast.

Coastal trading states acquired their slaves from the interior, purchasing them with both their local products such as salt, dried fish, kola nuts, and cotton textiles, as well as European goods. Unusual trade routes opened by Europeans were often developed as well by the Africans. The Dutch initiated direct oceanic trade on a systematic basis with small coastal yachts between the Gold Coast and the Slave Coast in the seventeenth century in order to purchase African products demanded on the Elmina markets for gold. Africans in ocean-going canoes soon followed and created a major new cabotage trade for regular African commercial goods between these two coasts for the first time. This in fact explains the origins of the settlement at Little Popo on the Slave Coast, which was a portage stopping point for Gold Coast canoes transferring from the ocean to the inland lagoon shipping channels.[38] And new American foods imported by Europeans for their own needs were soon cultivated by African producers. These imports included such fundamental crops as maize and sweet potatoes, along with manioc (casava), coffee, and cacao. Already by the 1680s the Slave Coast communities were supplying maize to the European slave ships. The Europeans also introduced pigs and such unfamiliar Asian products as citrus fruits. Many of these crops slowly replaced or supplemented traditional African foodstuffs, often permitting denser and healthier populations. Although some of these products were integrated into traditional food production arrangements, others became the basis for new local industries. There was even the case of imported European woollen cloths being unraveled for their thread and rewoven by Slave Coast weavers to produce new style cloth for consumption on the Gold Coast. This is aside from the well-known importation of Swedish bar iron that was used by African blacksmiths to produce agricultural instruments for clearing bush and planting crops. Nowhere on the coast were the Africans incapable of benefiting from European trade or the introduction of new products and using them to their own ends and needs.

A thriving market economy with specialization of tasks and production and a well-defined merchant class, in existence before the arrival of the Europeans in most areas of Africa, goes a long way toward explaining the rapidity and efficiency of the African response to European trade. Gold and slaves had been exported from Africa for centuries. From the Saharan caravan traders of the western Sudan to the stone cities and gold fairs organized by Swahili Islamic traders in East Africa, Africans were well accustomed to market economies and international trade well before the arrival of the Portuguese on the West African coast. This does not mean that these were necessarily full capitalist markets, since local monopolies, kinship, and religious constraints, along with state intervention, often created unequal access and restricted markets here as they still did in most of Europe in the fifteenth century. But in general, prices defined in whatever currencies, units of account or mixes of trade goods, fluctuated in response to supply and demand across the entire continent. Nor were traders reluctant to expand their markets or adopt new technologies.

To determine the price for which slaves were sold in Africa to Europeans is a very complex calculation. For the Europeans the price was the European cost of goods that they needed to purchase and offer in exchange for obtaining a slave. This European cost of goods was called the "prime" price. On the African coast, these goods were often doubled in value when sold to the Africans, and these prices in the trade were called the "trade" price. Almost all the European accounts when estimating their own costs used the "prime" price.[39] But even using the trade price for what the Africans had to pay is a complex calculation. The actual mix of goods used in any purchase was expressed in both European currencies as well as in African monetary accounts, which included such monies as cowry shells, copper wires, or even palm cloth, or was defined in such units as a trade "ounce"—which originally meant the value of an ounce of gold—"bundles," and other arbitrary units. These African currencies or units of account were not uniform across Africa. Cowry shells, for example, were a primary money used in the Gulf of Guinea, but not on the Congo-Angola coast. The ounce was very common in the Gold Coast and associated areas, but not used elsewhere, while a "bundle" of goods was common in the trade to the Loango and Angolan coast. In the Upper Guinea coast, in the seventeenth century, Spanish silver coins were used and Brazilian gold dust was accepted as payment in the Biafran coast in the eighteenth century. Moreover, there was no uniform price for all slaves in a given port at a given time. Prices varied quite widely, based on the age, sex, and health of each slave. Women were on average 20 percent cheaper then men, and children were even cheaper still. Moreover, prime age males were 20 percent more expensive than older males, and so on. Thus average prices varied depending on the sex, age, and health of the slaves purchased, though in general between half and two-thirds of any group of slaves carried off the coast were made up of so-called "prime" age adult males. Slave prices also varied depending on local trading competition and supply conditions and even varied over the time of trading for an individual voyage. Moreover, the actual prices paid for each slave varied quite considerably from this mean, since each purchase was made with a variety of goods of markedly differing costs for the Europeans. Thus one English observer calculated that though the average price per slave on a given group of slaves was £3 15s., those purchased with cowry shells cost £4 each, those with beads and iron bars cost only £2 15s, and those sold for pieces of Indian cotton goods were valued as high as £6.

Nor were the goods that were sold by the Europeans over several centuries of the trade of the same quality, quantity, or price. Nor were the products that the Africans wanted for their slaves the same over time and place. While beads and brass bracelets did well in the seventeenth century, the number needed to purchase slaves increased to such an extent that they were worthless in such exchanges in the eighteenth century, as African sellers no longer expressed interest. Nor were European goods, East Indian textiles, or Indian Ocean cowries the only products used by the Europeans to purchase slaves. There were also large quantities of American goods imported, including Brazilian tobacco and North American rum. Often African goods were also employed. Cotton textiles of African make, salt, dried fish, kola nuts, various woven cloths, and other local coastal products were also used to purchase slaves in the interior markets, along with European goods. On the Senegambia and Upper Guinean savanna coast north of the forest areas, there were even imports of horses brought by Portuguese vessels from Arab sources in North Africa, which were used to purchase slaves, gold, and ivory. In the Cross River zone of the Bight of Biafra it was necessary to buy local copper rods to be used to purchase slaves, and in the

Loango Bay north of the Zaire River and on the river itself, it was customary to purchase locally produced palm cloth of the Vili group of Loango Coast to pay for slaves in the early centuries of the trade from the Congo and Angolan regions. The Portuguese were required to import North African cloths in their Senegambia trade in the sixteenth and seventeenth centuries.[40]

Despite all these variations there were some long-term trends evident in the selling of African slaves to the Europeans. Almost everywhere slave prices remained low or even declined from the early period to the end of the seventeenth century, suggesting that the growing supply exceeded any increased demand in most cases. Though American demand for slaves was on the increase, especially after the middle of the sixteenth century, the steady level of African wars due to state expansion and the European exploitation of new areas of the coast more than satisfied American needs. Slave prices on the Slave Coast, for example, which only developed as a prime source for slaves in the early seventeenth century, saw average prices decline for most of the century until the 1690s, when they began a long but steady period of growth.

The origin and manner in which slaves were obtained for sale to the Europeans is one of the more difficult areas to fully detail. It is evident from most sources that coastal peoples were able to supply sufficient slaves from groups close to the sea for the first century or so of the trade. By the eighteenth century slaves were being drawn from interior groups far from the coast. But who these groups were and how far from the coast they were situated is an issue difficult to resolve. Much of this difficulty is due to the ignorance of the European traders. They had only the vaguest notions of the names of interior groups or of their placement and relative importance. No ship's manifest lists the ethnic origin of the slaves they carried to America, just their port of purchase.

Most commentators have suggested that the slaves taken to America in the first two centuries of the trade came from the coastal areas probably no further than a few days march from the sea. Densely populated regions along the Senegambian, Guinean, and Congo coasts were a major and constant source of slaves over very long periods of time. But it is assumed that slaves were coming from much further inland by the second half of the eighteenth century as the trade expanded and intensified. Some diminishing of coastal slave trading must have occurred as local groups either were incorporated into more powerful states and obtained protection from raiding and enslavement or migrated out of range of the slave hunters, thus forcing local traders ever deeper into the interior. But even in the late eighteenth and early nineteenth century the majority of the slaves in most zones were still coming from relatively close to the coast. Thus, for example, the major exporting ports of the Bight of Biafran coast—Bonny and

New Calabar (in the Niger Delta) and Calabar (in the Cross River Delta)—took the majority of their slaves from the Igbo and Ibibio language groups, who were densely settled between the Niger and the Cross Rivers quite close to these ports. Though Hausa, Nupe, and Kakanda peoples far to the north passed through these ports, they were definitely a minority.[41] What occurred in this major exporting zone was probably typical of what occurred in most regions. Traditional areas, if they were still exporting, were still the major source for slaves, and these were mostly located close to the coast. Only in the eighteenth century did the Loango and Angolan ports begin to develop major caravan routes that stretched several hundred miles into the interior.[42]

The ratio of total slaves leaving Africa as war captives is difficult to estimate. All studies agree that there were numerous ways to enslave peoples. Aside from captives taken in war, there was large-scale raiding for slaves along with more random individual kidnapping of individuals almost everywhere, especially on the poorly defined frontiers of the larger states. Common to most societies was the judicial enslavement for civil and religious crimes and indebtedness. Larger states often required dependent regions to provide tribute in slaves, which could then be shipped overseas. It is clear that there was no one dominant source of enslavement in any region, though force was ultimately the basic instrument used everywhere to obtain slaves. The fact that almost all African states recognized domestic slavery meant that enslavement was an accepted institution within the continent and that the cost of transportation and security of slaves was much less than otherwise would have been the case. The slave coffers in the market were respected by local peoples as long as their own members were not affected, and slaves were found in almost all internal markets as well as coastal ones.

It is evident that since force was required, and trading involved many communities and states, that the costs of entrance into the slave trading business were relatively high. Merchants had to organize porters, buy goods for trade, and have extensive personal, kin, or religious contacts over a wide area so as to guarantee peaceful passage, and they clearly needed soldiers or armed followers to protect their purchases from others or prevent their slaves from escaping. In turn, those who raided for slaves had to outfit well-armed and mobile groups that also needed to be able to resort to peaceful markets as well. All this meant that only relatively wealthy individuals, or well-defined associations of small merchants (found everywhere from Nigeria to Loango), could engage in the slave trade. These merchants had to be skilled in determining local market conditions and to be able to trade with the Europeans, along with using and obtaining credit from all their contacts. Though managed state trading existed in some places,

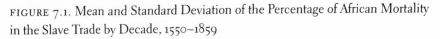

FIGURE 7.1. Mean and Standard Deviation of the Percentage of African Mortality in the Slave Trade by Decade, 1550–1859

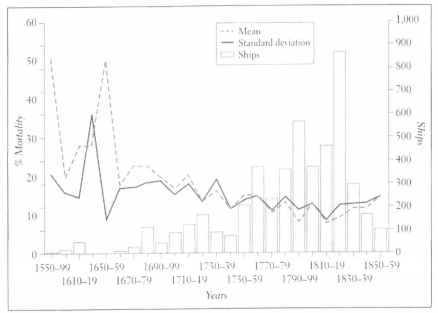

Source: David Eltis, Stephen D. Behrendt, David Richardson, and Herbert S. Klein, *The Transatlantic Slave Trade, 1562–1867: A Database* (Cambridge: Cambridge University Press, 2000).

and kings and royal officials traded on their own everywhere, the market for slaves was dominated by merchants, who are reported to have been major actors everywhere, even in such royally dominated trading kingdoms as that of Dahomey. Since Europeans were free to trade anywhere on the coast and often refused to trade in areas where prices were too high, there rarely developed any market domination on the part of Africans. In turn the Africans refused to be confined to any one trading nation and actively fostered competition among buyers. The result was that slave prices varied according to supply and demand and tended to be uniform across all the coastal regions of West Africa.

The slaves destined to America would cross the Atlantic in a journey that became known as the "Middle Passage." To put the Middle Passage in context, it should be recalled that the water crossing on average took a month from Africa to Brazil and two months from the West African coast to the Caribbean and North America. But for most slaves a minimum of six months passed between being captured and boarding European ships, with an average of three months spent waiting on the coast just to board.

If the purchase of slaves on the African coast was not a costless transaction,

FIGURE 7.2. Slave Mortality in the Middle Passage, 1550–1799

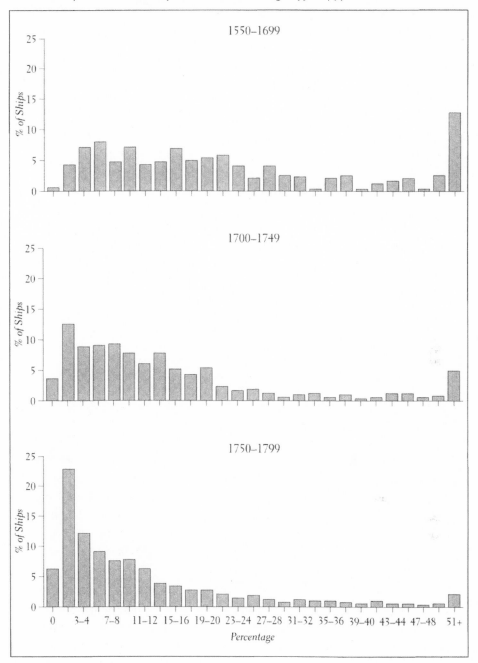

TABLE 7.1. Estimates of African Slave Arrivals, by Region, 1451–1650

Quarter	Europe	Atlantic Islands	São Tomé	Spanish America	Brazil
1451–75	19,396	2,500			
1476–1500	37,544	5,000	1,000		
1501–25	43,200	5,000	25,000		
1526–50	7,500	5,000	18,800	12,500	
1551–75	2,500	5,000	18,800	25,000	10,000
1576–1600	1,300	2,500	12,500	37,500	40,000
1601–25	300		12,500	93,850	150,000
1626–50	300		6,300	93,850	50,000
Total	112,040	25,000	94,900	262,700	250,000

Sources: Based on the tables in Philip Curtin, The Atlantic Slave Trade: A Census (Madison: University of Wisconsin Press, 1969); as revised in David Eltis, "The Volume and Structure of the Transatlantic Slave Trade: A Reassessment," William & Mary Quarterly, 3rd series, 58, no. 1 (January 2001): 44. Professor Eltis has generously provided the numbers he used to create this table and his most recent revisions of these numbers. These in turn will appear in David Eltis et al., The Transatlantic Slave Trade: A New Census (Cambridge: Cambridge University Press, forthcoming). The data for the 1451–1525 period was taken from Ivana Elbel, "The Volume of the Early Atlantic Slave Trade, 1450–1521," Journal of African History, 38, no. 1 (1997): 73.

Note: I have summed Elbl's annual figures and assigned them to Curtin's twenty-five-year periods. I have also assumed that 1522–25 averaged the same as the last years Elbl studied. Finally, since her data is on African shipping estimates, I have reduced her figures by 20 percent for mortality. Also, I have arbitrarily assigned all the increased numbers to Europe.

then any loss of slaves on route would directly affect the ultimate profitability of the voyage. In fact high slave mortality on the crossing resulted in financial loss on the trip. It has been estimated for the eighteenth-century French slave trade—which has the best information on costs—that each transatlantic slave death on a ship carrying 300 slaves would reduce profits by 0.67 percentage points. Thus a mortality rate of 15 percent could reduce trading profits by as much as 30 percent.[43] This fact explains the increasing efficiency and uniformity that all trades achieved by the eighteenth century. Over time Europeans learned to better transport slaves on specially built ships most adequate for African trading. By the end of the seventeenth century, the average size ship was approaching half the size of a normal cargo ship of the period. Various late-eighteenth-century calculations estimate a space of from five to seven square feet per slave.[44] But despite this very limited space, in no large collection of slave voyages currently available is there a correlation between the manner of carrying slaves (either using a tonnage or space indicator) and the mortality they suffered.

| British | West Indies | | | Total | Average Annual |
	French	Dutch	Danish		
				21,896	876
				43,544	1,742
				73,200	2,928
				43,800	1,752
				61,300	2,452
				93,800	3,752
				256,650	10,266
20,700	2,500			173,650	6,946
20,700	2,500			767,840	7,678

Moreover, at least in the seventeenth century, the 15 to 20 percent mortality experienced by slaves in the Atlantic crossing was probably similar to the experience of troop ships of the time in similar crossings.[45]

In dealing with mortality rates, most writers have defined it as the number of slaves who died on ship in the Middle Passage, either recorded directly in the contemporary records or calculated as the difference between slaves who boarded and slaves who landed divided by the number of slaves loaded on the African coast. The most significant pattern discerned in these death rates is the very wide distribution of mortality rates by voyage. This is found even when holding other features constant such as sailing times, ship sizes, African embarkation areas, and the age and sexual composition of slaves carried. There was a broad range of outcomes, with very many quite different experiences, even for the same captains or the same nationality of shippers. Very high mortality rates tend to be associated with unexpectedly long voyages, or to unusual outbreaks of disease, but in general it is the very broad range of outcomes rather than any bunching at specific mortality rates that has been the main characteristic of the transatlantic slave trade for most of its existence.

From the earliest recorded voyages in the late sixteenth and early seventeenth century when the death rates averaged 20 percent, slave shipboard mortality declined to less than one-half this level in the late eighteenth century. While the decline was relatively monotonic over time, there was an especially large decline in the last quarter of the eighteenth century. As sharp as was the decline in the mean of slave mortality rates, the median of slave ship mortality declined even more rapidly, with the entire distribution of mortality rates shifting down. There was a great increase in the share of voyages coming in at relatively low mortalities. Correspondingly, the percentage of ships with mortality rates above a selected threshold level fell, meaning that over time there were relatively fewer ships with very high mortality rates (see figures 7.1 and 7.2) and more and more ships were coming in at close to the mean mortality.

The general trend in ships sizes was to move from a wide range of ships with the majority being of very low tonnage, often carrying high ratios of slaves, to a middle range tonnage vessel carrying fewer slaves per ton. Moreover, this trend was common to all slave trades regardless of the local national tonnage measurements used. This meant that there was a progressive increase of slaves carried per ship, as average tonnage increased and a more steady ratio of slaves per ton was achieved.[46] Also, all such slaving ships were unique in their internal arrangements as well, usually using temporary decks to house the slaves, which were divided by bulkheads made of open grates with open latticed grates for deck hatches. On several of the ships for which designs exist (all from the eighteenth century), there were even opened up ventilation ports (with hatches to be closed in inclement weather) on the sides of the ships between the gun ports and above the platforms built over the lower deck, creating air flow across the platforms. These design features were unique to slavers and were specifically designed for bringing air into the sleeping quarters of the slaves. Late-eighteenth-century British vessel measurements show that they all divided their internal space in a common pattern: the men's rooms were on average three times the size of the boys', and twice the size of the women's and infants' quarters, with the boys' quarters separating the men's and women's rooms.

Though all these arrangements had still not been fully worked out by 1650, enough of the basic knowledge about markets in Africa and America and the transportation of slaves across the Atlantic had been developed that it can be said that the modern slave trade had now fully evolved in its basic structure. The only change in markets after this would be the incorporation of South East African slaves into the Atlantic slave trade in the early nineteenth century, or the more intensive development of New World markets. By 1650 a minimum of some 708,000 slaves had been delivered by the European trades to the European and Atlantic markets (see table 7.1). The trade was now moving some 7,000 slaves per annum, a figure that would rise to 24,000 per annum by the last quarter of the seventeenth century. Thus the trade was well able to keep pace with American demand as the century-long stability of prices in slaves has suggested. It was not until well into the eighteenth century that demand for slaves in America finally outpaced supply.

NOTES

1. Paul E. Lovejoy, *Transformations in Slavery: A History of Slavery in Africa* (Cambridge: Cambridge University Press, 1983), 24.

2. John Vogt, *Portuguese Rule on the Gold Coast, 1469–1682* (Athens: University of Georgia Press, 1979), 9.

3. A. C. de C. M. Sauders, *A Social History of Black Slaves and Freedmen in Portugal, 1441–1555* (Cambridge: Cambridge University Press, 1982), 55.

4. Ibid., 29.

5. On the Madeira sugar industry and the role of African slaves, see Alberto Vieira, *Os escravos no arquipélago da Madeira* (Funchal: Centro de Estudos de História do Atlântico, 1981).

6. By 1582 there were some 14,000 African slaves and 400 freedmen in these previously uninhabited islands, though most of the slaves were dedicated to local cloth production. Walter Rodney, *A History of the Upper Guinea Coast, 1545 to 1800* (Oxford: Clarendon Press, 1970), 72.

7. Tony Hodges and Malyn Newitt, *São Tomé and Príncipe: From Plantation Colony to Microstate* (Boulder, Colo.: Westview Press, 1988), 19–21.

8. Frederick P. Bowser, *The African Slave in Colonial Peru, 1524–1650* (Stanford: Stanford University Press, 1974), 75.

9. Ibid., 92ff.

10. This discussion is based on ibid., chaps. 4 and 6.

11. Ibid., 341.

12. Alberto Crespo R., *Esclavos negros en Bolivia* (La Paz: Academia Nacional de Ciencias de Bolivia, 1977), 31.

13. Colin A. Palmer, *Slaves of the White God: Blacks in Mexico, 1570–1650* (Cambridge: Harvard University Press, 1976), 76. For the experience of African slaves in the mines, see P. J. Bakewell, *Silver Mining and Society in Colonial Mexico, Zacatecas, 1546–1700* (Cambridge: Cambridge University Press, 1971), 122–24.

14. For an analysis of African slaves in Mexico the best survey is that of Palmer, *Slaves of the White God*. See also Gonzalo Aguirre Beltrán, *La población negra de México, 1519–1810: Estudio etnohistórico*, 2d ed. (Mexico City: Fondo de Cultura Económica, 1972).

15. Beltrán, *La población negra*, 219.

16. Stuart B. Schwartz, *Sugar Plantations in the Formation of Brazilian Society: Bahia, 1550–1835* (Cambridge: Cambridge University Press, 1985), 65.

17. In fact, the 15,000 Africans on the islands, compared to only 2,000 in British North America at this time, meant that the former were more populated than the latter. John McCusker and Russell R. Menard, *The Economy of British America, 1607–1789* (Chapel Hill: University of North Carolina Press, 1985), 54, table 3.1.

18. The density estimates have been converted from those given in square kilometers by Jean Meyer, *Histoire du sucre* (Paris: Ed. Desjonquères, 1989), 129.

19. For the demographic history of Barbados, see Richard S. Dunn, *Sugar and Slaves: The Rise of the Planter Class in the English West Indies, 1624–1713* (Chapel Hill: University of North Carolina Press, 1972). By 1700 there were an estimated 115,000 black slaves and only 33,000 whites in the British West Indies; see McCusker and Menard, *Economy of British America*, 154, table 7.2.

20. This estimate is taken from Christian Schnakenbourg, "Statistiques pour l'histoire de l'économie de plantation en Guadeloupe et Martinique (1635–1835)," *Bulletin de la Sociétié d'Historie de La Guadeloupe* 21 (1977): 44–47; David Watts, *The West Indies: Patterns of Development, Culture, and Environmental Change since 1492* (Cambridge: Cambridge University Press, 1987), 320, table 7.9; and Meyer, *Histoire du sucre*, 112.

21. The best survey of this early trade is found in Saunders, *Social History of Black Slaves*, chap. 1.

22. Ivana Elbl, "The Volume of the Early Atlantic Slave Trade, 1450–1521," *Journal of African History* 38, no. 1 (1997). The Portuguese by the first two decades of the sixteenth century were purchasing over 400 kilos per annum from inland traders, which was roughly one-fourth of West African gold production in that period.

23. For the comparative volume of the different external African trades over time, see Lovejoy, *Transformations in Slavery*, 45, table 3.1.

24. David Eltis and Lawrence C. Jennings, "Trade between Western Africa and the Atlantic World in the Pre-Colonial Era," *American Historical Review* 93 (1988): 936–59. There is also a recent debate about value of slaves in total trade before 1700 in Ernst van den Boogart, "The Trade between Western Africa and the Atlantic World, 1600–90," *Journal of African History* 33, no. 3 (1992): 369–85; and the reply of David Eltis, "The Relative Importance of Slaves and Commodities in the Atlantic Slave Trade of Seventeenth-Century Africa," *Journal of African History* 35, no. 2 (1994).

25. One of the best introductions to the *asiento* arrangements is found in Bowser, *African Slave*, 28. For studies of individual Asiento contract arrangements in the pre-1700 period, see Enriqueta Vila Vilar, *Hispanoamérica y el comercio de esclavos: Los asientos portugeses* (Seville: Escuela de Estudios Hispano-Americanos, 1977); and María Vega Franco, *El tráfico de esclavos con América . . . 1663–1674* (Seville, 1984). On the Portuguese experiment with localized monopoly companies, see Antonio Carreira, *As Companhias Pombalinas de navegação, comercio e tráfico de escravos entre a costa africana e o nordeste brasileiro* (Porto, 1969).

26. On the development of the French monopoly companies, see Abdoulaye Ly, *La Compagnie du Senégal* (Paris: Présence Africaine, 1958).

27. Johannes Menne Postma, *The Dutch in the Atlantic Slave Trade, 1600–1815* (Cambridge: Cambridge University Press, 1990), 54.

28. On the RAC see K. G. Davies, *The Royal African Company* (London: Longman Green, 1957), as well as the much earlier studies of George Frederick Zook, *The Company of Royal Adventurers Trading into Africa* (Lancaster: New Era Printing, 1919).

29. The discussion of the economics of the slave trading voyages is a summary based on the detailed studies of Postma, *Dutch in the Atlantic Slave Trade*, esp. chap. 11; Jean Meyer, *L'armement nantais dans le deuxième moitié du XVIII siècle* (Paris: SEVPEN, 1969); and Robert Louis Stein, *The French Slave Trade in the Eighteenth Century: An Old Regime Business* (Madison: University of Wisconsin Press, 1979).

30. Postma, *Dutch in the Atlantic Slave Trade*, 156. Nor did these figures change much over time. Of the 312 captains who made slaving voyages out of the port of Bordeaux in the seventeenth and eighteenth centuries, 71 percent made only one voyage and the average was 1.5 voyages for all the captains. Eric Sauger, *Bordeaux, port négrie: chronologie, économie, idéologie, XVIIe–XIXe siècles* (Paris: Kartalha, 1995), 225.

31. The average for 870 slavers leaving Nantes in the 1749–92 was thirty-four crewmen per vessel; see Meyer, *L'armement nantais dans la deuxième moitié*, 79–80, 83–86. Dutch slavers on average carried a crew of forty in the seventeenth century. See Postma, *Dutch in the Atlantic Slave Trade*, 153.

32. Numbers generated from the voyage data in David Eltis, Stephen D. Behrendt, David Richardson, and Herbert S. Klein, *The Transatlantic Slave Trade: 1562–1867: A Database* (CD-ROM) (Cambridge: Cambridge University Press, 1999).

33. This is the generally agreed upon estimate provided for the best-studied French trade by Meyer, *L'armement nantais dans la deuxième moitié*, 159; and Stein, *French Slave Trade*, 139.

34. For the Dutch data, see Postma, *Dutch in the Altantic Slave Trade*, 157.

35. Arquivo Historico Ultramarino (Lisbon), Angola, caixa no. 10. Printed law dated 18 March 1684.

36. Stein, *French Slave Trade*, 110, covers average French patterns; and for the English selling practices, see David W. Galenson, *Traders, Planters and Slaves: Market Behavior in Early English America* (Cambridge: Cambridge University Press, 1986). See also Trevor Burnard, "Who Bought Slaves in Early America? Purchases of Slaves from the Royal African Company in Jamaica, 1674–1708," *Slavery and Abolition* 17, no. 2 (1996).

37. Robin Law, *The Slave Coast of West Africa, 1550–1750: The Impact of the Atlantic Slave Trade on an African Society* (Oxford: Clarendon Press, 1991), 207.

38. Ibid., 148–50.

39. Ibid., 170.

40. Vogt, *Portuguese Rule on the Gold Coast*, 9, 66–69.

41. David Northrup, *Trade Without Rulers: Pre-Colonial Economic Development in South-Eastern Nigeria* (Oxford: Clarendon Press, 1978), 59ff.

42. For an analysis of these trades, see David Birmingham, *Trade and Conflict in Angola: The Mbundu and Their Neighbours under the Influence of the Portuguese, 1483–1790* (Oxford: Clarendon Press, 1966); and Phylis M. Martin, *The External Trade of the Loango Coast, 1576–1870* (Oxford: Clarendon Press, 1972).

43. Stein, *French Slave Trade*, 141–42.

44. For a detailed analysis of space aboard slave ships, see Charles Garland and Herbert S. Klein, "The Allotment of Space for African Slaves Aboard Eighteenth Century British Slave Ships," *William & Mary Quarterly* 42, no. 2 (1985).

45. For an analysis of slave mortality over time and its comparison to other shipboard mortality rates, see Herbert S. Klein and Stanley L. Engerman, "Long-Term Trends in African Mortality in the Transatlantic Slave Trade," in *Routes to Slavery: Direction, Ethnicity and Mortality in the Transatlantic Slave Trade*, ed. David Eltis and David Richardson (London: Frank Cass, 1997), 36–49.

46. For a detailed discussion of ships and their changing size and carrying capacity, see Herbert S. Klein, *The Middle Passage: Comparative Studies in the Atlantic Slave Trade* (Princeton: Princeton University Press, 1978).

CHAPTER EIGHT

The Expansion of the Sugar Market in Western Europe

Eddy Stols

At the palace of Brussels, on 18 November 1565, on the occasion of the festivities of the marriage of Alexander Farnese with Princess Maria of Portugal, a gallant company of great lords and ladies, surrounding the regent of the Netherlands, Margaret of Parma, crowded around a long table to admire crystallized fruits from around the world, from Spain, Portugal, Genoa, and Naples, from Africa and the marvelous Indies, laid out on dishes, in jars, in cups, and on plates, with matching and paring knives and napkins amidst chandeliers and candelabras.[1] Everything, except the cloth on the table, was made of sugar.

Even more impressive was the set in a neighboring room, four or five times larger, where on another, even longer table were set scenes of the voyage of the Portuguese princess. One saw at the start the Pillars of Hercules and the imperial eagle, the squadrons of ships, the unfurled sails marked with the arms of Portugal and Spain, the raging ocean with its whales, dolphins, and sea monsters, the wreck of one boat and another in flames, the passengers throwing themselves in the water or drowning, the arrival in Zeeland, the reception at Middelburg and then in Gand, the river with its barges, the celebrating people, and, on the road to Termonde, packed with cavaliers and carriages, the princess surrounded by her ladies and black slaves in livery—and all was this in a region where custom prohibited slavery. A great carriage led to the entry gate of the city of Brussels, which enclosed the city's churches and its towers, its roads and houses full of people, the palace with Her Highness the Regent, and an animal park, with lions, antelopes, and a herd of elephants ridden by Indians. The scene lacked neither card and dice players in taverns nor a theater of comedies. Behind some windows there were parakeets in cages, apes, and tiny cats.

There were more than three thousand pieces made from the finest sugar, and they looked so natural that they could fool some people. Nevertheless, the guests were not at all embarrassed to eat sugar and fill their pockets with more. Soon there remained nothing but the heaviest pieces that one hardly dared to touch. The people and horses weighed up to nine or ten pounds a piece. It required no fewer than four men to carry each of these cities, three feet high and six long. Never had the Italian observer Francesco de Marchi seen a spread in sugar as splendid as this, except perhaps in Naples in 1536, at the marriage of Margaret of Parma and Alexander de Medici.[2] These pieces were displayed by the magistrate of Antwerp and the cost was estimated at more than three thousand ducats.

As extravagant as these expenses might appear, they no doubt justified themselves to the head of these municipal authorities. They provided evidence of the abundance of sugar and established Antwerp as its principal European market, while they also brought the city closer to its principal Portuguese supplier. The display of such colossal quantities of sugar, which could probably be estimated at more than 6000 pounds, defies the imagination, yet became very familiar and beloved by these lands of plenty. In the past, sumptuous banquets with gigantic, soaring pieces were seen at the court of Burgundy, but they had been created from nearly inedible substances such as beef fat or wax. However, after 1530s, following the sugar sculptures invented in Italy for several wedding banquets of the Este, Sforza, Montefeltro, and Medici families between 1473 and 1539, the new hype in festivities in northern Europe swept to a similar conspicuous use of sugar. On 12 December 1531, celebrating in Brussels the birth of the infant Manuel, successor to the Portuguese throne, the Portuguese ambassador Pedro de Mascarenhas regaled his invitees, the Emperor Charles V himself, his sister, Queen Mary of Hungary, and the high-ranking nobility of the Low Countries with the new tempting delicacies, the sweets of Madeira. A rich midnight *"banc-quet de confitures et de succades"* closed on 26 October 1544, after the splendid feast and dance at the Brussels's Palace on the occasion of the visit of yet another sister of Charles V, Leonore, widow of King Manuel of Portugal and at that moment queen of France through her marriage to François I. At the end of August 1549, Queen Mary of Hungary offered her nephew, the hereditary Prince Philip II, on his maiden tour through the Low Countries, several days of splendid and memorable festivities at the castle of Binche. The apotheosis came with an astonishing banquet in the *Cámara encantada* (enchanted chamber), where with lightning, thunderbolts, and a hail of *comfits*, three tables descended successively from the ceiling, richly laded with all kinds of preserves. One carried a rock of candy sugar with five trees full of sugar fruit. Already on the outward journey, the prince had been treated to several *collazione de zucchero* in Barcelona and Milano.

Decorative display made of sugar for the wedding of Johann Wilhelm, heir to the ducal seat of Jülich-Kleve, from an engraving by Frans Hogenberg in Didederich Graminaeus, Beschreibung derer furstlichter Juelichscher Hochzeit *(Cologne, 1587). From Giuseppe Bertini,* Le nozze di Alessandro Farnese *(Milan, 1997).*

During the same period at Hampton Court Palace, King Henry VIII also succumbed to the costly new taste for sugar. His cooks furnished the royal table and receptions with all kinds of confectionery, spices coated in sugar, marmalades, marzipan, sugar plates, and subtleties such as figures of soldiers, saints, and even a St. George on horseback or a St. Paul's Cathedral. Under Elizabeth and James I, the sugar banquet evolved into a standard element in court entertainment. Other examples of sugar collations are registered in the city of Paris for the entry of Elizabeth of Austria in 1571 and at the marriage of the heir to the duchy of Jülich-Kleve in 1587.

From Sugar-Spice and Sugar-Medicine to Colonial Commodity

Many authors, including Fernand Braudel and Immanuel Wallerstein, still underestimate the importance of the sugar trade in the rapid expansion of large-scale capitalist commerce; instead they privilege the trade in pepper, grains, wool, and textiles. Like Sidney Mintz, they argue that the creation of a large sugar market and its mass consumption was closely linked to the spread of tea and other sugared drinks toward the end of the seventeenth century.[3] The story of the Brussels table of 1565 seems to suggest that one could advance the beginning of the sugar boom and that the overabundance of sugar was already a fact from the middle of the sixteenth century, or even earlier. Its consumption was no longer restricted to minute quantities for medical use or as a luxury spice, but rather it gradually became an entirely separate and important foodstuff, a veritable colonial commodity. André Thevet in his *Cosmographie de Levant* (1554) wrote and Abraham Ortelius in his *Theatrum Orbis terrarum* (1570) repeated "What in times passed was scarcely found but in Arabia Felix [Yemen] or India; and [which] the Ancients used only in medicines; today the confectioner knows well how to apply it to our use." This vulgarization, which had already occurred at the end of the Middle Ages in the Mediterranean basin, moved, from that point on, to the Atlantic coast of Europe and increased along an axis from south to north, from the Iberian peninsula to the Netherlands by way of France, to reach all of western and northern Europe.

The quantities of sugar increased considerably with its use in the conserving of fruits and jam making. This method of preserving fruits was admittedly not unknown in the Middle Ages, but it spread from royal and princely courts to the kitchens of more modest and more numerous social groups such as shopkeepers, artisans, and peasants. Later, the making of these preserves became a supplementary job, a rather important one for this new bourgeoisie, often of rural origin, that still possessed several acres of pleasure gardens and orchards at the gates of the city. Women, especially the wives of merchants, found in it a worthy occupation that could only enchant their husbands and visitors. The new cultural prestige of preserves was well summarized in the work of the French agronomist, Olivier des Serres: "Thus it will be here where the honorable lady will find pleasure, continuing the proof of the subtlety of her spirit. So she can secure pleasure and honor, when, on the unexpected arrival of her relatives and friends, she will cover the table for them with diverse jams carefully prepared."[4] Other women, more modest, widows or women deserted by some sailor who left for the Indies, or else servants, found the sale of these preserves a supplementary or compensatory income. On these grounds the feminine work of jam making can be compared to lace making, which also developed as a specialty of women's work in countries that were already well known for their jams and marmalades. Moreover, in Portugal, women accompanied their preserves with unique papers cut out in lace work. Although male *confeiteiros* (jam makers) in Lisbon organized themselves, very early in 1539, in their *Casa dos Vinte e Quatro* (House of Twenty-Four), under the banner of São Miguel, and with professional regulations in 1572, there were at least 200 women who publicly sold their products in 1644.

It was precisely in Portugal and Spain that Olivier de Serres voluntarily conceded the merit of the invention of new methods and recipes of confectionery. In his *Singularités de la France Antarctique* (1557), André Thévet reputes especially the people of Madeira "for the best and most delicious conserves shaped as men, women, lions, birds and fishes, beautiful to see and even better to taste." Without a doubt, the Mediterranean world, especially those countries under Islamic influence, had specialized for a long time in this practice of preservation.[5] What appears particular to Portugal is that the abundance of sugar permitted the use of those fruits and legumes that were heavy and bulky, inexpensive and bland, and did not lend themselves to the use of honey, much too expensive and difficult to use in such large quantity.[6] Thus, the Portuguese did not hesitate to conserve in syrup the omnipresent chestnuts, known as the fruit of the poor, or to cook in sugar the astringent quince or different varieties of squashes and gourds, the *cabaças* [calabashes], *jirimuns*, and *chila*. These, similar to the *doces de abóbora* (Brazilian sweet pumpkins) and relatives of the Mexican *camotes* (sweet potatoes), seem almost unique in Europe and do not appear as abundantly elsewhere. *Le cuisinier français* (1651) mentions only sugared pumpkins and *marrons glacés* (iced chestnuts).[7] In addition, in Portugal, sugar even served to salvage leftover rice as *arroz doce* (sweet rice), or slices of stale bread as *rabanadas* (French toast). Although the poor north of Portugal resigned itself at first to adopting cornmeal for breadmaking, judged inferior to wheat, it also dared to mix this almost flavorless grain with sugar to make tasty *broas de santos* (saint's bread). Because of their early contact with India, the Portuguese would have been able to develop techniques of conservation and preparation with vinegars and *achar* spices, but it remains obvious that it was sugar that was preferred as nowhere else.

More than others, the people of Portugal, Andalusia, and other Spanish regions developed and maintained a surprising ingeniousness and an almost disturbing creativity to vary and differentiate their sweets in all forms and colors and under the most evocative names, such as *toucinho celeste* (celestial lard), *tutano do céu* (heaven's marrow), *papos de anjo* (angelic Adam's Apples), and *barringuinhas de freira* (little nun's bellies). If it is impossible to date in a precise fashion the origin of all these wonders, one can perhaps attribute this sugary explosion to the beginning of the sixteenth century, the archetypal Portuguese golden age, with its society of plenty and leisure, more clement weather, more generous nature, the absence of strong barriers of social distinction, the omnipresence of black slaves and servants from the beginning of the sixteenth century on, or even the persistence of a more pagan religiosity fixed on the cult of fertility, that protected Portugal from the Christian fundamentalism of Protestants from the north. One could pretend that Portugal, together with southern Spain, became one of the first European regions in which sugar formed a part of the popular diet.

It is likely that convents, which were at that time multiplying in Portugal and Spain at a wild rate, while they were menaced or closed in England and the Netherlands, acted as the instrument of mediation *par excellence* of this descent of sugar to the lower levels of society.[8] These convents, lavishly provisioned with dozens of eggs and supplied with alms given in sugar, competed among themselves to welcome, with *mimos* and *tabuleiros de doces* (trays of sweets), royal or princely visitors who could dispense new favors. At the same time, they found in sugar confectionery not only their own material subsistence but also a form of redistribution and the appreciation of their fellow citizens. Simultaneously, several Portuguese religious writers developed a distinctive spirituality based on a symbolic and spiritual valorization of fruits and flowers.[9] Confronted with the Reformation spirit, traditional faith justified itself in a certain way by a veritable confectionery debauchery that could even combine with the famous *amor freirático*, that rather strange predilection for cloistered women cultivated by several Portuguese and Spanish kings and princes. Shortly thereafter, the convents of Puebla and other Mexican towns would develop this association of religion, the feminine, and sugar to the point of the paroxysm of a cultural *chrisme*.[10]

While Lisbon seemed to be the capital of this rapid expansion of the new art of preserving that was at once aristocratic and more quotidian, the first, more concrete indications of the art's economic and social importance were also found in the Portuguese capital.[11] In his inventory of the economic riches of the city in 1552, João Brandão counted no fewer than thirty *tendas de confeiteiros* (confectionery shops), each employing four to five people, amounting to a hundred fifty in total, including fifty women, making marmalade (*açúcar rosado e laranjadas*), which they sold to those going to the Indies or Guinea.[12] There were ten more *tendas de pastéis* (pastry shops), where more than thirty people busied themselves with making small pastries or morsels, often lightly sugared. Two weeks before Christmas, at the Ribeira and at Pelourinho Velho, some thirty women installed their table covered with a white cloth and filled with *gulodices* (delicacies) or sweets and preserves such as orange marmalade, *sidrada* (apple jelly), and *fartéis*. Their estimated sales were at least 2,000 *cruzados*, and together with the expenses paid in noble households for fruit preserves, they perhaps amounted to more than 20,000 *cruzados*. Fifty women sold the *arroz doce* (sweet rice) that nourished and warmed so well the bellies of children and made them cry for it as soon as they woke up.

The city also had at its disposal a *casa de refinar* (refinery) that employed more than twenty people. There were even forty carpenters who made cases for the various compotes and marmalades, each of whom made 1,000 to 2,000 pieces a year. As one case was already worth twenty to thirty *réis*, and was filled with 300 to 400 *réis* worth of merchandise, the whole was worth at least 11,500 *cruzados*. If included in that were cases of *rosado* (rosy) sugar, they were worth 23,000 *cruzados*. The total value of the sugar importations from Madeira, São Tomé, and Brazil, excluding the Canaries, was calculated by the *alfandega* (customs house of Lisbon) at 45,000 to 50,000 *cruzados*, while Brandão estimated it at 70,000.

This humming sugar activity was confirmed by the humanist Damião de Góis, who in his description of Lisbon called attention to the vast *terreiro* (debarkation square) that served as much as a market for fish as for preserves and where fishmongers, market gardeners, preserve sellers, bakers, and confectioners gathered to sell their wares.[13] During their passage through Lisbon in 1585, the Japanese pupils of Portuguese Jesuits were amazed to discover a street where sugared preserves were sold in sufficient abundance as to satisfy easily the needs of the people of Lisbon as well as to export to numerous cities in Europe.[14] Passengers embarking for the East Indies bought in large quantities *comer feito*, prepared meals often consisting in large part of *ovos moles*, egg yolks lightly cooked in sugar. Well-wrapped preserved foods of all kinds did very well on board the ships, relieved illness, and lasted long enough even to sell well in the Indies, where people did not seem familiar with cherry and plum marmalades.[15] Some brought more than 100 kilos.

In particular, the preparation of quince jelly had considerable economic impact on the sugar market because it required enormous quantities of sugar. A recipe in the cookbook that Maria of Portugal took with her to the Netherlands required three to four kilograms of sugar for an equal weight of quince. The accounts of Queen Catarina of Portugal, whose confectioner, Cornelio Izarte, was Flemish, include in an entry for 24 July 1554 an order of payment for no less than fifty-one *arrobas* and nine *arratéis* of conserves.[16] The taste for these marmalades widened in a manner characteristic for this period of perfumed gloves, gilded leather, jewels, clocks, glass, parrots, genre painting, and devotional images. Preserves were thus carted off in large quantities to the markets of northern Europe. King Phillip III, visiting his Portuguese kingdom in August 1619, had sent to his sister in Brussels, Archduchess Isabelle, two large shipments that included thirty-six cases of *conservas cubiertas* and twenty-six kegs of preserves in syrup.[17] French poet Vincent Voiture, returning in 1632 by sea from Lisbon, complained that the boat was so filled to the brim with sugared preserves that he feared being candied.

In the sixteenth century, the Spanish still visibly recognized the superiority of the Portuguese. Francisco Martínez Montiño, the author of *Arte de cocina, pastelería y vizcochería y conserveria,* a classic Spanish cookbook, undertook his apprenticeship in the service of Dona Juana of Austria, the sister of Phillip II, widow of the Portuguese heir Dom João (prematurely deceased in 1554), and regent of Spain from 1554 to 1559, and he borrowed several recipes from his Portuguese tutor.[18] In 1543, at the time of the marriage of the prince and future king, Phillip II, with his cousin Princess Maria Manuela of Portugal, her estate included a *confitero,* the Portuguese Francisco Machado.[19] Upon the death of the princess, he received 42,000 *maravédis* to abandon the court. It was probably the first time that such a function was mentioned in the Spanish court in the sixteenth century. At any rate, Phillip II continued to receive his daily ration of *pasteles ojaldrados* (sugared pears and peaches), and three times a week *manjar blanco* (white pudding).[20]

In fact the Spanish were themselves already experienced in sugaring and no less conscious of their *savoir-faire.* As for the Italians, since the fourteenth century, they had been referring to several handwritten Catalan manuscripts such as the *Libre de sent soví* and the *Libre de totes maneres de fer confits* (Book of the methods of making preserves). In the sixteenth century they added to it a rather extensive and varied printed bibliography with, among others, the *Libre del coch* (1520), by Ruperto de Nola, translated into Spanish as *Libro de guisados* (1525), and the *Libro del arte de cozina* (1599) (Book of the art of the kitchen) by Diego Granado, who was accused of plagiarism by Martínez Montiño.[21] In 1592 Miguel de Baeza published in Alcalá de Henares one of the first specialized cookbooks, *Los quarto libros del arte de la confietería* (Four volumes on the art of confectionery). He carefully described the process of sugar production and the different types of jams and preserves. He also explained the method of making sugar-candy, by which sugar was reduced to half its weight in round pitchers made in Seville. He classified the sugars of the Canaries, by quality, ahead of those of the coast of Granada and those of Gandia on the coast of Alicante. For preserves he preferred two-year-old sugarcane to one-year-old sugar (*alitas* in Santo Domingo).

Several Spanish treatises on health and medicine evince an appreciation for sugar, fruits, and preserves, as in the famous *Banquete de nobles caballeros,* by Luis Lobera de Avila (the surgeon to Charles V), who based his work on the principles of Galenus and preferred refined sugar to honey and, in particular, preferred small sugar pills, though he clearly warned against excess in case of fever or choleric temperament.[22]

Highly esteemed for their medical knowledge, the Portuguese, strangely, had to wait until 1680 for the publication of their first cookbook, *Arte de cozinha* (The art of cooking), by Domingos Rodrigues.[23] This delay should not be interpreted as a lack of interest in cooking, but rather as proof of the vitality of the practice and the oral, visual, and handwritten transmission of recipes. Moreover, there existed, in addition to the *Arte de cozinha*, another manuscript, drawn up by Alvaro Martins, chef of Dona Juana of Austria, mentioned by Barbosa Machado in the *Biblioteca lusitania*, but presumed lost. It is also necessary to recall that in France no new cookbook was published between the French translation of Platina's *De honesta voluptate* (1505) and *Le Cuisinier François* (The French chef) (1651), despite the definite progress and the growing prestige of French cooking throughout this period.

Meanwhile, the success of preserves spread to France, where jam making, according to Olivier de Serres, had remained "for a long time ignored in this kingdom, having been kept secret, as if a cabal." We must remember, however, that since the fourteenth century, the city of Bar in Lorraine was very famous for its jam made with currants from which women and young girls carefully removed the seeds with a feather before cooking them in sugar syrup.[24] Nevertheless, the secrets of the preparation of other kinds of preserves were disclosed rather early with the publication of the *Petit traicté contenant la manière pour faites toutes les confitures, compostez, vins* (Short treatise containing the method of making all preserves, compotes, wines) (1545), by Jehan Longis, and *Pratique de faire toutes confitures* (Practice of making preserves) in Lyons in 1555.[25] In the same year appeared the first edition of *Le vray et parfaict embellissement de la face . . . & la seconde partie contenant la façon et manière de faire toutes confitures* (The true and perfect embellishment of the face . . . and the second part containing the way and manner of making all kinds of preserves), by the celebrated physician and diviner Michael Nostradamus.

Many other treatises on home economics, agriculture, pharmacy, and chemistry address more closely food preservation with sugar, as in the very popular *L'agriculture et maison rustique* (Agriculture and the rural household), by Charles Estienne and Jean Liébault, with numerous editions after its first appearance in Paris in 1564, or *Les eléments de chymie* (Basics of chemistry), by Jean Béguin (1637). The great agronomist Olivier de Serres condescended to dedicate an entire chapter to the method of jam making, and, parsimonious like a true Frenchman, he even instructed his readers to reuse the sugar from old preserve to make new ones, though only for "the dark walnut and almond" preserves, and not for jellies.

Recipes are found scattered throughout the most diverse books. Thus Jacques Pons offered in his *Sommaire traité des melons* (Summary treatment of melons) (1583) details of the preparation of melon compote with brown sugar. In *Discours contenant la conférence de la pharmacie chymique* (Discourse containing a lecture on chemical pharmacy) (1671), Jacques Pascal devoted a chapter to the role of sugar in the preparation of *alkermès*, a cinnamon- and clove-based liqueur much valued at that time.

Conversely, certain historians of food who, somewhat shocked, ponder this strange absence of French cookbooks lose sight of the fact that in the sixteenth century many French books were printed outside France, mostly in Antwerp, which nevertheless could circulate within France without suffering censure. This was the case with the Nostradamus book printed in 1558, by Plantin, in Antwerp. The chef of the bishop-prince of Liège, Lancelot du Casteau, included in his *Ouverture de cuisine* (Work on cooking) (1604) several sugared desserts, *gaufre succrée, succades liquides, pastez de coing, marmelade en forme, grand biscuit succré* (sugared waffles, sweet liquids, quince paste, molded jelly, sugared cookies).[26]

It is, therefore, clear that a long period of experimentation supported *Le cuisinier français* when, in 1653, it distinguished very clearly among the different manners of cooking sugar, *"à lisse, à perle, à la plume et au brûlé"* (smooth, beaded, feathery, and burnt), or revealed the secret of the clarification of sugar.[27] The entirety of French knowledge on the matter attained the highest degree of perfection with the 140 pages dedicated to *Confiturier royal*, attributed to Massaliot, as part of the summary of gastronomy of the age of Louis XIV, *L'ecole parfaite des officiers de Bouche, contenant le vray maistre d'hotel; le grand Escuyer-Tranchant; le sommelier royal; le cuisinier royal et le patissier royal* (The perfect school of the officiers de bouche, including the actual maître d'hôtel, the escuyer-tranchant, the royal sommelier, the royal chef, and the royal pastry chef) (1662; 1676). It would be impossible to omit many other publications, often published in the Netherlands, such as *Le patissier françois* (1655), *Le jardinier françois, qui enseigne les arbres et herbes potagères avec la manière de conserver les fruicts et faire toutes sortes de confitures* (The gardener, who raises trees and edible herbs, including the method of preserving fruit and making all kinds of jams), by Nicolas de Bonnefons (1660), and *Traité de confiture, ou le nouveau et parfait confiturier* (Treatise on jam making, or, the new and perfected jam maker) (1698). In *Le nouveau recueil de curiositez rares et nouvelles des plus admirables effets de la nature et de l'art* (The new collection of rare curiosities and news of the most admirable effects of nature and art) (1685), Nicolas de Memery integrated the knowledge of candy making with the knowledge of the perfect *honnête homme*. It must be emphasized that in France these delicacies were no longer limited to the royal household. In 1662 F. P. de la Varenne published in Troyes the famous *bibliothèque bleue*, *Le patissier françois*, which spread throughout France on the backs of traveling book peddlers (*colporteurs*). That a physician of the poor, Philbert Guybert, devoted several pages of his *Les œuvres charitables* (Charitable works) (1630) to preserves, sugar, brown sugar, and syrups gives pause for thought.

Although Portuguese and Spanish influences are undeniable and perfectly possible by means of dynastic alliances (we must not forget that marriage of Queen Leonore, sister of Charles V and widow of Manuel, with François I), as well as increasingly regular commercial relations and the arrival of numerous New Christians at Bayonne, Bordeaux, Nantes, and Rouen, new refinements in French cooking are traditionally attributed to the Italians. Their cookbooks such as *Liber de coquina* introduced the custom of powdering dishes with sugar, substituting for the more traditional honey in the German kitchen, and the preference for a more sour taste in the French cuisine. Most notably the Italians figured as precursors in the matter of confectioneries, at that time still close to the arts of the apothecary, pills, and bonbons.[28] Quirico degli Augusti gathered in his *Lumen apothicariorum* (1504) no fewer than thirty-one notices on recipes using sugar and was the first to use the word "marzipan," while in the same year his fellow Italian Paolo Suardi gave even more recipes in his *Thesaurus aromatariorum*. In 1564 *L'empirie, et secrets*, by Alessio Piemontese, alias Girolamo Ruscelli, was published in Lyons; the Latin original appeared in 1555 in Venice, and with an English translation, *The Secretes of the Reverend Maister Alexis of Piedmont*, in 1562. In addition, the Italian princely courts, as much if not more than the Burgundian courts, set the tone for social events and public festivities. Thus Cristoforo da Messisburgo, chef at the court of the Duke of Ferrara, of Flemish origin and ennobled by Charles V, delivered in his *Banchetti, composizioni di vivvande e apparecchio generale* (1549) the model of a banquet offered to the counselors of the emperor, consisting of no fewer than six luxurious courses that ended with an apotheosis of sugary desserts.[29] His model, however, could be adapted to the purse of more modest lords who would expend a third less sugar and spices.

It is necessary to recall that this Italian influence extended also to Germany, above all the southern part, from Frankfurt to Augsburg, which still maintained very close commercial ties with northern Italy. These ties manifested themselves in the translation of the famous *De honesta voluptate*, by Bartolomeo Sacchia (alias Platina), Walter Ryff's *Von der eerlichen zimlichen auych erlaubten Wolust des leibs* (1542)—which inspired such masterpieces as the *ConfectBuch und hauss apoteck kunstlich zubereiten, einmachen und gebrauchen* (1544)—and Marx Rumpolt's *Ein neu Kochbuch* (*A New Cookbook*) (1581).

In terms of France, the historical record has always emphasized, and no doubt too exclusively, the decisive role of Catherine de Medici, who from the time of her marriage with Henri II in 1535 dominated the French court for almost half a century. Indeed, during her son Charles IX's journey through France in 1568, she accompanied him, followed by two pack animals intended to carry fruits and preserves.[30] Aside from the royal court, one can assume that after the wars of religion convents similarly functioned as veritable laboratories for the perfecting of new recipes. In this way, the Ursulines of Flavigny, borrowing a Benedictine recipe, developed the famous crystallized anise, an aniseed surrounded by sugar and scented with orange blossom water, still for sale today in a small, colorful box.[31] The sugar-crystallized stems of the *angélique* flower of the Sisters of the Visitation in Niort owed their reputation as a panacea against the plague most notably to the recommendations of Madame de Sévigné. According to *Le cuisinier françois* there were also *pets de putain* (whore's farts).

In Flanders, in the Spanish Netherlands, these were known under the name of *nonnescheten*, or *pets de nonne* (nun's farts). The term refers to the fact that, from the end of the sixteenth century on, after the excesses of the *gueux*, the Calvinist Protestants, the very dynamic agents of the Catholic Counter-Reformation indiscriminately covered the country with numerous new convents. Several were founded and populated by religious women from Spain, and thus they adopted the new Carmelite reforms of Teresa of Avila. Several indications allow the supposition that these nuns similarly introduced to the Netherlands the art and practice of the *dulcerías conventuales* or, at the very least, enriched the existing traditions of the *Béguines* (religious women who resided communally in *béguinages* without taking vows).[32]

In the main cities of the southern Netherlands, the *béguinages* repopulated themselves with dozens, or even hundreds, of *Béguines*. Although devotional literature constructed for them an aura of great abnegation and alimentary sacrifice, the popular voice saw them, instead, as both lazy and *gourmandes*. Even amongst themselves a legend presented a *Béguine*, Beatrice of Brussels, who tarried too long in the chapel before supper, but who in returning to her kitchen found at the table a handsome young man, none other than Jesus himself, stirring the soup with a spoon.[33] In their naive imagination heaven became the place where one ate *rijstpap*, or rice pudding, with golden spoons. In fact, these religious women had to contribute to the cost of their maintenance, and hence busied themselves with all kinds of work, especially embroidery, cutting out *découpage* figures from lace, or making artificial flowers or small dolls. Nothing could be more natural for those women who invented, in addition to waffles, other cookies and sweets, all the more so as they gladly combined them with the meticulous observation of religious festivals such as the Saint-Martin, the Feast of the Kings, and *Den graaf van half vasten*, or *Mi-carême* (a festival held more or less at the midpoint of Lent).[34] The *Béguines* of Antwerp distinguished between *"crakelingen, weggen, marsepijn, spans suyker, amandelen, bacades, mostasollen, muskesletteren, busquit."* They ate them communally, reserved them for the ill in the infirmary, or threw them at random to children. At the *béguinage* of Diest the laywomen distributed at funerals so much *lijkmikken*, or bread lightly dusted with sugar, that the bakers of the city took offense and protested this disloyal competition with their livelihood. Although sugary comestibles lent themselves marvelously to rites of distribution, the religious women could not claim to monopolize them. Previously, in the thick of the religious wars, Spanish soldiers distributed fruits and sweets during their carnival festivities.[35]

In effect, outside this small world of religious Flemish women, sugar products had long since acquired the status of a commercial commodity and became an important market product in the epicenter of the Netherlands. In 1561, in Antwerp, they already took center stage in the allegorical plays presented at the time of *Landjuweel* or the theater festival of the *Rederijkerskamers* or *Chambres de rhétorique*. Their characters frequently undertook promotion of sugar products: *Préparez moi pour le banquet de douces succades / des conserves, des sirops et des marmelades / des savouereuses gelées / des vins de Romanie bien sucrés* ("Prepare me for the banquet of sweet drinks / of conserves, syrups and marmalades / savory jellies / sweetened Romani wines").[36] Aside from the book of Nostradamus, confectioners in the Spanish Netherlands could also make use of the recipe books, in Flemish translations, of Alessio Piemontese, *Die secreten* (1558); of Charles Estienne, *De landtwinninge ende hoeve van M. Kaerle Stevens* (1566); or, better yet, the *Secreet-boeck* (1600), by Carolus Battus, a surgeon in Antwerp, exiled in 1585. The professionalization—or rather, in contrast to the female predominance of the practice in Portugal, the *masculinization* of the *suiker-bakker*, or candy maker, and *pasteibakker*, or confectioner—seems obvious and appears in the public acts of the period, even though those working with sugar lacked their own guild; this is how the sugar economy appears in the Antwerp chronicle of Godevaert van Haecht with his *"Peer de suyckerbakker."*[37] It is probable that this new profession based itself on relatively simple products and the almost obligatory bourgeois consumption of sweets. An Italian collection, written between 1585 and 1625, refers to a recipe for *confetto di Fiandra*, a sugar-based paste, molded with *gomma adragante* (a kind of gum), musk, and cinnamon, that served as table decoration.[38] From that point on, bookkeeping of all kinds of feasts in the milieu of the *Chambres de Rhétorique*, such as the *Brabantse Olijftak*, and among merchant families and *Béguines* reserved considerable sums for confectioners' expenses. Even in inns, in Flanders every meal ended with sweets, and in particular small candies, according to one French traveler, the Jansenist Charles Lemaître.[39]

During this invasion of preserves and sweets, sugar did not disappear as a spice from the main dishes of Flemish cuisine. To the contrary, it seemed to impose itself more and more on all sorts of preparations of meat and fish, as one last resurgence of medieval tastes before the arrival in the second half of the seventeenth century of the new French cuisine and its stricter separation of sweet, salty, and bitter flavors. One finds this proliferation of sugar in all cookbooks, of which the Spanish Netherlands had no lack, either in copies or new editions, since the premier of Thomas Vander Noot's *Een notable boexken van cokerijen* (1510), and the subsequent *Eenen nyeuwen coock boek* (1560) by Gheeraert Vorselman, Carolus Battus's *Eenen seer schonen excellenten gheexperimenteerden nieuwen coc-boec* (1593), *L'ouverture du cuisine* by Lancelot du Casteau, and *Koock-boeck ofte familieren Keuken-boeck* (1612; 1655) by Antonius Magirus.[40] Similarly, a manuscript from Antwerp, from the end of the sixteenth century, recommends sugar as well for roast rabbit, veal pâté, mutton, or carp as for beef tongue, *poivrade de lièvre* (hare in pepper sauce), or dressing a capon.[41] Lancelot du Casteau used sugar most notably in a dish of minced carp and in a *tourte de Portugal* (Portuguese pie) with veal.[42] Sugar appears similarly in one of the most favorite dishes of the period, the *blanc-manger, witmoes op zijn Catalaans*, or *manjar blanco* of the Spanish, the white meat of chicken breasts served with rice creamed with the milk of crushed almonds. There was even sugar in another highly appreciated sauce, the *sopa dorada* or *vergulde soep* that accompanied roasts and fish. In each case the recipe did not skimp on the quantities of sugar used, often reaching or even easily surpassing a pound of sugar. Frequently, sugar was mixed with cinnamon, almonds, and oranges, all of which were also imported in large quantities. Popular drinks such as *l'hypocras*, a kind of mulled wine flavored with cinnamon, vanilla, and cloves, also contained much sugar. Moreover, even new wine, often of poor quality, was drunk sugared.

The popularization of sugar and its spread into more common cooking remains to be investigated. According to Pierre Belon, the naturalist specializing in ichthyology, sugar served to improve the bad taste of large fish obviously destined for the masses, such as tuna or dolphin, whose gustatory qualities he promoted in his *L'histoire naturelle des estranges poissons marins* (The natural history of strange marine fishes) (1551). Even today in Flanders as in many regions of northern Europe one willingly puts sugar in such very popular dishes as *boudin* (blood sausage), *carbonnade* (a beef and onion stew popular in Ghent and northern France), red cabbage, or apple compote. Sugar mixes more easily with dairy foods such as butter or cream cheese than does honey and similarly improves such rather insipid and heavy pasta and stews such as *mastellen*, fruit tarts, or rice puddings. In Holland, *wentelteefje*, similar to the American but misnamed "French" toast, or the Portuguese *rabanada*, has become very popular, as have *poffertjes*, fritters sprinkled with powdered sugar (similar to the *beignet* of New Orleans). These are probably identical to the *snoeperije en bancketsuycker*, sweet baked goods of this type that were sold, according to the Antwerp chronicle of Godevaert van Haecht, from sheds erected next to the frozen Scheldt at the end of December 1565.

However, it is very difficult to find precise information on the purchase and use of sugar among more common people. Accounts very often mention *stroop*, a molasses-like by-product of sugar refining that was often of dubious quality. It could replace honey, whose production remained very limited, or compete with or supplement honeycombs. An appellation that appeared rather early was that of *broodsuiker* or *pain de sucre* (sugar molded into a large, round shape similar to that of a loaf of bread, whence the name "sugarloaf"). The term may indicate both the spherical shape and the coarse, inferior quality for its most common use with bread. The slices of bread sprinkled with brown sugar still eaten in Flanders may also date from this time period, just as the traditional gingerbread, the so-called *pain à la grecque*, is an old specialty of the Brussels bakeries. For the preparation of such edibles, one can picture a practice similar to that of a merchant in Segovia, Juan de Cuellar, in which a sugarloaf was hung in the kitchen and flattened with a warm glass or plate, then garnished with several dribbles of melted sugar.[43] Such a sugarloaf appeared in *La visite à la ferme* (Visit to the farm), by Pieter Breughel the Elder, and was recaptured in the paintings and engravings of Jan Breughel the Elder and Pieter Breughel the Younger around 1597–1625. In the images are visitors, probably bourgeois landowners, giving as a gift to their tenant farmer a large sugarloaf, wrapped and tied with paper, just as they are still sold today. Some years later, the Antwerp Franciscans received two sugar loaves as a New Year's present.

The Cultural Promotion of Sugar

The diffusion of sugar was not only a question of alimentary innovation; it also appealed to the pleasures of the senses, especially sight. Sugar, because it is so easy to manipulate, color, carve, and file, lends itself marvelously to all sorts of decorative fantasies in the typical taste of the Renaissance and the early Baroque for arabesques, grotesques, and fantastic creatures that inhabit and animate the borders of carpets, paintings, and the wainscoting of the period.[44] Sugar was perfectly suitable to developing this ephemeral art, and confectionery emerged as the major branch of architecture, as Antonin Carême would define it two centuries later. Better than papier-mâché or calcified materials, sugar served to articulate fantastic constructions in miniature, temples, and arcs de triomphe that exalted the power and glory of antiquity. It facilitated an almost encyclo-pedic miniaturization at once admiring and possessive of the surrounding world. In this aspect, sugar had the advantage of being edible, and thus that it could potentially satisfy cannibalistic fantasies, common in the imagination of the period.[45] Furthermore, the sumptuous displays remained fashionable for a long time, like the one offered in 1640 by Cardinal Borja (president of the Supreme Tribunal of Aragon), which featured a château of marzipan and sugar worked in filigree, with a remarkable likeness of the cardinal in the portico of the entry.[46] In 1667 in Amsterdam, the grand duke of Tuscany, Cosimo de Medici, during his visit to the Netherlands, was presented by a delegation of the community of the Portuguese Jews with "a triumph of sugar, representing a ship, finely worked with its decks and inner rooms in fullest detail. There was a mausoleum made in the grotesque style with many little statues, and a bowl of a pastiche of ambers, and one of little pieces of chocolate . . . Four kinds of Portuguese-style confec-tions".[47] One may note both the rather precocious association of sugar with chocolate and the persistence of Portuguese specialties.

Sugar, more than bread, appealed to the popular imagination and produced figures of all kinds—people, animals, and even devils, sugared and easy to crunch.[48] It is not astonishing that a victim of the Inquisition in Goa, Charles Dellon, set up a strange comparison between sugarloaves and the sanbenitos worn by the condemned: "Paper hats rising like a sugar loaf covered with devils and flames."[49] One can also write one's name or initials with the famous lettres d'Hollande (Dutch letters) or lettergebak (an almond and sugar pastry). Such smaller letter cookies were apparently used for fostering literacy among children.

Although it remains almost impossible to trace the origin of the first sugary fantasies, one can easily establish the chronology of their appearance in scien-tific or artistic iconography. Initially, the fabrication of sugar, along with gold and silver mines, represented colonial technology and wealth as much as the exploitation of slaves. It was largely divulged through the engravings of Théo-dore de Bry from 1590 onward and did not cease to haunt the imagination through numerous imitations and reinterpretations, as well as the drawings and paintings of Jan Van der Straeten (Stradanus), Crispijn van de Passe (see his *Hortus floridus* [1914]), and Frans Post, among other witnesses of the Dutch occupation of Brazil.[50] Such works, painted on the cabinets of the Antwerp furniture workshops, depicted the painful job of cutting sugarcane.[51] It seems unlikely that at this point in time this evocation of hard labor and slavery could have provoked the strong consumption anxieties that were later addressed in the eighteenth century.[52] In the *Schat der Gesontheyt* (1636) by Johan van Beverwyck, however, next to an engraving of sugar extraction, there appears an early commentary in this vein by the poet, moralist, and polymath Jacob Cats: "What suffering of fierce blows today in torrid Brazil / To harvest the fruits in this distant land."[53] In contrast, neither the *Temptations of Saint Anthony* by Jerome Bosch, nor the Breughel allegories of coarse and scant food, nor the numerous Flemish *quermesses*, nor the *arcimboldique* fantasies, include depictions of sugar or confectioneries.[54] From the engravings of Pieter Breughel the Elder and his son's, one could perhaps with effort mention the *mise-en-scène* of the *Lutte entre le Carnaval et le Carême* (Struggle between Carnaval and Lent), in which Lent is coiffed with a hive of live bees, while Carnaval seems to wear sugar tarts. To view this tableau as a representation of the confrontation between traditional honey and imported sugar is a far stretch. The hive appeared again in *L'âne à l'école* (The donkey at school), in which a child sticks his head in an overturned hive, and in *L'espérance* (Hope), in the "Seven Virtues" series, which presented a virgin crowned with a hive. In the drawing *La prudence* (1559), preserves arranged in pots figure as a symbol of domestic foresight.

Even stranger is the absence of sugar products in the exuberant scenes of the market, the kitchen, the table, and genre paintings, made popular in the sixteenth century by Flemish painters such as Joachim de Beuckelaer and Pieter Aertsen.[55] Of the latter's work, however, there survives the *Wafelbakster* (1560), a painting of a woman selling waffles with butter, but without the visible syrup or sugar of the actual *lacquemant* vaffle. In the festive meals painted by Jerome Francken (1540–1610), several sweets can just barely be made out. It is probable that, in contrast to fruits and vegetables or bloody meats and chops, heavily charged with symbolic significations and erotic suggestion, the painting of confectioneries and preserves, much more innocent and without their own symbolic value, fascinated and gratified the eye of the spectator to a much lesser extent. They lacked the beauty of form, a strong color, and a very pictorial substance. Additionally, it must be remembered that in the Netherlands, confectionery had become a man's business, while women sold fruits and vegetables.

Not until the end of the sixteenth and beginning of the seventeenth century did confectioneries make a more marked appearance in still-life paintings. It is not surprising that the precursors in this record were Italian painters such as Vincenzo Campi, Ludovico di Susio, and Jacopo Chimenti da Empoli, followed by Giovanna Garzoni and Bartolomeo Arbotoni around 1660.[56] To the north of the Alps, a student of Lucas van Valckenborch, Georg Flegel, who was active in the same city of Frankfurt as the De Bry family and where the first sugar refiners suddenly appeared with the influx of Flemish immigrants, appears to have been one of the first to appreciate confectioneries and to give them a place in his *Schauessen* or still-life paintings. With their bizarre forms and their white color, sticks of spun sugar and candied nuts contrasted very well with plates of olives, of preserved fruits and nuts, glasses, shellfish, or butterflies. Almost simultaneously, around 1610–20, the theme surfaced in Antwerp, in Jan Breughel the Elder's allegories of taste, but above all in the works of Osias Beert, an expert in this matter. Amsterdam followed immediately with David Vinckboons, Pieter Claesz, and Clara Peeters, originally from Antwerp, who introduced sugar in the *bancketjes*. Shortly thereafter, around 1620, confectioneries appeared in Spain in the *bodegones* of Juan Sánchez Cotán, Juan van der Hamen y León, Antonio de Pereda, and Francisco de Palacios. Van der Hamen willingly added the round though austere forms of *cajas de dulces*, while Pedro de Medina even showed a piece of cane sugar. Confections also appeared in the works of French painters such as Lubin Baugin in 1635. It is not surprising that Portuguese painter Josefa de Ayala de Obidos's still lifes from 1660–80 were so realistic that they tempted the viewer with the most varied confectioneries.[57]

This genre of paintings played a role comparable to modern publicity and remained popular, at least, until the end of the seventeenth century. One could perhaps object that this pictorial exaltation of confectionery concerned only an elite minority, relatively wealthy to be able to permit themselves the purchase of such tableaux. By way of response it must be stated that many Flemish and Dutch paintings served as a matrix for engravings printed by the hundred and widely distributed through the lively trade in images, or else in books.[58] They inspired a whole imagery, such as that of Abraham Bosse.

Sweets, and above all *galette* cakes, perfectly expressed the fragility of temporal things and of existence; they appeared mostly in portraits of children. They held the sweets in their hands or had them just out of reach, as they did with other attributes of innocent carelessness such as flowers, small birds, or *sjiboleths*.[59] At the same time, these sweets served to entice or reward children, turning them into inveterate consumers. By 1490 the *cortes* of Évora complained of the *alfeloeiros*, who came from Castille to sell these caramels of twisted sugar that made children cry in front of their parents to obtain the money necessary to buy them.[60] Shortly thereafter Dom Manuel forbade their sale by men, reserving it

Sugar and sweets became regular elements in bodegones *(still-lifes). A considerable market for these existed in Spain. This example was painted by Juan van der Hamen or one of his students in the seventeenth century. Courtesy of the Museum of the Royal Academy of Fine Arts of San Fernando.*

for women, widows, and children. Michel Montaigne advised to "sweeten with sugar the meats healthy for children and make bilious those harmful to them."

The childhood feast-days par excellence, those of Saint Thomas, Saint Martin, and the Three Kings, persisted with ease in the period of the Catholic Counter-Reformation, while even the most recalcitrant Calvinist Dutch could not bring themselves to strike off the calendar the feast of good Saint Nicholas, who brought to good children their candies and to recalcitrant ones the blows of *père fouettard* (similar to the American "boogeyman," that is, a mythical character used to frighten misbehaving children; in the French version, he carries a whip, "fouet," thus the name). Nothing manifests this better than the famous painting of Jan Steen.[61]

More secretly, many adults, melancholic because of religious conflict and incessant wars, sought and found consolation in a bonbon. Amorous discourse and relations often borrowed references from sugar, as in the poems of Jan van der Noot or the *"Mijn life . . . mijn suyckerdoos"* ("My love . . . my box of sugar") in *De gecroonde leers* (verse 466) of Michiel de Swaen. Cristóvão Godinho's book *Poderes de amor em geral, e obras de conservaçam particular* ("The procuration of love in general, and how to maintain it") (1657) discussed at length the links between the erotic and sugar.[62] In the comedy *De Spaanse Brabander* (1617), by the successful Dutch author Gerbrand Adriaensz Bredero, the main character, Jerolimo from Antwerp, a poor wretch with the look of a great lord, presents his mother as "the wife of a poor confectioner, but she knows how to bring the tarts and marzipans right quick to captains, colonels, and grands pagadores."[63] Thus, a whole body of literature contributed, even before and certainly following Rabelais in the *Quart Livre de Pantagruel*, to the enthroning of confectioneries at the height of extreme happiness in the mythic land of plenty, where the snow and hail are made of powdered sugar and sugared almonds and tarts cover all the roofs.[64] Finally the emblematic imagination took hold of sugar: in *Menselijk Bedrijf*, his survey of human action, Jan Luiken associated the *Suikerbakker* (confectioner) with "the divine sweetness of the blood of Christ" and advised that "those who would vanquish acrid sourness / must not begin with aqua-fortis / but rather sugar is the proper sword / o my God, how you have given / to bitter life your greatest sweetness / and thus prevented the Great Fall."[65]

Even cities began to identify themselves with all kinds of sweets, which became a new emblem, such as the *calissons d'Aix* (marzipan petit-four) or the *bêtises de Cambrai* (literally, "idiocies of Cambrai": rectangular, mint-flavored candies).[66] The latter owed their name to poorly made candies that Marguerite of Burgundy had distributed "to the common folk by her confectioner" on the occasion of her marriage.[67] As way of punishment he was paraded in a *carnavalesque* procession that was repeated each year, with a shower of candies. The municipality of Orléans offered to the king its *cotignac*, small cubes of quince paste (no fewer than thirty-eight dozens in 1576), Verdun its candies with musk and anise, and Metz its preserved *mirabelle* plums, to the point of arousing the jealousy and imitation of Nancy.[68] Already in 1565 in Toulouse the meal offered by the *capitouls* (municipal magistrates of Toulouse) consisted of "fifteen badges of the king, with collars of the Order and fifteen Fleur de Lys." Bruges was famous for its *Brugse mokken*, and Antwerp for the *Antwerpse handjes*, which refers to the hands that the mythical giant Antigone cut off shipmasters reluctant to pay their taxes and then threw into the River Escaut. Political connotations of certain sweets soon emerged: during the Fronde, the duc of Praslin mollified the most belligerent of the *jurat* (municipal magistrates of the Ancien Régime, especially in the Midi) of Bordeaux with toasted almonds. According to humanist and Protestant critics, sites of pilgrimage seemed equally associated with one or another candy, which the devoted would buy there to recover their strength or to bring back as a souvenir of their proceedings. The painting *Les trois sens* (The three senses) (1620) of Jan Breughel the Elder, at the Prado in Madrid, illustrates the part dedicated to taste with an pile of sweets and cookies garnished with small pilgrimage flags on top.

Like pots of wine and other gifts, sugar confections served perfectly to support familial and friendly relationships and as recompense for favors. Thus, in 1554 in Venice, Flemish merchant Maarten de Hane, in his will, left to his sister Catharina, a nun in the convent of Woutersbrakel, an annual income of fifteen ducats, devoted "partially to good wine and sugar and spices according to her practice."[69] Almost a half-century later, one of the successors of his firm, Antwerp merchant Jan della Faille, facilitated the registration of his purchase of a lordship by distributing sweets to competent functionaries, no fewer than "4 brootsuyckers to councilor Grysperre."

The many new uses of sugar required the creation of new, appropriate utensils that incorporated confectionery more visibly in daily life and assured it a place among the domestic equipment and in the familial patrimony. Inventories of kitchens included stoves and copper saucepans, bells for cooking fruits, skimmers, graters for sugar, pie pans and cookie molds, while on the tables of dining rooms appeared sugar pots, shakers, boxes of sugar, and, most likely later on, the sugar spoon and tongs.[70] It appears that these objects of luxury were made first of silver, gold-plated silver, or earthenware and were not produced in fine porcelain until later.[71] It is worth noting that Chinese porcelain made its way to Europe through the same route as sugar, from Lisbon to Antwerp, and subsequently to Amsterdam. There were also goblets filled with small fruit-and-seed conserves that were left on the tables at the disposal of visitors, and candy purses or pocket boxes that could be worn on a belt, which would later evolve into the candy tin.

Finally, the increasing familiarity with sugar manifested itself in toponymy with roads that bore the name of sugar almost as often as the older *rues au beurre* (butter streets, that is, the streets on which butter was sold) or *rues aux harengs* (herring streets). Thus in Antwerp, since at least 1565, a *Suyckerroije* or *Suikerrui* (sugar street) was very well situated on a small stream recently covered over, very close to the new Hôtel de Ville under construction.[72] In Gand there is a *Suikersteeg*, and in Amsterdam a *Suyckerhuys*. Lisbon has its *Rua dos Confeiteiros*.

It seemed as though nothing could hold back the triumphant ascent of this colonial commodity. One finds few warnings or critical preoccupations concerning the excess of marzipan and preserves in *Miroir universel des arts et sciences en générale* (The universal mirror of arts and sciences in general) (Paris,

1584) by Leonard Fioravanti, though nothing comparable to the numerous and often vehement diatribes against alcohol and tobacco. For those with good manners, Erasmus advised in his *De civilitate morum puerilium* (1535) not to let children lick sugar or other sweets attached to a plate or dish: "Such is the behavior of a cat, not of a human." Religious prescriptions and restrictions during Lent, abstention, and sobriety addressed sugar only rarely, as many still considered it to be, in the tradition of Galenus and of Thomas Aquinas, a medication, not food. In *Il vitto quaresimale* (1637), Paulo Zacchia explicitly mentioned cakes and preserves as substitutes for meat and bacon during Lent. Certain physicians began, however, to denounce the ill effects of an excessive consumption of sugar. According to Henry IV's physician, Joseph du Chesne, in his *Le pourctrait de la santé* (The picture of health) (1606), candies and preserves heated and burned the blood, and rotted and blackened the teeth.

Some warnings against excessive expenditure on luxury items (silk, lace, and silver dishes) targeted the consumption of sugar. Already in Portugal, royal orders like those of João III (3 July 1535), Cardinal Henrique (8 June 1560), and King Sebastião (28 April 1570) sought to limit expense in accordance with resources, or to forbid *manjar blanco* or *bolos de rodilla*.[73] Madrid, too, attempted to forbid the *figones* or caterers from selling in public "neither manjar blanco, nor tortadas, nor pastellitos nor other sweet things."[74] It is evident that such prohibitions had little effect and impelled instead a more highly valued consumption. In the Netherlands, since the beginning of the seventeenth century, critics had denounced the growing mania for buying expensive sugar in order to show them to visitors and citizens.[75] Several moralists, including the poet Jacob Westerbaen in his *Minnedichten* (1633), and the painter Joseph de Bray, promoted good national dishes, simple and plain, such as cheese or pickled herring, attacked delicacies of foreign origin and, thus, indirectly sugar. However, burgomaster Tulp of Amsterdam, who promulgated the restrictive decrees, did not respect them himself.

The Great Antwerp Sugar Market

How can we evaluate this unprecedented ascent and valorization of sugar in the context of the extraordinary commercial expansion of the sixteenth century? As Antwerp became the driving force in commercial capitalism, overtaking Venice, according to the Braudelian thesis, the role of sugar in the rapid expansion of the great northern market must be investigated. Did sugar attain the rank of important commodity, or that of a prime necessity, in the transactions of the period?

The first observation is that sugar by no means found a prominent place in the still prevailing classical works on the growth of the Antwerp market. Although Herman Van der Wee published quite a long list of sugar prices, he did not spend much time on it in his analysis of the rise of the market, nor on the effects of the increase in sugar prices.[76] It is true that there were no other quantitative data comparable to the data set on pepper. Thus, the calculations were perforce based on very approximate and debatable estimations. In 1560, according to one of the most prominent specialists of the economic history of Antwerp, Wilfrid Brulez, sugar imports reached 15,200 chests annually, with a value of 250,000 florins (guilders), as he based his calculations on Ludovico Guicciardini's description of the Netherlands and a general depiction of the economy undertaken by Gerard Gramaye.[77] This is truly a modest sum in comparison to the two million guilder value of the spice trade, and is almost insignificant in relation to the ensemble of commercial trade in the Low Countries, of which sugar represented less than 2 percent of the total imports. It is worth noting that, in general, Brulez tended to underestimate the role of colonial products in the commerce of Antwerp, while at the same time he emphasized the importance of more traditional commodities such as textiles and grains. Although his argument appears valid for the spice trade, he underestimated the role of other primary overseas materials, the rich trades, most notably pearls and precious stones, clandestine merchandise though they were, yet nevertheless decisive for the fortunes of a great commercial city and its luxury crafts. Likewise, it appears now that sugar occupied a much more important place among commercial transactions and the riches accumulated in Anvers, and that it even constituted an essential pivot.

However, Portuguese data, ignored by these Flemish historians (as is too often the case), suggested a much higher value. The sugar from Madeira reserved for Flanders, by Dom Manuel, reached 40,000 *arrobas* (nearly 460,000 kilograms) out of a total of 1,080,000 *arrobas*, of which almost half was destined for the Italian ports.[78] Almost definitely, these Portuguese sugar exports increased considerably, at least until the unleashing of hostilities in 1570 and the closing of the Scheldt in 1585. The fourth of June 1564 saw in Antwerp ten or eleven Portuguese ships filled with sugar. Aside from sugar, the rather significant quantities of conserves must also be considered. In 1517 Diogo de Medina, *confeiteiro* (confectioner) in Madeira, sent annually, by royal order, twelve *arrobas* of nothing but conserves to the *feitor*, the Portuguese factor in Flanders.[79] Later, after the crisis of 1566 and the departure of so many merchants, the solidity of the sugar trade played a greater role in the recuperation, however incomplete, of the economic prosperity of the city, in its "long Indian summer," lasting until the end of the seventeenth century, and in the survival of a rather numerous and wealthy colony of Portuguese merchants.[80] Between 1590 and 1629 the latter imported almost 75,000 chests.

It is true that a more precise estimate of the sugar trade in Antwerp becomes all the more difficult for the sixteenth century, as the sources of supply increasingly diversified and continuously evolved. In contrast to spices, alum, Spanish wool, English cloth, and Baltic grains, sugar was not subject to any staple right, required warehousing, or monopoly in the matter of refining. Sugar chests were not negotiated as bulk goods on one or more ships, but could be imported as packaged goods in smaller or bigger amounts or even as personal luggage. However, because of its excessive weight and the difficulties of storage and conservation, it was always preferred to transport it directly to its destination for refining. Fortunately, the sugar trade was well suited for the new mercantile practices of double-entry bookkeeping, the commitment of information and regular epistolary correspondence between partners, the sharing of interests, and the mutual insurance. Thanks to their familiarity with the *Dispositionshandel*, the new type of merchants could negotiate their sugar chests without passing through Antwerp, or even through other ports of the Netherlands, but rather dispatch those directly to other ports on the Atlantic Coast, the North Sea, the Baltic Sea, and even the Mediterranean.

Jan Materné has tried to quantify the origin of the sugar traded in Antwerp.[81] Hence, he proposed for the years 1552–53 figures of 51 percent from São Tomé, 20 percent from Madeira, 10 percent from the Antilles, 9 percent from the Canaries, 6 percent from North Africa, and 4 percent of unknown origin. By 1570, São Tomé would furnish 70 percent, Brazil 15 percent, and North Africa 5 percent, with the remaining 10 percent of unknown origin. For 1590–99, he attributed 86 percent to Brazil and barely 2 percent to São Tomé, without being able to identify the remaining 12 percent. Although there appear in these statistics general trends, it is necessary to nuance this distribution and to complete it.

First, while writing a history of sugar supply in northern Europe, one should recall the precursory role of Bruges, where from the Middle Ages on, Italian, Andalusian, Catalan, and German merchants brought sugar from the Mediterranean, especially from Damascus, Egypt, Venice, and Málaga.[82] They provided for a local market in the middle of a comparatively wealthy region, but also reexported their sugar to England and Germany. The merchants of the Hanse and particularly the *Grosse Ravensburger Gesellschaft* may have opened the path to a more active participation of Flemish merchants in the commerce and production of sugar.[83] The access of the Bruges merchants to the exploitation of the new plantations of Madeira and to the production of new Portuguese sugars visibly resulted in interaction among Italians, Portuguese, Germans, and Flemish. In parallel, the political and dynastic ties between the kings of Portugal and the counts of Flanders, and particularly the marriage of Isabella, sister of Henry the Navigator, with Philip the Good facilitated the establishment of a Portuguese colony in Bruges and a Flemish one in Lisbon, both greatly privileged.[84] Once Madeiran sugars arrived at the market in Bruges, nothing was more natural for the merchants native to the city or the neighboring region of the Artois, as well as those from Tournai (such as the Despars; the Nieulant or da Terra; the sons of Maarten Lam or Leme; Jean Esmenault or João Esmeraldo; and João Lombardo), than to engage themselves in the depth of this traffic to the point of acquiring land and constructing *engenhos* in Madeira.[85] Paradoxically, the growing difficulties and decadence of the Bruges market forced its merchants to take greater risks outside their homeport. At the same time, they also hoped to assure themselves a place in the new market of Antwerp. Shortly thereafter the Bruges merchants were also very active in the Canaries.

It was probably in Antwerp that Canary sugar achieved its entry into the supply of northern Europe. Again, as in the case of Madeira, there was interaction among German merchants from Cologne, Augsburg, or Ulm, and Italians, Spaniards, and Flemish through either synergy or competition. In 1509 a representative of Welser purchased and developed the first *ingenios* in Tazacorte, which subsequently passed in 1520 to the hands of Johann Bies of Cologne and Jakob Groenenberg of Antwerp.[86] The latter became the ancestor of a large family, Monteverde, in the Canaries. Other Antwerpers, such as the Van Dale family, followed them.[87] It is important, however, to emphasize the simultaneous, if not prior, presence of Brugeois such as Lieven van Ooghe, Gilis Dhane, Juan Jaques, and, above all, Thomas and Jorge Vandewalle or Bendoval.[88] According to Emmanuel van Meeteren, the first Canary Islands sugar arrived in Antwerp in 1508. There it acquired an *appellation d'origine* and a reputation of high quality, as recipes explicitly prescribed *Canariesuycker* and its price appears to have been higher than those of other sugars. Note that these transactions still remained very important during the difficult decade of the 1580s.[89]

On the trade route from the Canaries, and hence often serving as a port of call, lay the *Cabo de Guer*, a name that covered in effect the entire Atlantic coast of Morocco, frequented by merchants in particular for the purchase of Moroccan sugar. The latter certainly made up a significant portion of the sugar market.[90] Aside from the Antwerpers, the French were equally active there, and in 1561 King Charles IX sought to obtain a monopoly from the Moroccan sovereign, while in 1570 a society of merchants from Rouen attempted to organize plantations.[91]

It was again experience acquired in the Canaries that impelled commerce-minded men such as the Welsers to develop an interest in the sugar plantations on the island of Santo Domingo, between 1530 and 1556, and their product passed steadily through the Antwerp markets and elsewhere in western Europe.[92]

Meanwhile, Portuguese sugars from Madeira remained in Antwerp, but they were substantially reinforced with arrivals from other Portuguese possessions, from São Tomé and Brazil. The relations between Antwerp and the third new important supplier of sugar, São Tomé, are clearly less well known or studied. There is, nevertheless, an occasional indication, such as a certain Antoinette Raes, wife of the merchant Louis le Candele, who claimed that by 1611 she had lived for eighteen years in São Tomé and knew of another Flemish man, Jan de Clercq.[93] It appears that Portuguese merchants, principally cristãos novos (New Christians), controlled São Tomé's sugar production.

In contrast, the Antwerp involvement in the tapping of the Brazilian sugar vein is better known. It revolved around a merchant of great skill, Erasmus Schetz, and his sons Gaspar, Melchior, Baltasar, and Conrad, and their links of family, business, and trust with a varied group of merchants of diverse origins, all active and established, in one way or another, along the Lisbon-Antwerp axis.[94] They found German sponsors through Erasmus's father-in-law, Lucas van Rechtergem (originally from Aachen), and German factors and servants in Lisbon, such as Guillermo del Reno or del Rey and Hans Ingelbertus. This explains why German soldiers, such as Hans Staden or Ulrich Schmidl, who had difficulties in Brazil, were welcomed in 1553–54 at the Schetz property in São Vicente. Through its other interests in metal, spices, gems, and even tapestries, as well as their loans to the king of Portugal, the Schetz family positioned itself at least at the level of the Höchstetter family, and not far behind the Fugger or Welser families. Afterward came Flemish or Antwerp relations and associates with the brother-in-law and nephew João van Hilst or Venist de Hasselt, the son-in-law Jan Vleminck, the Wernaert's, the Pruenen's, and Van Stralen. There were probably also overtures made to the wealthy merchants of French-speaking Flanders, of which Pedro Rouzée d'Arras would be their representative in São Vicente. The Schetz family's relation with the great Italian merchants was consolidated by the marriage of Baltasar Schetz with the widow of Jean-Charles Affaitadi, Lucretia. Later on, they would have in their service in Brazil a Jean-Baptiste Maglio and a Jeronimo Maya. We must not forget the relationships between Erasmus and the converso merchant of Burgos, Francisco de Valle, who became his brother-in-law, and his protection of the cristãos novos Gabriel de

Nigro and Diego Mendes, who had found refuge in Antwerp but were again threatened there in 1532.

Finally, the insertion of these merchants into the highest political and intellectual realms manifested itself in the lodging in 1549 of Charles V and his heir Philip in their splendid estate in Antwerp, *Huis van Aaken*; by the purchase of lordships and the concession of noble titles such as the lords of Grobbendonk, Wezemaal, and Hoboken; by the nomination of Gaspar as financial factor of King Philip II at Antwerp in 1555; through relationships and literary exchanges with humanists such as Erasmus of Rotterdam or the German Heliodorus Eobanus; and Melchior Schetz's patronage of the *Landjuweel* in Antwerp in 1561. Although the Schetz and their familiars certainly associated with influential people and followers of the Reformation, and shared somewhat heterodox ideas, they chose loyalty to the king and did not flee during the religious turmoil. Gaspar Schetz established good relations with the Jesuits, to whom he sold the family's *Huis van Aaken* and who visibly reciprocated through their spiritual assistance in controlling the behavior of their representatives in Brazil. Thus, in 1578 he even received a letter from Father José de Anchieta. The Schetz family also sent the Jesuits supplies, paintings, images, and a small harpsichord. It is not astonishing that, from that point on, they passed as partisans of the king of Spain, a reputation that undoubtedly instigated the Dutch, led by Joris Van Spillbergen (from Antwerp himself, but a rebel), to burn the *Engenho dos Erasmos* at the time of their passage in 1615.

The example of and rivalry with other notable merchants in Bruges and Antwerp, who established a pattern with the acquisition of estates on the Atlantic islands and who flaunted their royal titles, most probably impelled Erasmus Schetz to purchase, from his nephew João Veniste of Lisbon, a significant share in a new *engenho* in São Vicente in 1535. A short time later, the three other shareholders, Martim Afonso, Vicente Gonçalves, and Francisco Lobo, ceded their shares as well. From the 1540s on, Schetz sought to develop his *engenho* through direct management of his factor, probably Pedro Rouzée. Their first results may have incited other German merchants active in the Lisbon-Antwerp-Upper Germany axis to work through Brazilian channels. It was, most notably, Sebald Lins of Ulm and the de Holanda and Hoelscher families who established their *engenhos* at Pernambuco and Bahia.

Although from the beginning of the sixteenth century all these various Atlantic sources of sugar dominated the western European sugar supply, we cannot ignore the importance of the Mediterranean sugar, which maintained its hold over part of the market. For example, in 1589, the della Failles received molasses shipments from Palermo in Zeeland and in Amsterdam.[95] The Andalusian coast, along with Motril, remained in the seventeenth century a rather significant supplier to both Antwerp and Marseilles.[96] The port of Marseilles even continued to receive, on occasion, sugar from the Levant and from Alexandria in Egypt. It is significant that the *pepercoeckbackers*, or the gingerbread artisans, protested in the 1670s against an increase in taxes on syrups imported from Motril, which would affect the lower classes' consumption.[97] In Portugal, sugarcane, planted at Algarve, seemed to move up toward the north at Coimbra and Lisbon, where the Milanese financier João Batista de Rovelasca introduced it in his estate at Alcantara.[98] Moreover, attempts were made to extend sugarcane plantations to other regions of Europe, most notably in the Midi of France with an attempt in Hyères starting in 1551 and ending in failure in 1584.[99] According to botanist Mathieu de l'Obel, who noted another attempt in the Low Countries, the failure of the project was due to the extremely cold weather.[100]

In addition to this diversification in supply, the volume of transactions and its larger radius of resale also distinguished the Antwerp market from that of Bruges. Clearly, Antwerp was already a larger city, approaching 100,000 inhabitants, in which the local consumption of sugar was more considerable because of its relatively high standard of living. The fact that the daily salary for a ship's master was equivalent to 400 grams of sugar leads to the supposition that people of more modest means could occasionally afford to buy some sugar.[101] In the immediate vicinity of Antwerp was an exceptionally dense demographic area that enjoyed a rather high purchasing power or else benefited from a system of donations and occasional redistributions. The inventories of spice sellers in Tournai and Saint-Armand in 1568 attested to the diffusion of sugar as an ordinary commodity in the more average cities of the Low Countries.[102] It is not surprising that the Van der Meulen family had confections sent from Antwerp to Haarlem for a wedding.[103] The *tonlieu* (shipping tax) of Lith on the Scheldt for the period 1622 to 1630 was rather high for taxes paid for the passage of sugar upstream the Meuse.[104] Hans Pohl estimated the capacity of the Antwerp sugar market at an annual average of more than 2,000 chests for the period 1609–21, but a contemporary writer, Manuel Lopes Sueiro, reports a much higher amount of 6,000 during the war years before.[105] Later on, political and economic difficulties only moderately diminished these needs and imports of sugar into the Antwerp market for thirteen months starting in June 1655; in 1656 imports rose to almost 2,026 cases.[106]

After 1500 the Rhineland at Cologne opened up for the Antwerp merchants a very important market for local consumption and resale elsewhere in Germany.[107] The Antwerp merchants, however, also exported directly to more distant destinations in Germany, to Frankfurt, Ulm, Augsburg, and Breslau. In his *Livre d'arithmeticque* (1587), Michel Coignet offered arithmetic exercises concerning the sugars sent to Nuremberg.[108] A second important market was situated in France, particularly in the northeast of the country.[109] Thus in 1572 Jehan de Boisy registered a debt of more than sixty-four Flemish pounds for sugars sent to the late Jehan Barlet in Arbois in Burgundy.[110] In contrast, due to their extensive Atlantic coastline, the numerous ports in the west favored direct trade with producing countries, or, at the least, with Lisbon, to such an extent that French consumption witnessed a strong increase. According to Henri Lapeyre, in 1550 France purchased 250,000 pounds of Portuguese sugar and 50,000 pounds of Spanish sugar, while the only market of Rouen absorbed in 1565 no fewer than 3,000 chests from Madeira and the Canaries as well as São Tomé and Barbary.[111] Nevertheless, a port such as La Rochelle continued to receive sugar from Antwerp, although it is true that it also occasionally shipped sugar directly to the northern port.[112]

If the Antwerp merchants maintained for a period of time an advantage in redistribution over their French and German competitors, it was because they profited from a more highly developed commercial infrastructure. Antwerp had at its disposal in the immediate vicinity a compound of ports frequented by hundreds of boats (Spanish and Portuguese as well as Breton and Dutch) and a bourse where the latest information, opportunities for credit, and insurance could readily be found daily. The city's merchants corresponded with a number of factors, associates, or individual participants throughout Europe and even overseas, and could always include sugar in their other more numerous and heterogeneous commercial operations. Thus merchants of a modest but solid scope such as Maarten and Jan della Faille sent sugar chests both on one sole ship to Narva in the Baltic, as they did in 1565, and in hundreds of ships to Venice from Cádiz, as they did with Santo Domingo sugar in 1585.[113] In this way, they practiced *dispositionshandel*, that is to say, long-distance commercial dealings between two cities distant from the merchants' base, well before the 1585 seizure of the city by the Spanish army, under Alexander Farnese, and the closing of the Escaut.

This political and religious crisis, with the obligatory imposition of Catholicism as the sole faith, provoked the departure of several hundred merchants and artisans. They sought refuge temporarily or permanently in Holland, England, and Germany. Obviously, such an exodus considerably weakened the sugar market in Antwerp. However, paradoxically, it was also sugar that contributed to the reestablishment (modest in comparison to the city's previous stature, but nonetheless rapid) of a not negligible prosperity and Baroque splendor in the city on the Scheldt, which subsequently enjoyed the "long Indian summer." After 1590 the political situation gradually normalized, particular under the reign of the Archdukes Albert and Isabelle. During the Twelve Years' Truce, from 1609 to 1621, the sugar market recovered not only in Antwerp but also in all the other cities of western and southern Europe. One could even contend that the great sugar boom took place from 1590 to 1630. It remained in the hands of the Portuguese, but included numerous Flemish merchants. The latter formed part of the so-called diaspora of political, economic, and religious refugees in Middelburg, Amsterdam, Emden, Hamburg, Frankfurt, and London. Many Flemish families settled in Rouen, Nantes, and Bordeaux; in San Sebastian, Viana do Castelo, Porto, and Lisbon; in Sanlúcar, Cádiz, Seville, and Málaga; in the Canaries and Madeira; on the Brazilian coast at Pernambuco and Bahia; in Safi, Morocco; in Algiers; and in the Italian cities of Naples, Livorno, and Venice. There were fewer political and religious refugees than adventurers, pawns, agents, and emissaries of an immense network that extended to four continents.[114] Some worked hard to represent the great Antwerp merchant houses that were reestablishing themselves, while others labored alone and hastily amassed a small fortune, especially in Brazil, in order to return to Europe, to Amsterdam or Seville, with considerable capital. Take, for example, the case of the young Jasper Basiliers from Antwerp, who in 1600 engaged himself for five years in Bahia in Brazil in the service of a group of nine merchants from Antwerp, Amsterdam, Rotterdam, and Lisbon, each of whom invested 10,000 guilders in the business; he himself contributed 2,000 additional guilders to the project.[115] If a Pedro Clarisse in Lisbon handled only 100 chests a year, others such as the De Groots could extend their purchases to 500 chests or even more. Whether they resided in Antwerp or in the Iberian Peninsula, merchants could export just as easily to northern Europe or to Italy.

A large portion of this commerce took place in total legality following the normalization of relations of the Spanish crown with France and England after 1598 and 1604. However, in principle, the Dutch enemy remained excluded, and foreign merchants who shipped sugar to the north were required to produce *testimonios* or certificates stating that the sugar had been disembarked in friendly areas—La Rochelle, Rouen, Calais, London, or Hamburg.[116] Several examples illustrate the importance of this trade: Jorge Benson, a London merchant, presented on 24 December 1605 a *testimonio* for a total of fifty-four chests of sugar that arrived on three boats between May and December of that year; in La Rochelle, Michel Reau and Joseph attested on 20 June 1606 to the arrival of thirty-seven chests of partially refined sugar from Pernambuco and Bahia, received from Jacques Godin in Lisbon on the Marie de St. Gilles. On his return voyage from the East Indies in 1610 the French traveler François Pyrard de Laval wondered how in the imbroglio of that sugar trade Flemish and Dutch merchants and shipmasters, having their residence as well in Lisbon as in Bahia, enjoying Portuguese nationality, owning a well-armed Dunkirk hulk, managed to associate themselves with New Christians in a loading valued at 500.000 *escudos* and shipped to Bayona de Galicia, a port on the frontier between Portugal and Spain well known for its smuggling facilities.

All these prohibitions, requirements, and precautions did not prevent Amsterdam from becoming one of the most important sugar markets of northern Europe in 1600. This port was able to consolidate its position because of a highly diverse supply of sugar, from the Canaries, Madeira, São Tomé, Santo Domingo, and Brazil, combining purchase with theft and confiscation. We must not exclude collaboration with Barbary corsairs, among whom were often found Dutch renegades. In 1626 the denizens of Lisbon complained of the loss of 60,000 chests of sugar as a result of the seizure of not fewer than 120 ships over three years.[117] After 1616, Asian sugar from China, Formosa, and Siam, negotiated by the V.O.C. (Verenigde Oost-Indische Compagnie or Dutch East India Company), was added. The latter did not hesitate to create in 1637 its own production system on its plantations, first in Bantam and later in the surrounding area. It is worth noting that the English were also interested in the possibilities of Chinese sugar; Asian sugar could compensate for the insufficiencies and failures of the Brazilian *engenhos*, which after the 1630 conquest of Pernambuco by the W.I.C. (West-Indische Compagnie) no longer delivered all the expected cases. After the restoration of the independence of Portugal in 1640, Brazilian sugar arrived again by way of Lisbon in Amsterdam, which shortly thereafter would vie with sugar from the Dutch islands conquered from the Spanish in the Caribbean.

Antwerp, despite the rise of Amsterdam and the Dutch conquest of Brazil, still maintained a significant commerce in sugar that arrived either by way of the Flemish coast and the ports of Ostende and Dunkirk or from an intermediate stop in Middelburg or Amsterdam, having paid if necessary the *licenten* or rights of trade with the enemy, at least until the conclusion in 1648 of the Treaty of Munster between Spain and Holland. Thus, from 1648 to 1660, the Moretus family of the powerful printing house Plantin did not hesitate to receive cases of sugar through Amsterdam in the return for their book shipments.[118]

The Art and Technique of Refining

The continuity and longevity of the Antwerp market cannot be explained solely by its central position for commerce and the resistance of the Spanish power and supremacy in this part of northern Europe, but rather by its early mastery of the secrets of sugar refining and its technical perfection. Although the art and techniques of refining could not be safeguarded or monopolized indefinitely, they clearly ensured a certain lead as long as the consumption of refined sugar did not soar.

Although sugar did not lend itself to commercial monopoly and permitted the participation of a large number of merchants and shopkeepers, this social mobility of sugar was not realized as easily at the level of refineries and confectioneries.[119] Thus, in Lisbon between 1533 and 1545, the merchants solicited King João III to forbid or restrain the use of São Tomé sugar by the confectioners or refiners of the city.[120] They advanced as pretext the risks of exhausting the supply of firewood and of the adulteration of sugar destined for export. Moreover, it is necessary distinguish between the numerous *suyckerbackers*, the keepers of simple booths, or small confectionary workshops, and the refineries of a proto-industrial character. The refineries themselves demanded during their initial phase considerable investments, equal to if not surpassing those of breweries: a relatively spacious workshop not too distant from the docks (in order to limit the very high costs of transportation), several ovens, large copper basins, and a collection of pottery sufficient for hundreds of sugarloafs. It is true that copper was rather readily available in Antwerp due to the brassworks of the Meuse and to imports from central Europe and Sweden, which, in turn, were exported to Africa in large quantities through the mediation of the Portuguese. It was especially important to guarantee a sufficient, regular, and inexpensive supply of sugar. As for professional and technological know-how, the merchants and artisans of northern Italy, of Cremona and Venice, were rather numerous and in all probability introduced the secrets of refining, just as they had done for glass making, double-entry bookkeeping, and other financial innovations. Sugar was boiled up to three times, skimmed carefully, and purified and clarified with lime, egg white, or even ox blood. Many aspects of sugar refining were similar goldsmithery, an art trade that employed many craftsmen in Antwerp and other Flemish cities during this period. So it is not surprising that a Spanish craftsman at Malines, Bernardin Maroufle, accepted in 1535 an apprentice, Nicolas Saillot, to educate him for twenty guilders in the "art of preparing conserves, candied fruits and other preserves," in the "art of drawing gold and silver wires and fashioning buttons, chains, trimmings, and other baubles," and in the art of perfumery. It was most important to gain mastery of the boiling point and the precautions to protect the sugar from moisture. Including the rather long drying period, the entire refining process could last as long as nine months and could easily employ several workers.

Antwerp welcomed foreign refiners to register as *bourgeois* and encouraged them to join the guild of those associated with the food trades. Thus the records of the bourgeoisie registered twelve new refiners for the period 1525–39, and seventeen more for the periods 1560–64 and 1565–69. These numbers indicate the scale of the activity and the success of this profession. While several authors such as Fernand Donnet and Hans Pohl estimated the number of refiners at nineteen in 1556, the more precise research of Alfons K. Thijs was able to establish a list of twenty-five names for that same year and proposed another twenty-eight names for 1575. A research on the social and labor structure of the city, by Jan Van Roey, collected eighty-eight *suyckerbackers*, but probably included many *pâtissiers* and confectioners, who did not refine their raw material.[121] Furthermore, many of these refineries were no more than simple *suyckersieders* or boiling houses. Others worked only small quantities of sugar and only the most prosperous employed four to five workers and several servants. Nevertheless, according to the calculations of Thijs, they succeeded in refining more than a third of the 15,200 chests imported to Antwerp in 1560.

The refinery that was constructed on the *Korte Raapstraat* around 1545–48 by the merchant from Lucca, Jan Balbani, and his associates Vincent and Baltasar Guinisy, who employed several dozen workers, was so large that other refiners feared that it would lead to monopolization in Antwerp.[122] In 1550 their *suikerhuis* would be sold to Jean-Charles de Affaitadi. Although the other *suyckerbakkers* or *banketbakkers* consisted of many modest artisans, some middle-class workers made a fortune; in 1658 the brothers Karel and Willem van den Eynde left a legacy of at least 30,000 guilders to the almoners of the city.[123]

This dominant position of the Antwerp refineries would obviously suffer from the closing of the Scheldt in 1585, as the Antwerp sugar market itself did, and would be damaged by the departure of several refiners. However, these dramatic events must not be considered as the sole determining factors in the diffusion of refining throughout northern Europe. Although the secrets and techniques of refining took their time to become familiar in western Europe, sugar gradually became a commodity almost as essential as salt, cheese, and grains, and every city of a certain size had to provide for its supply through its own merchants and sugar refiners, very similar to how breweries operated. Even without the political-religious crises of 1567–85, the Antwerp refining industry would have had to let go of its ascendancy and see its exclusive or predominant position break up and expand toward the great cities of consumption in northern Europe. Wage costs, the more capitalist organization of work, and the technological specialization of the refinery could hardly slow down or limit this proliferation of sugar refining. Only the volume and weight of these thousands of chests of sugar and the difficulties of conservation and storing called for rapid processing and distribution. Quality was a relative term for consumers who were still uninformed and relatively easy to please, and less fine sugars, such as brown sugars and molasses, were disposed of more easily or often underwent a second transformation.

On the other hand, after 1585 Antwerp recovered a part of its refining capacity corresponding to the reestablishment of both its sugar market and local and regional consumption. The city even reexported part of this refined sugar, due to its persisting reputation as a specialized refiner. According to the assertions of the *cahier d'apprentissage* (apprenticeship notebook) of the Van Colen-De Groots in 1643, Antwerp's refineries produced a refined sugar of a higher quality than their competitors in Amsterdam and Hamburg, who used saltier water.[124] This qualitative specialization corresponded somewhat to the renewed manufacturing activity through the seventeenth century, particularly in luxury items such as silk, furnishings, musical instruments, painting, and printing, which produced a great purchasing power and stimulated the consumption of sweets. In 1676, the city still had approximately sixteen refineries, thanks to its exclusive rights to the market in the southern Low Countries and a considerable mass consumption. Recall the aforementioned protest of the *pepercoeckbackers*, who made gingerbread as a protest against a new tax on syrups from Motril and Málaga, under the pretext of a general inflation of prices that affected modest people's consumption habits.[125]

The dispersion of sugar refining began well before 1585. Aside from Antwerp, one would expect that refining would develop earlier in the large French ports, given their direct, early relations with Brazil. In 1546, according to a letter from Luis de Góis to João III, seven to eight French ships annually frequented the Brazilian coast between Cabo Frio and Rio de Janeiro. In fact, in the middle of the seventeenth century Jean de Léry was astonished that "we French had not yet, when I was there, the appropriate people or the necessary instruments to export sugar (as the Portuguese had in the areas they possessed). We only infused water with sugar to sweeten it, or else those who desired could suck and eat the [cane] pulp."[126] They knew at most how to extract a sort of liqueur from old, moldy canes. It is easy to comprehend why in 1556 a Venetian settled in Antwerp proposed to go to France to make sugar.[127] Already in 1548, in Rouen, a Spanish refiner volunteered to teach this art to an apothecary, and later on similar efforts were made (such as that of the Hollé-Seigneur company in 1570) before successful refineries started operating in 1611.[128] In La Rochelle the refinery began at the end of the sixteenth century, after the arrival of Flemish refiners such as Joseph Baertz in 1598 and Gillis Tsermarttyns de Malines in 1599; however, another Flemish refiner, Brisson, went bankrupt there in 1605.[129]

Two attempts to establish a refinery in Marseilles failed.[130] The first, in 1547, was a proposal by the Italian Jean-Baptiste des Aspectat, who claimed to have experience in the industry in Antwerp, and who could probably be identified with the aforementioned Jean-Baptiste de Affaitadi. In 1574 a second refinery was set up by François de Corbie and Pierre Hostagier, businessmen from Marseilles, but their company lasted only for a few months and made a modest number of sugarloaves.

Among the plausible explanations for the absence of refineries, there is no need to invoke the infamous "French economic backwardness," particularly in colonial matters. In fact, one could argue that the French already had sufficient maritime enterprises with their cod and other fishing industries, and especially with trade in Brazilian wood. The ships they outfitted in La Rochelle in 1561 came back loaded with Brazilian wood rather than sugar.[131] Moreover, as a result of the increasingly fierce Portuguese defense of their monopoly from the 1540s on, the French did not succeed in securing a regular supply through the possession of plantations and mills. On the other hand, for an internal market as vast as that of France, with a smaller population density and a lower degree of urbanization, and for a market that could develop the very broad variety of its immense natural resources, refined sugar remained too strange and too expensive in comparison with fruit syrups and honey, which were produced there in greater abundance, especially in the Midi of France. Finally, commerce presupposed exchange, and if French commercial interests wanted to sell their wines, salt, and textiles, sugar became a more suitable import product.

England also attempted to free itself from the mediation of the Antwerp market, starting in 1544, with an attempt by Sir William Chester, followed by others, all still without the ability to supplant the role of Antwerp as distributor.[132] Although from 1650 to 1670 the large English ports had at their disposal large-scale refiners such as the Sugar House Close in Liverpool and the East Sugar House in Glasgow, the English continued to direct a large part of their sugar from the Caribbean islands to Dutch ports.

In Germany in 1547, a refiner in Nuremberg found himself in trouble because of the debt from his expensive equipment.[133] The Augsburg refinery, founded in 1573 on the initiative of Konrad Rot, son-in-law of Bartholomäus Welser, who passed as much time in Antwerp as in Venice and Lisbon, was a little more successful. Familiar with the monopolistic contracts for sugar and copper awarded by the Portuguese crown, he attempted to obtain similar exclusive rights from municipal authorities.[134] However, a dozen years later, a second refinery appeared in Leipzig under the control of Hieronimus Rauscher. It appeared that the great Hanseatic ports offered a better location for the development of a powerful refinery because, after the 1590s, ships from Hamburg had engaged in the Brazil trade routes and brought back Brazilian sugar directly, while at the same time numerous New Christian and Flemish merchants established themselves in Hamburg.

If Antwerp finally ceded its monopoly in the matter of sugar refining, it was to the profit of the United Provinces, especially Amsterdam. The initiative for the installation of the first refinery in Holland is generally attributed to the brothers Pieter and Jasper Morimont, merchants from Antwerp who, after a trip to England, settled in Leyden in 1577, but it is doubtful that their refinery functioned effectively.[135] In any case, in 1585 the Van der Meulens, emigrants to Holland, had to request a chest of *banket* (good-quality pastries) from a *pâtissier* in Antwerp on the occasion of a marriage in Haarlem.[136] Shortly thereafter, the marked increase in sugar prices may have encouraged the establishment of new refineries, which permitted more substantial profits through the rapid transformation of material initially contested.

At the same time, the pirate activity of Zeeland, Dutch, and English captains against Portuguese and Spanish boats increased dangerously and inundated the sugar market of Amsterdam, although it was true that not all the cargo of stolen sugar was immediately destined for Dutch ports.[137] As the real ownership often belonged to *cristãos novos* settled in Amsterdam but associated with Flemish merchants in Antwerp, or settled in Brazilian or Portuguese ports, it was of importance for the pirates to avoid an easy, rapid judicial confiscation of their booty. Thus they obscured their tracks by putting large quantities up for sale in the ports of the Barbary Coast or England. There they may have had covert arrangements with insurers, whereas the Spanish authorities considered the ready surrender of Portuguese crews as treason. In February 1607, an informant notified the Spanish authorities of 800 chests sold in Barbary in May 1606 and of 4,000 chests available for purchase at Plymouth in September 1606.[138] These predators obviously had an interest in delivering their loot as rapidly as possible to refiners, and Amsterdam offered the greatest possibilities of camouflage, despite the respect for the law of certain groups among the merchants. Although during the Twelve Years' Truce sugar commerce temporarily surged back to normal levels, after the foundation of the West Indies Company in 1621 these extra-legal import deals resumed in full force. According to Johan De Laet, the company managed to confiscate, through the seizure of 547 Portuguese and Spanish boats during the first thirteen years of activity, no fewer than 40,000 chests of sugar, of which a good part was finally unloaded in the ports of Holland and Zeeland.

This was all that was necessary for sugar refining to impose itself as one of the most dynamic sectors of industrial growth and capitalist accumulation. The first refinery in Amsterdam was reported in 1597.[139] From three refineries in 1605, the number increased to twenty-five in 1622, forty in 1650, and fifty or sixty in 1661. Each refinery could process nearly 1,500 chests per year, and could have stocks in reserve that were worth two tons of gold. At the time of the fire at the Nuyts refinery in 1660, the sugar burned was worth three tons of gold.

The refineries were installed in old convents and brand new buildings of five to six floors. They housed thousands of pots necessary for drying sugarloaves. The refineries stimulated the development of pottery in surrounding areas. On the other hand, industrial activity in the center of the city—with continuous smoke coming out of the numerous chimneys (the first of the truly modern type, according to Jan De Vries and Ad Van der Woude)—caused serious pollution problems, after which coal heating was prohibited in the summer.[140] Despite the fact that the number of workers involved was relatively modest—scarcely 1,500, or 1.5 percent of the manual workers employed by the principal Dutch industries for the period 1672–1700—municipal authorities showed themselves to be rather understanding of the refiners, indicating also their economic weight and political influence. It has been claimed that a fifth of the *waaggelden* (municipal taxes) came from sugar and that its commerce at any time maintained at least one hundred ships en route.

Among the entrepreneurs were several immigrants from Antwerp, such as Abraham and Hans Pelt, Adam and Hans Nijs, and Cornelis Nuyts. One refinery even called itself *De stad Antwerpen*, another quite simply *Suyckerbackery*, where the painter Rembrandt came to live in 1639. Of the thirty-one refineries set up before 1670, no fewer than twenty-one would maintain their name until the nineteenth century. It appears that it was only in 1655 that the first Portuguese New Christians, Abraham and Isaac de Pereira, received the authorization to set up a refinery, on the condition that they only sold wholesale.

Obviously it was important to safeguard the local trade and internal market of the United Provinces, whose population witnessed a substantial growth from 940,000 in 1500 to 1,900,000 in 1650. The ability to purchase and consume sugar, like Chinese porcelain or other luxury objects, was no longer limited to the bourgeoisie because seamen, artisans, and even workers in the United Provinces earned some of the best salaries in Europe. A majority of the production was exported, not only to the ports of the Baltic and Germany but also to France and Italy.[141] In 1645 the United Provinces sold to France sugar that was worth 1,885,150 livres, or 8.75 percent of the total exports to that country. The steady penetration of Dutch ships into the Mediterranean derived in large part from the success of Dutch sugar. In Tuscany, Dutch sugarloaves were preferred for their whiter, more brilliant consistency, harder, sweeter, and silkier than Venetian sugar.[142]

According to the somewhat debatable hypothesis that Amsterdam had forcefully taken from Antwerp the dominant position in the refined sugar trade, its new preeminence also crumbled more rapidly. In 1700, the number of refineries diminished to thirty-five. Already within the United Provinces itself, refineries had arisen earlier in the other important cities. Thus was the case in 1627 in Middelburg with the creation of a refinery by Daniel Dierckens, who brought his experience as a refiner in Rouen.[143] Other refiners, nearly a dozen, established themselves in Rotterdam, Delft, and Gouda. Moreover, this internalization of refining activity occurred in the southern Low Countries and began to threaten the regional monopoly of Antwerp, after the opening of a refinery in Brussels in 1650 by two brothers from that very same city on the Scheldt.[144] Later in the eighteenth century, there would be more in Ypres, Ghent, Mons, and Liège.

In France, the refineries rapidly regained ground after the 1664 proclamation, by Colbert, of a new tariff that intended to attract the sugar of the French Caribbean islands to French ports and to exclude Dutch refined sugar. Duties rose from fifteen to 22.05 *livres*. Between twenty and thirty refineries were created in Rouen, Nantes, Orléans, Bordeaux, and Bayonne.[145] Very rapidly, the city of Nantes gained a reputation for its refined Antillean sugar, and its production increased from 5,400 tons in 1674 to 9,300 tons in 1683. None of this prevented Dutch sugar from entering as contraband. In turn, Sweden and Denmark similarly enacted protectionist measures against Dutch imports.

In Italy, this Dutch explosion rapidly reached its limits. According to the Van Colen–De Groot's *cahier d'apprentissage*, refined sugar exports were perforce limited to Messina, Naples, and Genoa, as Venice and Livorno had at their disposal their own refineries.[146] The latter city saw a rapid expansion from 1620, with the influx of Flemish and Portuguese Jewish merchants, and thus became a Mediterranean center for all kinds of legal and illegal business revolving around Brazilian sugar, often purchased from Algerian corsairs. Moreover, its first refinery was the work of a Dutchman, Bernard Jansz Van Ens, a merchant and refiner from Hoorn.[147] In 1624, together with his neighbor Theodoor Reiniers, he obtained from the grand duke of Tuscany a monopoly for ten years. Upon his death in 1626, the business passed into the hands of two other Flemish businessmen, Daniel Bevers and Paris Gautier. Nicolas Du Gardin, a large-scale merchant form Amsterdam deeply involved with the West Indies Company, functioned as the source of financing. When, at the beginning of 1630, business began decline, it appeared that the plague was not the sole cause; confectioners refined their own sugar and competed with refiners. As for Venice, it defended its regional monopoly against the creation of refineries in its hinterland of Veneto.[148]

Clearly, sugar trade and refining developed much earlier and had an economic and cultural importance that was much greater than is generally admitted in the majority of works that synthesize the European economy during the period 1500–1650. Too often sugar is ranked behind the other colonial commodities, especially spices. However, among the "rich trades," it was the only one to have continued and expanded so remarkably, successively enriching the various European economies, from Italy and the Iberian Peninsula to Flanders, from the United Provinces to France and England. On these grounds, the sugar economy would certainly merit more detailed, targeted research, as much on its place in creating commercial fortunes as in industrial investments. Authors such as Jan De Vries and Ad Van der Woude have somewhat underestimated the importance of sugar as the engine of economic growth, sugar refining in relation to the other *trafieken*, or processing industries, of raw materials such as tobacco. Although the processing of tobacco may have employed more workers, sugar refineries not only generated employment in other industries (pottery, ceramics, and silversmiths) but also transformed the work of *pâtissiers*, confectioners, and women. This applies not only to the so-called "first modern economy" of the United Provinces but also to the purportedly more traditional economies of Italy, the Iberian Peninsula, the Spanish Netherlands, and France.

In addition, sugar encourages the questioning of the negative relationship between high prices and mass consumption and the verification of a hypothesis that would, rather, posit an increase in consumption preceding the depression of prices since the middle of the seventeenth century. There are numerous indications that the prestige of sugar in European society from 1500 to 1650, characterized by a marked urbanization and modernization, was such that it took on symbolic value of social integration and promotion, and that its high price incited consumption, rather than abstention, among the middle or lower urban classes. Like drugs today in poor neighborhoods, the high cost of sugar was not prohibitive, and many by-products such as syrup and molasses were cheaper. One should not forget that sugar once figured among the drugs of the "Orient."

NOTES

1. Giuseppe Bertini, *Le nozze di Alessandro Farnese, feste alle corti di Lisbona e Bruxelles* (Milan: Skira, 1997); A. Castan, *Les noces d'Alexandre Farnèse et de Marie de Portugal, narration faite au cardinal de Granvelle par un cousin germain Pierre Bordey* (Brussels: Hayez, 1888). With warm thanks to my colleagues Giuseppe Bertini, who brought this text to my

attention, and to Bart De Prins, who helped me with the revision of this essay. On sugar banquets, see also Roy Strong, *Feast: A History of Grand Eating* (London: Jonathan Cape, 2002), 194–95, 198–201; Krista De Jonge, "Rencontres portugaises, L'art de la fête au Portugal et aux Pays-Bas méridionaux au XVIe et au début du XVIIe siècle," in *Portugal et Flandre, Visions de l'Europe*, exhibition catalog (Brussels: Musées royaux des Beaux-Arts de Belgique, Europalia Portugal, 1991), 84–101; and Ghislaine De Boom, *Archiduchesse Eléonore (1498–1558), Reine de France—Sœur de Charles Quint* (Brussels: Le Cri, 2003), 147. See also Juan Christóval Calvete de Estrella, *El felicíssimo viaje del muy alto y muy poderoso Príncipe don Phelippe*, ed. Paloma Cuenca (Madrid: Sociedad Estatal para la Conmemoración de los Centenarios de Felipe II y Carlos V, 2001), 352–53; and Peter Brears, *All the King's Cooks: The Tudor Kitchens of King Henry VIII at Hampton Court Palace* (London: Souvenir Press, 1999), 64–85.

2. Francesco De' Marchi, *Narratione particolare delle gran Feste, e Trionfi fatti in Portogallo et in Fiandra* (Bologna, 1566). On the soaring pieces of sugar, see F. Yates, "L'entrata di Carlo IX e della sua regina in Parigi nel 1571," in *Astrea, L'idea di Impero nel Cinquecento* (Turin: Einaudi, 1975), 173; and K. J. Watson, "Sugar Sculpture for Grand Ducal Weddings from the Giambologna Workshop," *The Connoisseur* 199, no. 799 (1978). After the second half of the fifteenth century, the *colazioni di zucchero* were already highly appreciated at the Court of the Estes in Ferrara; see T. Truohy, *Herculean Ferrara, Ercole d'Este, 1471–1501, and the Invention of a Ducal Capital* (Cambridge: Cambridge University Press, 1996), 172–76.

3. Sidney W. Mintz, *Sweetness and Power* (New York: Viking, 1985); Jean Meyer, *Les Européens et les autres* (Paris: Armand Colin, 1975), 217, proposes the years 1730–40 as the period in which sugar became part of the daily diet, and ceased to be a medicine that was given to desperate patients at the hospital of Saint-Yves in Rennes, as well as in Parisian hospitals. See also H. J. Teuteberg, "Der Beitrag des Rübenzuckers zur 'Ernährungsrevolution' des 19. Jahrhunderts," in *Unsere tägliche Kost, Geschiche und Regionale Prägung*, ed. H. J. Teuteberg and G. Wiegelmann (Munster: Coppenrath, 1986); and Martin Bruegel, "A Bourgeois Good? Sugar, Norms of Consumption and the Labouring Classes in Nineteenth-Century France," in *Food, Drink and Identity, Cooking, Eating and Drinking in Europe Since the Middle Ages*, ed. Peter Scholliers (Oxford and New York: Berg, 2001).

4. Olivier de Serres, *Le théâtre d'agriculture et mesnage des champs* (Arles: Actes Sud, 1996). One of the pioneering books on sugar in Portugal is Emanuel Ribeiro, *O doce nunca amargou..., doçaria portuguesa, história, decoraçao, receituário* (Sintra: Colares, 1997), with the first edition appearing in 1928. On the papers cut from lace, the *rosas*, and *periquitos*, see Eurico Gama, "A Arte do Papel Recortado," *Revista de Etnografia* 7, no. 2 (1966); and *O doce nunca amargou*, exhibition catalog (Lisbon: Museu de Arte Popular, 1977). On women and sweets in Spain, see María Angeles, "Los recetarios de mujeres y para mujeres, sobre la conservación y transmisión de los saberes domésticos en la época moderna," *Cuadernos de Historia Moderna* 19 (1997).

5. Lucie Bolens, *La cuisine andalouse, un art de vivre, Xie–XIIIe siècle* (Paris: Albin Michel, 1999); Manuel Urban Pérez Ortega, *Viaje por la mesa del alto Guadalquivir* (Jaén: Diputación Provincial de Jaén, 1993). On the use of honey and sugar for the preservation of food, see Lucie Bolens, "Sciences Humaines et Histoire d'alimentation: Conservation des

aliments et associations des saveurs culturelles (de l'Andalousie à la Suisse Romande)," in *Alimentazione e nutrizione, secc. XII–XVIII* (Prato: Instituto Datini, 1997).

6. Maria de Lourdes Modesto, Afonso Praça, and Nuno Calvet, *Festas e comeres do povo português*, 2 vols. (Lisbon: Verbo, 1999); Alfredo Saramago and António Monteiro, *Cozinha transmontana, Enquadramento histórico e receitas* (Lisbon: Assírio & Alvim, 1999).

7. Jean-Louis Flandrin and Philip and Mary Hyman, eds., *Le cuisinier françois* (Paris: Bibliothèque Bleue, Montalba, 1983). Note, however, that *confites largos de cidra* and the *calbazate de Valencia* are found in displays during commencement ceremonies at the University of Salamanca, according to a 1619 document. Luis Enrique Rodríguez-San Pedro Bezares, *La Universidad Salmantina del Barroco, período 1598–1625*, 3 vols. (Salamanca, 1986), 910–12; and María de los Angeles Pérez Samper, *La alimentacíon en la España del Siglo de Oro, Domingo Hernández de Maceras "Libro del Arte de Cocina"* (Huesca: La Val de Onsera, 1998), 123.

8. Alfredo Saramago, *Doçaria conventual do Alentejo, as receitas e seu enquadramento histórico* (Sintra: Colares, 1997); Alfredo Saramago, *Doçaria conventual do norte, história e alquimia da farinha* (Sintra: Colares, 1997); Alfredo Saramago, *A tradição conventual na doçaria de Lisboa* (Sintra: Colares, 1998); Alfredo Saramago and Maneul Fialho, *Doçaria dos conventos de Portugal* (Lisbon: Assírio & Alvim, 1997); Maria Isabel de Vasconscelos Cabral, *O livro das receitas da última freira de Odivelas* (Lisbon: Verbo, 2000). On Spanish convents and confectioneries, see Luis San Valentín, *La cocina de las monjas* (Madrid: Alianza, 1989); and María Luisa Fraga Ibarne, *Gula de dulces de los conventos sevillanos de clausura* (Córdoba, 1988).

9. Frei Isidoro Barreira, *Tractado das significações das plantas, flores e fructas* (Lisbon, 1622); Sóror Maria do Céu, *Significações das frutas moralizadas em estylo singello* (Lisbon, 1735).

10. Carlos Zolla, *Elogio del dulce, ensayo sobre la dulcería mexicana* (Mexico City: Fondo de Cultura Económica, 1988). Un *chrisme* is a Christ symbol, formed by the Greek letters X (*chi*) and P (*rho*).

11. On Lisbon during this period, see in particular the appraisals of two Flemish visitors, the pilgrim Jan Taccoen and the humanist Nicolas Clenardus. Eddy Stols, "A repercussão das viagens e das conquistas portuguesas nas Indias orientais na vida cultural da Flandres no século XVI," in *Vasco da Gama, Homens, Viagens e Culturas*, Actas do Congresso Internacional, Lisboa, 4–7 November 1998, ed. Joaquim Romero Magalhães, 2 vols. (Lisbon: Comissão Nacional para as Comemorações dos Descobrimentos Portugueses, 2001), 2:11–38. It is astonishing that Joaquim Romero Magalhaes, who specializes in the economic history of the sixteenth and seventeenth centuries, devotes little attention to the refining and working of sugar in his chapters in the *História de Portugal*, ed. José Mattoso, 3 vols. (Lisbon, 1993), vol. 3.

12. João Brandão, *Grandeza e abastança de Lisboa em 1552*, ed. José de Felicidade Alves (Lisbon: Livros Horizonte, 1990), 32–33, 59–60, 72, 210–14. In his *Sumário* of Lisbon in 1551, Cristóvão Rodrígues de Oliveira offers slightly less significant figures, with only eight sugarmasters, twenty-three *alfeloeiros*, thirteen candymakers, forty-three pastrymakers, but also sixty-six women who made sugared fruits, twenty-three women who made *alféloas*, and thirty

jam makers. Cristóvão Rodrígues de Oliveira, *Lisboa em 1551, summário em que brevemente se contém algumas coisas assim eclesiásticas como seculares que há na cidade de Lisboa* (1551), ed. José de Felicidade Alves (Lisbon: Livros Horizonte, 1987), 97–99. On the economic wealth of several confectioners in Lisbon in 1565, see António Borges Coelho, *Quadros para uma viagem a Portugal no século XVI* (Lisbon: Editorial Caminho, 1986), 63, 126, 358.

13. Damião de Góis, *Descrição da cidade de Lisboa*, ed. Raúl Machado (1554), 59.

14. D. De Sande, *Diálogo sobre a missão dos embaixadores japoneses à curia romana*, ed. A. da Costa Ramalho (Macao: Fundação Oriente, 1997).

15. Artur Teodoro de Matos, " 'Quem vai ao mar em terra se avia,' preparativos e recomendaçoes aos passageiros da carreira da India no século XVII," in *A Carreira da India e as rotas dos estreitos*, ed. Artur Teodoro de Matos and Luís Filipe F. Reis Thomaz (Angra do Heroismo: Actas do VIII Seminário Internacional de História Indo-Portuguesa, 1998), 385–86. It is probable that the Portuguese Indies were also supplied by way of China, where sugar production had witnessed a dramatic expansion. See Sucheta Mazumdar, *Sugar and Society in China: Peasants, Technology, and the World Market* (Cambridge and London: Harvard University Press, 1998).

16. Giacinto Manupella, ed., *Libro de cozinha da infanta D. Maria* (Lisbon: Imprensa Nacional, Casa da Moeda, 1987); E. Newman, *A Critical Edition of an Early Portuguese Cook Book* (Chapel Hill: University of North Carolina Press, 1964).

17. Gerónimo Gascón de Torquemada, *Gaçeta y nuevas de la corte de España desde el año 1600 en adelante*, ed. Alfonso de Ceballos-Escalera y Gila (Madrid: Real Academia Matritense de Heráldica y Genealogía, 1991), 70.

18. María de los Angeles Pérez Samper, *La alimentación en la España*, 26–29.

19. José Martínez Millán, ed., *La corte de Carlos V*, 4 vols. (Madrid: Sociedad Estatal para la Conmemoración de los Centenarios de Felipe II y Carlos V, 2000), 2:132, 4:233.

20. José Luis Gonzalo Sánchez-Molero, *El aprendizaje cortesano de Felipe II (1527–1546), La formación de un príncipe del Renacimiento* (Madrid: Sociedad Estatal para la Conmemoración de los Centenarios de Felipe II y Carlos V, 1999), 117.

21. María del Carmen Simón Palmer, *Libros antiguos de cultura alimentaria (Siglo XV–1900)* (Córdoba: Imprenta Provincial, 1994); Manuel Martínez Llopis, *La dulcería española, Recetarios historico y popular* (Madrid: Alianza, 1999).

22. Luis Lobera de Avila, *El Banquete de nobles caballeros* (San Sebastián: R & B Ediciones, 1996).

23. Manuela Rêgo, ed., *Livros portugueses de cozinha* (Lisbon: Biblioteca Nacional, 1998).

24. Guy Cabourdin, *La vie quotidienne en Lorraine aux XVIIe et XVIIIe siecle* (Paris: Hachette, 1985).

25. Gérard Oberlé, *Les Fastes de Bacchus et de comus ou histoire du boire et du manger en Europe, de l'Antiquité à nos jours, à travers les livres* (Paris: Belfond, 1989).

26. Lancelot de Casteau, *Ouverture de cuisine* (Liège, 1604); reedition by Léo Moulin et al. (Antwerp and Brussels: De Schutter, 1983).

27. Jean-Louis Flandrin and Philip and Mary Hyman, eds., *Le cuisinier françois*; Françoise Sabban and Silvano Serventi, *La gastronomie au grand siècle, 100 recettes de France et d'Italie* (Paris: Stock, 1998).

28. Alberto Capatti and Massimo Montanari, *La cucina italiana: Storia di una cultura* (Rome and Bari: Giulio Laterza, 1999); Lord Westbury, *Handlist of Italian Cooking Books* (Firenza, 1963); O. Bagnasco, ed., *Catalogo del fondo italiano e latino delle opere di gastronomia, sec. XIV–XIX* (Sorengo, 1994), 3 vols.; Giampiero Negri et al., *Coquatur ponendo, cultura della cucine e della tavola in Europa tra medioevo ed età moderna* (Prato: Francesco Datini, 1996).

29. Cristoforo da Messisburgo, *Banchetti, composizioni di vivvande e apparecchio generale,* ed. F. Bandini (Vicenza, 1992), 103–4, in José Martínez Millán, *La corte de Carlos V,* 4:11.

30. Jean Boutier, Alain Dwerpe, and Daniel Nordman, *Un tour de France royal, le voyage de Charles IX (1564–1566)* (Paris: Aubier, 1984), 159–60.

31. Marie-Laure and Jacques Verroust, *Friandises d'hier et d'aujourd'hui* (Toulouse: Berger-Levrault, 1979); A. Perrier-Robert, *Bonbons et friandises* (Paris: Hatier, 1995); Catherine Amor, *Les bonbons* (Paris: Éditions du Chêne, 1998). On similar recipes for mixture of medical plants and sugar and actual confections in France and Belgium, see *Petite anthologie de la réglisse* (Barbentane: Éditions Équinoxe, 2002); and Bernard Dubrulle, *Petit futé, biscuits, confiseries de nos régions* (Brussels: Neocity, 2002).

32. In his otherwise innovative exploration of Flemish convents, Paul Vandenbroeck, ed., *Le jardin clos de l'âme,* exhibition catalog (Brussels, 1994), has ignored this sweeter aspect of monastic life in favor of suffering and the macabre.

33. *Het leven van de seer Edele Doorluchtighste H. Begga, Hertoginne van Brabant, stichteresse der Begijnen met een cort gegrip van de levens der salige, godtvruchtighe en Lofweerdighe Begijntjens der vermaerde en hoogh-gepresen Begijnhoven, bij een vergaedert door eenen onbekenden dienaer Godts* (Antwerp, 1711), 115.

34. W. A. Olyslager, *750 Jaar begijnen te Antwerpen* (Kapellen: Pelckmans, 1990); Johan Verberckmoes, *Laughter, Jestbooks, and Society in the Spanish Netherlands* (New York: Macmillan, 1999), 157.

35. Marcus Van Vaernewijck, *Van die beroerlicke tijden in die Nederlanden en voornamelyk in Ghendt (1566–1568),* ed. Ferdinand Vanderhaeghe (Gand, 1872–81).

36. "Bereyt my ten banckette soete sucaden / Conserven, Syropen ende Myrmiladen / . . . Soetmondighe smaken van geleyen / . . . Romenijen, wel ghesuyckert om drincken . . . "; "Ick moet ter mertwaerts, wat baet dat ickt hele En coopen daer vlaeyen, en suycker koeken" (C. Kruyskamp, *Het Anwerpse Landjuweel van 1561* [Antwerp, 1962], 29, 104); Elizabeth Alice Honig, *Painting and the Market in Early Modern Antwerp* (New Haven: Yale University Press, 1998), 219. See also Eddy Stols, "O açúcar na literatura e na pintura flamenga e holandesa (séculos XVI e XVII)," in *História do Açúcar, Rotas e mercados,* ed. Alberto Vieira (Funchal: Centro de Estudos de História do Atlântico, 2002), 221–35.

37. *Antwerps Archievenblad,* 23:241, 249, 24:170, 194; R. Van Roosbroeck, ed., *De kroniek van Godevaert Van Haecht over de troebelen van 1565 tot 1574 te Antwerpen en elders,* 2 vols. (Antwerp, 1929).

38. "Ricette diverse e segreti di Pietro Paolo di Carlo Beccuti Scala," in *Coquatur Ponendo,* ed. Giampiero Negri, 399.

39. Charles Le Maitre, *Relation de mon voiage de Flandre, de Hollande, et de Zélande fait en mil six cent quatre vint et un,* ed. Gilbert Van de Louw (Paris: Belles Lettres, 1978), 335–36.

40. Ria Jansen-Sieben and Marleen van der Molen-Willebrands, eds., *Een notabel Boecxken van Cokerijen door Thomas Vandernoot* (Amsterdam: De Kan, 1994); Elly Cockx-Indestege, ed., *Eenen nyeuwen coock boeck door Gheeraert Vorselman* (Wiesbaden: Gido Pressler, 1971); Elly Cockx-Indestege and Claude Lemaire, eds., *Een Secreet-Boeck uit de zeventiende eeuw over perfumeren, konfijten en koken* (Antwerp: De Schutter, 1983); J. V. A. Collen, ed., "Het Kock-Boeck van d. Carolum Battum, uit de zestiende eeuw," in *Academie voor de streekgebonden gastronomie*, no. 37 (1991); J. Witteveen, "450 jaar kookboeken in Nederland 1510–1960," in *Kookboeken door de eeuwen heen*, exhibition catalog (The Hague: Koninklijke Bibliotheek, 1991); John Landwehr, *Het Nederlands Koookboek, 1510–1945, een bibliographisch overzicht* (Utrecht: HES, 1995); Gillian Riley, *Kunst en koken: Recepten uit de Gouden Eeuw* (Bussum, 1994); Jozien Jobse-Van Putten, *Eenvoudig maar voedzaam, cultuurgeschiedenis van de dagelijkse maaltijd in Nederland* (Nijmegen: Sun, 1995); Lizet Kruyf and Judith Schuyf, *Twintig eeuwen koken, op zoek naar de eetcultuur van onze voorouders* (Utrecht: Kosmos-Z&K Uitgevers, 1997).

41. Walter L. Braekman, ed., *Een Antwerps kookboek voor "Leckertonghen"* (Antwerp: Municipal Library, 1995).

42. Lancelot du Casteau, *Ouverture de cuisine*, 24, 35.

43. R. Ródenas Vilar, *Vida cotidiana y negocio en la Segovia del siglo de oro. El mercader Juan de Cuéllar* (Salamanca: Junta de Castilla y León, 1990), 166.

44. On the forms of pastry, see Friedrich August Zuckerbäcker, *Die europäische conditorei in ihren ganzen umfang* (Heilbronn, 1837); and Georg Christian Neunhöfer, *Das neueste der conditoreikunst, in getreuen, mit iluminirten abbildungen von tafel und laden confekturen*, 2 vols. (Stuttgart, 1844–48). See also *El arte efímero en el mundo hispánico* (Mexico City: Instituto de Investigaciones Estéticas, UNAM, 1983); and Roy Strong, *Art and Power: Renaissance Festivals, 1450–1650* (Woolbridge: Boydell Press, 1995).

45. Frank Lestringant, *Le cannibale, grandeur et décadence* (Paris: Perrin, 1994). See also Piero Camporesi, *La terra e la luna, alimentazione, folclore, societá* (Milan: Mondadori, 1989), and *L'officine des sens, une anthropologie baroque* (Paris: Hachette, 1989).

46. Manuel Martínez Llopis, *Historia de la gastronomia española* (Huesca: La Val de Onsera, 1995), 265.

47. G. J. Hoogewerff, ed., *De twee reizen van cosimo de' Medici prins van Toscane door de Nederlanden (1667–1669)* (Amsterdam, 1919), 46; Zonnevylle-Heyning, "Achttiende eeuwse tafeldecoraties van suiker," in *Achttiend-eeuwse Kunst in de Nederlanden, Leids Kunsthistorisch Jaarboek 1985* (Delft, 1987), 437–53.

48. Christine Armengaud, *Le diable sucré, gâteaux, cannibalisme, mort et fécondité* (Paris: Éditions de La Martinière, 2000).

49. Charles Amiel and Anne Lima, eds., *L'Inquisition de Goa, la relation de Charles Dellon (1687)* (Paris: Chandeigne, 1997), 203.

50. Bernadette Bucher, *Icon and Conquest: A Structural Analysis of the Illustrations of De Bry's Grands Voyages* (Chicago: University of Chicago Press, 1981); Michèle Duchet, ed., *L'Amérique de Théodore de Bry, une collection de voyages protestante du XVIe siècle, Quatre études d'iconographie* (Paris: CNRS, 1987).

51. Ria Fabri, *De 17de-eeuwse en vroeg 18de-eeuwse Antwerpse kunstkast* (Brussels: Koninklijke Vlaamse Academie van België, 1993).

52. Charlotte Sussmann, *Consuming Anxieties: Consumer Protest, Gender, and British Slavery, 1711–1833* (Stanford: Stanford University Press, 2000).

53. Julie Berger Hochstrasser, "Feasting the Eye: Painting and Reality in the Seventeenth Century," in *Still-Life Paintings from the Netherlands, 1550–1720*, ed. Alan Chong and Wouter Kloek, exhibition catalog (Amsterdam, Cleveland, and Zwolle: Waanders Publishers, 1999), 80.

54. *The Arcimboldo Effect: Transformations of the Face from the Sixteenth to the Twentieth Centurty* (Milan: Editoriale Fabbri, Bompiani, 1987); Louis Lebeer, *Catalogue raisonné des estampes de Bruegel l'Ancien* (Brussels: Bibliothèque Royale, 1969); *Bruegel, Une dynastie de peintres*, exhibition catalog (Brussels, 1980); *Pieter Breughel der Jüngere-Jan Brueghel der Ältere, Flämische Malerei um 1600, Tradition und Fortschritt*, exhibition catalog, Villa Hügel, Essen (Lingen: Luca Verlag, 1997); Nadine M. Orenstein, ed., *Pieter Bruegel the Elder, Drawing and Prints*, exhibition catalog, Metropolitan Museum of Art, New York (New Haven: Yale University Press, 2001).

55. Joachim Beuckelaer, *Het markt- en keukenstuk in de Nederlanden, 1550–1650*, exhibition catalog (Ghent: Gemeentekrediet, 1986); Gerard Th. Lemmens and Wouter Kloek, eds., *Pieter Aertsen*, in *Nederlands Kunsthistorisch Jaarboek*, 40 vols. (The Hague: Gary Schwartz-SDU Uitgeverij, 1990); Francine-C. Legrand, *Les peintres flamands de genre au XVIIe siècle* (Brussels, 1963); Marie-Louise Heirs, *Les peintres flamands de nature morte au XVIIe siècle* (Brussels, 1963); Gillian Riley, *Painters and Food—The Dutch Table: Gastronomy in the Golden Age of the Netherlands* (San Francisco: Pomegranate Artbooks, 1994); Christa Nitze-Ertz et al., eds., *Das Flämische Stilleben, 1550–1680*, exhibition catalog, Villa Hügel, Essen (Lingen: Luca Verlag, 2002). One of the oldest paintings with sweets belongs to Frans Francken the Elder (1542–1616), *Les noces de Canaan*, in the collection of the OCMW in Antwerp.

56. Pamela Hibbs Decoteau, *Clara Peeters, 1594–ca. 1640, and the Development of Still-Life Painting in Northern Europe* (Lingen: Luca Verlag, 1992); Kurt Wettengl, ed., *George Flegel, 1566–1638, Stilleben* (Stuttgart: Gerd Hatje, 1999), 61–62; Claus Grimm, *Stilleben, die niederländische und deutsche Meister* (Stuttgart and Zurich: Belser, 1997), 38, 108, 114, 204; Claus Grimm, *Stilleben, die italienische, spanische und französische Meister* (Stuttgart and Zurich: Belser, 1995), 89, 137–39, 147, 175; Alan Chong and Wouter Kloek, *Still-Life Paintings from the Netherlands*, 126–27; Peter Cherry, *Arte y naturaleza, el bodegón Español en el siglo de oro* (Madrid: Ediciones Doce Calles, 1999), 158–66. For a very interesting depiction, see Lanfranco Ravelli, *Bartolomeo Arbotoni, Piacenza, 1594–1676* (Bergamo: Grafica & Arte, 2000), 50.

57. Vitor Serrão, ed., *Josefa de Obidos e o tempo barroco* (Lisbon: Instituto Português do Património Cultural, 1991), 131, 152, 169, 203, 246; Vitor Serrão et al., *Rouge et or, Trésors du Portugal Baroque*, exhibition catalog (Paris: Musée Jacquemart-André, 2002).

58. Eddy de Jongh and Ger Luijten, *Spiegel van alledag, Nederlandse genreprenten, 1550–1700*, exhibition catalog (Amsterdam and Gand: Rijksmuseum-Snoeck-Ducaju, 1997), 160–62.

59. See especially the portraits of Cornelis de Vos and of Govaert Flinck from 1620 to 1640; Jan Baptist Bedaux and Rudi Ekkart, eds., *Kinderen op hun mooist, het kinderportret in de Nederlanden, 1500–1700* (Gand and Amsterdam: Ludion, 2000), 142, 168.

60. Marques da Cruz, *A mesa com Luís Vaz de Camões ou o romance da cozinha no Portugal dos descobrimentos* (Sintra: Colares, 1998), 206.

61. H. Perry Chapman, Wouter Th. Kloek, and Arthur K. Wheelock, eds., *Jan Steen: Painter and Storyteller* (New Haven: Yale University Press, 1996), 198–99; Simon Schama, *The Embarrassment of Riches: An Interpretation of Dutch Culture in the Golden Age* (New York: Alfred A. Knopf, 1987), 182–85. On Saint Nicolas and sweets, see Isabelle Wanson, *Un Saint-Délice: Pain d'épice et Nicolas*, exhibition catalog (Brussels: Bibliothèque Royale de Belgique, 2002).

62. Paulino Mota Tavares, *Mesa, doces e amores no séc. XVII português* (Sintra: Colares, 1999).

63. E. K. Grootes, ed., *Gerbrand A. Bredero's Moortje en Spaanschen Brabander* (Amsterdam: Athneum-Polak & Van Gennep, 1999), vv. 1606–10.

64. Herman Pleij, *Dromen van Cocagne, Middeleeuwse fantasieën over het volmaakte leven* (Amsterdam: Prometheus, 1997), 43, 55, 207, 321.

65. Jan Luiken, *Spieghel van 't menselijk bedrijf* (Amsterdam, 1694). On symbolism, see D. R. Barnes, *The Butcher, the Baker, the Candlestick Maker: Jan Luyken's Mirrors of 17th-Century Dutch Daily Life* (New York: Hofstra Museum, 1995); and Karel Porteman, "De nationale benadering van het emblemo," in *Niederlandistik und Germanistik*, ed. H. Hipp (Frankfurt: Lang, 1992), 176–96. Warm thanks to my colleagues Karel Porteman and Marc Van Vaeck.

66. Jacques and Marie-Laure Verroust, *Friandises d'hier et d'aujourd'hui* (Paris, 1974); *Sucre d'art*, exhibition catalog (Paris: Musée des Arts Décoratifs, 1978); Claude Lebey, ed., *L'inventaire du patrimoine culinaire de France* (Paris: Albin Michel/CNAC, 1992–96); Catherine Amor, *Les bonbons*.

67. Maguelonne Toussaint-Samat, *Histoire naturelle & morale de la nourriture* (Paris: Bordas, 1987), 418.

68. Guy Cabourdin, *La vie quotidienne en Lorraine*, 220–22; Jean Boutier et al., *Un tour de France royal*.

69. Wilfrid Brulez, *De firma della Faille en de internationale handel van Vlaamse firma's in de 16e eeuw* (Brussels: Koninklijke Vlaamse Academie van België, 1959), 4, 204.

70. Françoise Lehous, *Le cadre de vie des médecins parisiens aux XVIe et XVIIe siècle* (Paris: Picard, 1976), 170; *Le sucre luxe d'autrefois*, exhibition catalog (La Rochelle: Musée du Nouveau Monde, 1991); J. Witteveen, "Kookboeken over kookgerei, het kookgerei van de middeleeuwen tot de twintigste eeuw," in *Quintessens, Wetenswaardigheden over acht eeuwen kookgerei* (Rotterdam: Museum Boymans-van Beuningen, 1992), 14–32.

71. In Antwerp in 1617, in the household inventory of Isabela de Veiga (wife of Manuel Ximenes, who already had at her disposal an entire room of porcelain) were found several sugar and syrup pots, a platter for marmalades, a sieve for sugar, and a silver shaker. Eric Duverger, *Antwerpse kunstinventarissen uit de zeventiende eeuw* (Brussels: Koninklijke Vlaamse Academie van België, 1984), vol. 1, inventory of 13–28 June 1617. The oldest silver

shaker preserved in Belgium at the Sterckxhof Museum of Antwerp dates, however, from approximately 1665. Anne D. Janssens and Arnould de Charette, *L'orfèvrerie et le sucre de XVIIe au XXe siècle en France et en Belgique* (1993); *Le sucre luxe d'autrefois*; Johanna Maria van Winter, "Keramiek in Nederlandse kookboeken van de vijftiende tot en met de achttiende eeuw," *Vormen uit vuur* 148 (1993); J. Witteveen, "Of Sugar and Porcelain: Table Decoration in the Netherlands in the 18th Century," in *Feasting and Fasting* (Proceedings of the Oxford Symposium on Food and Cooking, 1990).

72. Robert Vande Weghe, *Geschiedenis van de Antwerpse straatnamen* (Antwerp: Mercurius, 1977), 460–61.

73. Ribeiro, *O doce nunca amargou*, 40.

74. Matilde Santamaría Arnaíz, "La alimentacíon," in *La vida cotidiana en la España de Velázquez*, ed. José Alcalá-Zamora y Queipo de Llano (Madrid, 1994).

75. Schama, *Embarrassment of Riches*, 164–65; Jozien Jobse-Van Putten, *Eenvoudig maar voedzaam*, 173.

76. Herman Van der Wee, *The Growth of the Antwerp Market and the European Economy* (The Hague: Martin Nijhoff, 1963), 1:318–24.

77. Wilfrid Brulez, "De handelsbalans der Nederlanden in het midden van de 16e eeuw," *Bijdragen voor de Geschiedenis der Nederlanden* 21 (1966–67): 287; Wilfrid Brulez, "De handel," in *Antwerpen in de XVIe eeuw* (Antwerp, 1975), 123.

78. Vitorino Magalhães Godinho, *Os descobrimentos e a economia mundial* (Lisbon: Editorial Presença, 1983), 4:679; A. A. Marques de Almeida, *Capitais e capitalistas no comércio da especiaria, o eixo Lisboa-Antuérpia (1501–1549)* (Lisbon: Edições Cosmos, 1993).

79. Fernando Jasmins Pereira and José Pereira da Costa, eds., *Libros de contas da ilha da Madeira, 1504–1537* (Coimbra: Universidade de Coimbra, 1959), 1:181.

80. Eddy Stols, *De Spaanse Brabanders of de handelsbetrekkingen van de zuidelijke Nederlanden met de Iberische Wereld, 1598–1648* (Brussels: Koninklijke Vlaamse Academie van België, 1971); Roland Baetens, *De Nazomer van Antwerpens Welvaart, de diaspora en het handelshuis de groote tijdens de eerste helft der 17de eeuw*, 2 vols. (Brussels: Gemeentekrediet, 1976); Hans Pohl, *Die Portugiesen in Antwerpen (1567–1648), zur geschichte einer Minderheit* (Wiesbaden: Franz Steiner Verlag, 1977).

81. Jan Materné, "Anvers comme centre de distribution et d'affinage d'épices et de sucre depuis la fin du XVème jusqu'au XVIIème siècle," in *L'Europe à table, une exploration de notre univers gastronomique*, exhibition catalog (Antwerp: MIM, 1993), 57.

82. Jan A. van Houtte, *De geschiedenis van Brugge* (Tielt: Lannoo, 1982). For a general, if elementary, survey, see André Vandewalle, "Bruges et la Péninsule ibérique," in *Bruges et l'Europe*, ed. Valentin Vermeersch (Antwerp: Fonds Mercator, 1992).

83. Aloys Schulte, *Geschichte der grossen ravensburger handelsgesellschaft 1380–1530* (Wiesbaden: Steiner, 1964), 2:172–80; Eberhard Schmitt, ed., *Die mittelalterlichen ursprünge der europäischen expansion* (Munich: Verlag C. H. Beck, 1986), 141–48, 169–71, 177–79.

84. John Everaert and Eddy Stols, eds., *Flandre et Portugal, au confluent de deux cultures* (Antwerp: Fonds Mercator, 1991); Jacques Paviot, "Portugal et Bourgogne au XVe siècle, essai de synthèse," *Arquivos do Centro Cultural Português* 26 (Paris: Fundaçao Gulbenkian, 1989): 121–43.

85. Jan A. van Houtte and Eddy Stols, "Les Pays-Bas et la 'Méditerranée Atlantique' au XVIe siècle," in *Mélanges en l'honneur de Fernand Braudel* (Toulouse), 1:645–59; Eddy Stols, "Os mercadores flamengos em Portugal e no Brasil antes das conquistas holandesas," *Anais de História* 5 (1973); John Everaert, "Les Lem, alias Leme, une dynastie marchande d'origine flamande au service de l'expansion portugaise," in *Actas III Colóquio Iternacional de História da Madeira* (Funchal: Centro de Estudos de História do Atlântico, 1993).

86. Konrad Haebler, *Die überseeischen unternehmungen der Welser und ihrer Gesell-schafter* (Leipzig: Hirschfeld, 1903); Theodor Gustav Werner, "Die anfänge der deutschen zuckerindustrie und die augsburger zuckerraffinerie von 1573," *Scripta Mercaturae* (1972): 1–2, 166–86, (1973): 1, 88–103.

87. Fernand Donnet, *Histoire de l'établissement des Anversois aux Canaries au XVIe siècle* (Antwerp: De Backer, 1985).

88. Eddy Stols, "Les Canaries et l'expansion coloniale des Pays-Bas méridionaux et de la Belgique vers 1900," in *IV Colóquio de historia canario-americana (1980)*, ed. Francisco Morales Padrón (Grand Canary, 1982), 2:903–33; Manuela Marrero Rodríguez, "Mercaderes flamencos en Tenerife durante la primera mitad del siglo XVI," in *IV Colóquio de historia canario-americana (1980)*, ed. Francisco Morales Padrón (Grand Canary, 1982), 1:599–614; C. Negrín Delgado, "Jácome de Monteverde y las ermitas de su hacienda de Tazacorte en La Palma," *Anuario de Estudios Atlánticos* 34 (1988); Ana Viña Brito, "El azúcar canario y la cultura flamenca. Un viaje de ida y vuelta," in *España y las 17 Provincias de los Países Bajos, Una revisión historiográfica (XVI–XVII)*, ed. Ana Crespo Solana and Manuel Herrero Sánchez (Córdoba: Universidad de Córdoba, 2002), 2:615–37.

89. On 26 November 1582, Carlos de Santa Cruz insured for no less than 1,141 *livres* a cargo of more than 1,200 *arrobas* of various sugars from various *ingenios* in Tenerife (Orotava, Señor Capitán, Maria de las Cuevas et Diego de Mesa) and Grand Canary (Pedro de Soso) on the *Bon espoir*, Captain Pierre Pontu, departing from Croset for Middelburg, and, if possible, on to Antwerp. See the Archives Générales du Royaume, Brussels, Famille Comte Mercy-Argenteau, II, 14, 8 October 1588. Thus at the height of the crisis following the fall of Antwerp, Paul van Dale sold to the merchants Gillis Houtappel, Nicolas Ripet, and others the entire output of his *ingenios* in Tazacorte and Argual for the period November 1588–September 1589, for no less than 72,000 *reales*. Municipales Archives, Antwerp, Notaris L. Van Roekergem, 8-X-1588.

90. Jan Denucé, *L'Afrique au XVIe siècle et le commerce Anversois* (Antwerp: De Sikkel, 1937).

91. Joseph Billioud, *Histoire du commerce de Marseille*, ed. Gaston Rambert, 4 vols. (Paris: Plon, 1949–56), 3:284.

92. Enrique Otte, "Die Welser in Santo Domingo," in *Festschrift für Johannes Vincke* (Madrid, 1962–63).

93. Baetens, *De nazomer van Antwerpens welvaart*, 1:155; Pohl, *Die Portugiesen in Antwerpen*, 155–59.

94. Eddy Stols and Eduardo Dargent-Chamot, "Aventuriers des Pays-Bas en Amérique hispano-portugaise," in *Flandre et Amérique latine, 500 ans de confrontation et métissage*, ed.

Eddy Stols and Rudi Bleys (Antwerp: Fonds Mercator, 1993), 62–66; Eddy Stols, "Convivências luso-flamengas na rota do açúcar," *Ler História* 32 (1997): 119–45; Eddy Stols, "Humanistas y jesuitas en los negocios brasileños de los Schetz, grandes negociantes de Amberes y banqueros de Carlos V," in *Carlos V y la quiebra del humanismo político en Europa (1530–1558)*, ed. José Martínez Millán, 4 vols. (Madrid: Sociedad Estatal para la Conmemoración de los Centenarios de Felipe II y Carlos V, 2001), 4:29–47.

95. Brulez, *De firma della Faille*, 139–44.

96. Billioud, *Histoire du commerce de Marseille*, 4:147. On Andalusian sugar, see esp. J. Pérez Vidal, *La cultura de la caña de azúcar en el Levante español* (Madrid: Instituto Miguel de Cervantes, 1973); Margarita M. Birriel Salcedo, "La producción azucarera de la Andalucía Mediterránea, 1500–1750," in *Producción y comercio del azúcar de caña en época preindustrial: Actas del tercer seminario international, Motril, 23–27 Septiembre 1991* (Granada: Diputación Provincial, 1993); and Purificación Ruiz García and Antonio Parejo, *La axarquía, tierra de azúcar, cincuenta y dos documents históricos* (Vélez and Málaga: Asukaría Mediterránea, 2000).

97. Fernand Donnet, *Notice historique et statistique sur le raffinage et les raffineries de sucre à Anvers* (Antwerp: De Backer, 1895), 18–19.

98. António Borges Coelho, *Quadros*, 95, 268.

99. J. Fournier, "L'introduction et la culture de la canne à sucre en France au XVIe siècle," *Bulletin de géographie historique et descriptive* (1903).

100. Mathieu Lobelius, *Nova stirpium adversaria* (Antwerp, 1576), f. 19–20.

101. Jan Craeybeckx, *Un grand commerce d'importation: les vins de France aux anciens Pays-Bas (XIIIe–XVIe siècle)* (Paris: SEVPEN, 1958), 42.

102. H. A. Enno van Gelder, "Gegevens betreffende roerend en onroerend bezit in de Nederlanden in de 16ᵉ eeuw," in *Rijks Geschiedkundige Publicatiën*, Grote Serie 140 (The Hague, 1972), 1:524–25.

103. Gisèle Jongbloet-Van Houtte, ed., "Brieven en andere bescheiden betreffende Daniel Van der Meulen, 1584–1600," in *Rijks Geschiedkundige Publicatiën*, Grote Serie 196 (The Hague, 1986), 1:111.

104. M. Celis, *De Tol van Lith en de Maashandel* (Mémoire de licence, Katholieke Universteit Leuven, 1971), 106.

105. Hans Pohl, "Die zuckereinfuhr nach Antwerpen durch portugiesische Kaufleute während des 8 jährigen Krieges," *Jahrbuch für Geschichte von Staat, Wirtschaft und Gesellschaft Lateinamerikas* 4 (1967); Biblioteca Nacional Lisbon, Fundo Geral, 8457, f. 74–97.

106. Baetens, *De nazomer van Antwerpens welvaart*, 1:108–9.

107. Renée Doehaerd, *Études Antwerpoies, documents sur le commerce international à Anvers* (Paris: SEVPEN, 1962–63); Hans Pohl, "Köln und Antwerpen um 1500," *Mitteilungen aus dem Stadtarchiv von Köln* 60 (1970): 498–99; G. J. Gramulla, *Handelsbeziehungen Kölner Kaufleute zwischen 1500 und 1650* (Cologne and Vienna: Böhlau, 1972).

108. Municipal Archives, Antwerp, Insolvente Boedelskamer, 2126.

109. Emile Coornaert, *Les Français et le commerce international à Anvers*, 2 vols. (Paris: SEVPEN, 1962).

110. *Antwerps Archievenblad*, 23:417.

111. Henri Lapeyre, *Une famille de marchands: Les Ruiz* (Paris: SEVPEN, 1955), 589; Jacques Bottin, "La redistribution des produits américains par les réseaux marchands rouennais (1550–1620)," in *Dans le sillage de Colomb*, ed. Jean-Pierre Sanchez (Rennes: Presses Universitaires de Rennes, 1995).

112. Etienne Trocmé et Marcel Delafosse, *Le commerce Rochelais de la fin du XVe siècle au début du XVIIe* (Paris: SEVPEN, 1952), 125–26.

113. Brulez, *De firma della Faille*.

114. Hermann Kellenbenz, *Unternehmerkräfte im Hamburger Portugal und Spanienhandel 1590–1620* (Hamburg, 1954); Wilfrid Brulez, *Marchands flamands à Venise* (Brussels and Rome: Institut Historique Belge à Rome, 1965), 1:1568–1605; E. M. Koen, "Notarial Records in Amsterdam Relating to the Portuguese Jews in That Town up to 1639," *Studia Rosenthaliana* 3 (1969); Victor Enthoven, *Zeeland en de opkomst van de republiek, Handel en strijd in de Scheldedelta, c. 1550–1621* (Leyden, 1996); Marie-Christine Engels, *Merchants, Interlopers, Seamen and Corsairs: The "Flemish" Community in Livorno and Genoa (1615–1635)* (Hilversum: Verloren, 1997); Veronika Joukes, "Os Flamengos no Noroeste de Portugal (1620–1670)" (Master's thesis, Universidade do Porto, 1999).

115. Oscar Gelderblom, *Zuid-Nederlandse kooplieden en de opkomst van de Amsterdamse stapelmarkt (1578–1630)* (Hilversum: Verloren, 2000), 181.

116. Arquivo da Torre do Tombo, Lisbon, Corpo Cronológico, Maço 308, 309, 310.

117. Xavier de Castro, ed., *Voyage de Pyrard de Laval aux Indes orientales (1601–1611)* (Paris: Chandeigne, 1998), 2:815; J. J. Reesse, *De suikerhandel van Amsterdam van het begin der 17de eeuw tot 1813* (Haarlem: J. L. Kleynenberg, 1908); György Nováky, "On Trade, Production and Relations of Production: The Sugar Refineries of Seventeenth-Century Amsterdam," *Tijdschrift voor Sociale Geschiedenis* 23 (1997); Virgínia Rau and Maria Fernanda Gomes da Silva, *Os Manuscritos do Arquivo da Casa do Cadaval respeitantes ao Brasil* (Coimbra: Universidade Coimbra, 1956), 1:1, 23 July 1626.

118. Archives Plantiniennes, Plantin-Moretus Museum, Antwerp, 532, 1214–15.

119. Donnet, *Notice historique*; Alfons K. Thijs, "De geschiedenis van de suikernijverheid te Antwerpen (16de–19de eeuw): een terreinverkenning," *Bijdragen tot de geschiedenis* 62 (1979); D. De Mets and H. Houtman-De Smedt, "Antwerpen: een suikerstad? Apecten van een archeologisch-historisch onderzoek," in *Antwerpen* 4 (1984).

120. *Fontes para a história do antigo ultramar portugués*, vol. 2, *São Tomé e Príncipe* (Lisbon, 1982), 236–37. At any rate, in a letter dated 1 July 1577, the king authorized Nicolao Petro Cochino to refine São Tomé sugar in Lisbon. See Virgínia Rau, "O açúcar de S. Tome no segundo quartel do século XVI," in *Elementos de história da ilha de S. Tomé* (Lisbon, 1971), doc. 9; and Raymond Fagel, *De Hispano-Vlaamse Wereld, de contacten tussen spanjaarden en Nederlanders, 1496–1555* (Brussels: Archives et Bibliothèques de Belgique, 1996), 120.

121. Jan Van Roey, "De sociale structuur en de godsdienstige gezindheid van de Antwerpse bevolking op de vooravond van de reconciliatie met Farnese" (Ph.D. diss., University of Ghent, 1963), 72–73.

122. Jan Denucé, *Inventaire des affaitadi, banquiers italiens à Anvers, de l'année 1568*

(Antwerp: De Sikkel, 1934); Jan Denucé, *Italiaansche koopmansgeslachten te Antwerpen in de XVIe–XVIIIe eeuwe* (Malines and Amsterdam: Het Kompas—De Spiegel, n.d.).

123. Baetens, *De Nazomer van Antwerpens Welvaart*, 1:108–9.

124. Jan Denucé, *Koopmansleerboeken van de XVIe en XVIIe eeuwen in handschrift* (Antwerp and Brussels: Koninklijke Vlaamse Academie van België, 1941), 53–58.

125. Donnet, *Notice historique*, 18–19.

126. Jean de Léry, *Histoire d'un voyage faict en la terre du Brésil*, ed. Frank Lestringant (Paris: Le Livre de Poche, 1994), 323.

127. Luis de Matos, *Les Portugais en France au XVIe siècle* (Coimbra: Universidade de Coimbra, 1982), 88.

128. Bondois, "Les raffineries de sucre à Rouen," in *Compte rendu du Congrès de l'Union industrielle à Rouen*, 1922.

129. Étienne Trocmé and Marcel Delafosse, *Le commerce rochelais de la fin du XVe siècle au début du XVIIe* (Paris: A, Colin, 1952), 125–26.

130. Billioud, *Histoire du commerce de Marseille*, 434–37.

131. Trocmé and Delafosse, *Le commerce*, 100–101.

132. Edmund O. von Lippmann, *Geschichte des Zuckers, seiner Darstellung und Verwendung* (Leipzig: Hesse, 1890), 274; Noël Deerr, *The History of Sugar*, 2 vols. (London: Chapman and Hall, 1949–50), 458; Charles Wilson, *England's Apprenticeship 1603–1763* (London: Longman, 1979), 200–201.

133. Bob Scribner, ed., *Germany: A New Social and Economic History* (London and New York: Anold, 1996), 1:202.

134. Werner, "Die anfänge."

135. K. Ratelband, ed., "Reizen naar West-Afrika van Pieter van den Broecke, 1605–1614," in *Werken Linschoten-Vereniging*, vol. 52 (The Hague, 1950), 28; Reesse, *Suikerhandel*, 105.

136. Jongbloet-van Houtte, "Brieven en andere."

137. Stols, *Convivências*; Kenneth R. Andrews, *Elizabethan Privateering: English Privateering during the Spanish War, 1585–1603* (Cambridge: Cambridge University Press, 1964); Kenneth R. Andrews, *Trade, Plunder and Settlement: Maritime Enterprise and the Genesis of the British Empire, 1480–1630* (Cambridge: Cambridge University Press, 1986), 250.

138. Archivo General de Simancas, Secretarias provinciales, libro 1476, fol. 246 and 251.

139. Reesse, *Suikerhandel*; H. Brugmans, *Geschiedenis van Amsterdam* (1930; reprint, Utrecht and Amsterdam: Spectrum, 1973), 3:31–32, 105–7, 190.

140. Jan De Vries and Ad Van der Woude, *The First Modern Economy: Success, Failiure, and the Perseverance of the Dutch Economy, 1500–1815* (Cambridge: Cambridge University Press, 1997), 328–29; Robert S. Duplessis, *Transitions to Capitalism in Early Modern Europe* (Cambridge: Cambridge University Press, 1997), 116, 131.

141. Jonathan I. Israel, *Dutch Primacy in World Trade, 1585–1740* (Oxford: Oxford University Press, 1989), 100–101, 286–87, 356, 390–91.

142. Pietro Camporesi, *Le chocolat, L'art de vivre au siècle des Lumières* (Paris: Grasset, 1992), 262.

143. Reesse, *Suikerhandel*, 147.

144. Roger De Peuter, "Industrial Development and De-Industrialization in Pre-Modern

Towns: Brussels from the Sixteenth to the Eighteenth Century: A Provisional Survey," in *The Rise and Decline of Urban Industries in Italy and the Low Countries*, ed. Herman Van der Wee (Leuven: Leuven University Press, 1988), 227.

145. Jean Meyer, *Histoire du sucre* (Paris: Éditions Desjonquères, 1989); Jean-Yves Barzic, *L'hermine et le soleil, les Bretons au temps de Louis XIV* (Spezet: Coop Breizh, 1985), 208–9; Peter Voss, "Les raffineurs de sucre allemands à Bordeaux au XVIIe siècle," in *Dans le sillage de Colomb*, ed. Jean-Pierre Sanchez, 237–46.

146. Jan Denucé, *Koopmansleeerboeken*.

147. Engels, *Merchants*, 78, 93–94, 143–49, 159–60, 218.

148. Domenico Sella, *Commerci e industrie a Venezia nel secolo XVII* (Venice: Instituto per la collaborazione culturale, 1961).

CHAPTER NINE

The Sugar Industry in the Seventeenth Century

A New Perspective on the Barbadian "Sugar Revolution"

John J. McCusker and Russell R. Menard

The seventeenth century witnessed two important developments in the sugar industry. The first was a shift in the center of sugar production, away from the Spanish and Portuguese Atlantic Islands and Brazil to the British Caribbean, as the tiny island of Barbados became, for a while, the world's leading sugar producer. The second was the emergence, in Barbados, of a new way of organizing production that was eventually to change—but not revolutionize—the sugar industry. For more than two and a half centuries, the customary story of the rise of sugar production at Barbados has been told in terms that many have labeled the Barbadian "sugar revolution."[1] The sugar boom in Barbados is usually explained as the result of a conjuncture between a failure in the Barbadian economy and the interests of the Dutch. Around 1640 the island economy was thoroughly depressed as bad times in the tobacco industry hit the low-grade leaf grown in Barbados especially hard and as cotton failed to live up to its initial promise as an alternative crop. Barbadians, so the argument runs, either had to find a profitable export or abandon the pursuit of riches and accept a future much like that of New England's as small farmers at the edge of empire. According to conventional wisdom, the Dutch, after their control of the sugar industry in Pernambuco was threatened, proved to be the island's salvation. They taught the Barbadians how to grow, harvest, and process sugarcane; they loaned them the capital to build plantations; sold them the slaves to do the work; shipped the product across the Atlantic; and marketed it in the major European trading centers.[2]

As the tale is usually told, the results of this intervention were swift and sure, awesome and dreadful. The so-called Barbadian "sugar revolution" transformed the island in the decades surrounding 1650: sugar monoculture drove out diversified farming; large plantations replaced small farms; blacks arrived by the thousands and whites deserted the island; destructive demographic patterns took root among both whites and blacks; the island began to import food and fuel; and the great planters rose to wealth and power—all within the two decades after 1640. Regularly quoted in this regard is the striking testimony of Barbadian planter Richard Vines, in a 1647 letter to John Winthrop: "Men are so intent upon planting sugar that they had rather buy foode at very deare rates than produce it by labour, soe infinite is the profitt of sugar workes after once accomplished."[3] While what he wrote was surely true, nothing in his words equates with "sugar revolution."

The critical changes in this long list were the rapid growth of African slavery and the rise of great plantations. The traditional explanation for them seems straightforward. Sugar, because of its substantial scale economies and handsome profits, was most efficiently produced on big units that required large numbers of workers. That greatly increased the demand for labor. Increasing demand for labor pushed up wages and the price for indentured servants. That, in turn, stretched the capacity of the servant trade to the breaking point and forced planters to look elsewhere for workers. African slaves, available in large quantities through the century-old Atlantic trade, were the most attractive alternative. Other islands in the British and French West Indies, patterning themselves on Barbados, experienced similar transformations in the size of plantations and the composition of the labor force as sugar came to dominate their economies, too. Yet nowhere did the "sugar revolution" strike with the speed and power apparent at Barbados—or so we are told.

This is a compelling story, but, besides rushing things along, it is misleading in many of it details. In particular, it understates the performance of the economy just prior to the beginning of the sugar boom; it distorts the relationship between sugar and slavery in the crucial, early years of the transformation of the workforce; and it exaggerates the importance of the Dutch in introducing sugar to Barbados and, thereby, underestimates the role of the English. We think that the restructuring of the Barbadian sugar industry described in this essay came about slowly over time, that it came about through trial and error ("learning by doing"), that it involved a variety of social and economic factors, especially the nature of the labor supply, the character of the population, and the organization of the workforce, and that it contributed mightily to the reconfiguring of the social, economic, and political landscape. Most certainly, it did not exhibit any undue reticence on the part of plantation owners to invest their own human and financial capital in exploring change. Space will not allow us to explore each of

Map of Barbados showing the concentration of sugar estates on the western and southern coast in the 1630s. *From Richard Ligon,* A True and Exact History of the Island of Barbados *(London, 1657).*

these points in detail. We focus on two of them: the relationship between sugar and slavery when the Barbadian labor force was undergoing transformation; and the relative role of the English and Dutch in bringing sugar to the island.[4]

Before proceeding, it is necessary to establish the chronology. When, precisely, did sugar emerge as the dominant export crop in Barbados? When, precisely, did slavery emerge as the dominant mode of labor on the island? Unfortunately, the data required to answer the first question—annual series of Barbadian production and export figures—have not survived. Even though sugar was introduced into the colony soon after its settlement in 1627 and small amounts of it were grown from that time onward, it only became a major crop in the mid-1640s.[5] In the absence of production and export data, evidence from transactions on the island can serve as a proxy for determining sugar's growing relative importance (see table 9.1).[6] Just as colonists did elsewhere, the Barbadians used their major exports as forms of commodity money by assigning them set values in Barbados money currency.[7] By assuming that the frequency with which islanders referred to different commodities as money in transactions, we can establish a rough

TABLE 9.1. Commodity Monies Stipulated in Various Transactions, Barbados, 1639–1652

Year	Cotton	Tobacco	Indigo	Sugar
1639	43%	57%		
1640	79	21		
1641	74	26		
1642	72	28		
1643	43	47	6%	
1644	26	43	23	8%
1645	16	64	5	16
1646	22	47	4	27
1647	12	47		41
1648	8	32		60
1649				100
1650	10	10		80
1651				100
1652				100

Source: Recopied Deeds Books, 3/1–3/3, Barbados Department of Archives, Black Rock, St. James, Barbados.

guide to their importance in the economy. It is clear on that basis that sugar was not a major crop until the end of the 1640s and it was not until the 1650s that sugar fully dominated the Barbadian economy. In 1640 planters devoted more of their time and energy to tobacco and cotton, what are usually thought of as the minor staples, than to sugar.[8]

As for the second question, although demographic evidence for the island in the seventeenth century leaves much to be desired, from scattered references we estimate that at least 1,000 slaves were delivered to Barbados from 1627 to 1639 and at least 23,000 slaves in the 1640s.[9] By mid-century, the slave population is thought to have reached 12,800 or 30 percent of the total population, before sugar had fully established itself as the dominant crop on the island (see table 9.2). Further, the evidence indicates that the population as a whole increased much more rapidly in the 1630s than in later decades, largely because of migration from England and the importation of indentured servants.[10] The reason for the shift from servants to slaves is clear in the calculation behind what George Downing told his cousin John Winthrop Jr., in a letter written in August 1645. Barbadians had bought "no lesse than a thousand Negroes" that year. "The more they buie, the better able are they to buye, for in a year and halfe they will earne

TABLE 9.2. Estimated Population of Barbados, 1630–1690 (in Thousands)

Year	Whites	Blacks	Total
1630	0.9	0.1	1.0
1640	13.5	0.5	14.0
1650	30.0	12.8	42.8
1660	26.2	27.1	53.3
1670	22.4	40.4	62.8
1680	20.5	44.9	65.4
1690	17.9	47.8	65.7

Sources: John J. McCusker and Russell R. Menard, *The Economy of British America, 1607–1789*, 2d ed. (Chapel Hill: Published for the Institute of Early American History and Culture, Williamsburg, Virginia, by the University of North Carolina Press, 1991), p. 153, revised on the basis of additional research. The data were originally compiled and presented in John J. McCusker, *Rum and the American Revolution: The Rum Trade and the Balance of Payments of the Thirteen Continental Colonies, 1650–1775*, 2 vols. (New York: Garland Publishing, 1989), 2:644–45, 699. Compare P[eter] F. Campbell, "Barbados: The Early Years," *Journal of the Barbados Museum and Historical Society* 35 (no. 3, 1977): 155–77; Campbell, "Aspects of Barbados Land Tenure, 1627–1663," *Journal of the Barbados Museum and Historical Society* 37 (no. 2, 1984): 112–58; and Campbell, *Some Early Barbadian History* ([St. Michael, Barbados: Privately Published, 1993]), pp. 84–89.

(with gods blessing) as much as they cost." If you are "able to doe something upon a plantation, [you will] in a short tim be able with good husbandry to procure Negroes (the life of this place) out of the increase of your owne planta- tion."[11] The cost of indentured servants who would work for five or six years no longer made sense given the profits to be made in the sugar boom. Under the new circumstance, the larger investment in a slave could be recouped in eigh- teen months—and slaves could be worked a lifetime long.[12]

These estimates are supported by evidence that Richard Dunn and Hilary Beckles have gathered about the workforce on private estates (see table 9.3). The most important change described by these data is in the composition of the unfree labor force. In the late 1630s, slaves were still relatively rare on the island and most unfree workers were indentured servants. The slave population in- creased rapidly over the next decade, however, until, just before mid-century and before sugar had become the dominant commercial product, slaves out- numbered servants on island plantations by about two to one.[13] It is important to note that the servant population did not decline with the growth of slavery. Indeed, the number of servants per estate actually increased in the initial stages of slavery's expansion and it remained above the level of the late 1630s until the late 1650s. Slaves became a majority of the unfree workforce not because of a decline in the number of servants. The servant population actually increased after 1640; the slave population simply grew faster.

TABLE 9.3. Proportions of Servants and Slaves per Estate in Barbados in Different Periods, 1635–1670

Years	No. of Estates	Servants per Estate	Slaves per Estate	Ratio of Servants per Slave
1635–40	8	15.4	0.1	154.0
1639–43	15	12.9	3.4	3.8
1641–43	9	12.0	5.7	2.1
1646–49	6	19.5	11.1	1.8
1650–57	7	18.4	24.0	0.8
1658–70	10	3.1	111.1	0.0

Sources: Richard S. Dunn, Sugar and Slaves: The Rise of the Planter Class in the English West Indies, 1624–1713 (Chapel Hill: Published for the Institute of Early American History and Culture, Williamsburg, Virginia, by the University of North Carolina Press, 1972), pp. 54, 68; Hilary Beckles, "Plantation Production and White 'Proto-Slavery': White Indentured Servants and the Colonisation of the English West Indies, 1624–1645," *Americas: A Quarterly Review of Inter-American Cultural History* 41 (January 1985): 34; Hilary Beckles and Andrew Downes, "The Economics of the Transition to the Black Labor System in Barbados, 1630–1680," *Journal of Interdisciplinary History* 18 (Autumn 1987): 228.

These data cast doubt on the usual story of how Barbados became a sugar island and a slave society. Very importantly, they raise questions about the relationship between sugar and slavery. Slaves arrived in Barbados in substantial numbers while tobacco and cotton were still the major crops. Barbados gradually became a slave society in response to opportunities in cotton, tobacco, and indigo. Sugar did not bring slavery to Barbados.[14] It sped up and intensified a process already underway, which is why we speak of a sugar boom rather than a "sugar revolution."

In addition, these data make one wonder if the Barbadian economy was exceptionally depressed in the immediate, pre-sugar era. The late 1630s and early 1640s were years of decline in both the English economy and the English colonial economy in North America, but things had begun to improve by 1644.[15] Just as Barbados seems to have participated in the contraction phase of the cycle, it appears to have shared in the subsequent expansion. Otherwise, if their economy had continued to be so depressed, where did Bajans find the resources to pay for all those slaves? Given the evidence of slave purchases and the behavior of the prices of the island's other exports in the Atlantic World, one could in fact build a case that sugar came into its own in Barbados as part of a diversified production and export boom marking the end of a depression—rather than in its midst.

Traditionally, historians have credited the Dutch for the introduction of sugar to Barbados.[16] Perhaps the most extensive expression of these ideas occurs in a

1663 memorandum sent by Sir Robert Harley, former member of Parliament, island planter, and the Chancellor and Keeper of the Great Seal of Barbados: "The Earle of Carlisle sent a Governor to Barbadoes who ... granted Land as did others before to severall personns. Heitherto the Collonies did not thrive, but were like to bee extinguish for want of provision untill it happen'd that the Duch loosing Brasille, many duch and Jews repairring to Barbadoes began the planting and making of sugar, which caused the Duch with shipping to releive them and [supply] Credit when they were ready to perish, Likewise the Duch being ingaged on the coast of Giney in Affrick for negroe Slaves having Lost brasille not knowing where to vent them they trusted them to Barbadoes. This was the first rise of that plantation that makes it able to subsiste and traficke."[17]

One reason to quote this statement at length is the apparent authority of its source; another is its closeness in time to the events it describes. It may well have been the origin of other contemporary expressions of similar sentiments.[18] Despite its many confusions, distortions, and simple inaccuracies, Harley's account nevertheless conveys the flavor of the stories told then, in contrast to which we offer our own reading of the relevant evidence. A critical omission in Harley's account is any role for the English!

We have no interest in denying the contribution of Dutch merchants to the Barbadian sugar boom. Especially during the years of the Civil Wars, Dutch shipping was heavily engaged in freighting goods to and from all the English colonies.[19] Nevertheless, we think the traditional emphasis on the contribution of the Dutch has caused historians to miss the comparable importance of the role the English, both as investors who supplied capital to finance the conversion to sugar and as merchants who supplied the island with slaves to do the work.

Many English business firms sought direct investment in the Barbados sugar boom (see table 9.4). Our examination of the surviving deed books in the Barbados Department of Archives has identified seventy-five English merchants who invested in plantations on Barbados between 1640 and 1650; there are no references to Dutch merchants.[20] Many of them are known to have been prominent in colonial trade during the first half of the seventeenth century and are already familiar to all students of England's Atlantic economy, especially Maurice Thompson, William Penoyer, Thomas Andrews, and Richard Bateson. Some had long been active in West Indian trade, especially in tobacco and cotton. Others were obscure men with no previous Barbadian experience who saw an opportunity to make a big strike in sugar on the island and took the chance. In a few cases—Martin Noell is surely the best known—the results were spectacularly successful.[21]

TABLE 9.4. Purchases of Barbadian Land by English Merchants, 1639–1650

Years	Average No. of Tracts Purchased per Year
1639–42	1–2
1643–46	4–5
1647	30
1648	11
1649–50	2–3

Source: Recopied Deeds Books, Barbados Department of Archives, Black Rock, St. James, Barbados.

As table 9.4 shows, while English merchants were active in buying land and developing plantations on the island from the 1640s to the 1660s, their activity rose to a powerful crescendo in 1647 when they purchased thirty tracts.[22] In that one year alone there were twenty-two English merchants, acting alone or in partnership, all but two of them Londoners, all but one of them men. The sole woman was the widow Beatrice Odiarne, who seems to have taken over her husband Thomas's business. These merchants purchased more than ten thousand acres in Barbados, from March to December 1647, almost 10 percent of all land on the island. Some of the tracts were working plantations, with servants, slaves, housing, livestock, tools, and crops already in the ground. Merchant capital transformed these plantations, adding more workers and livestock, increasing the acreage, and purchasing the expensive equipment needed to process the cane and to make sugar.[23] Even though not all the recorded deeds included prices, based on the prices of those that were recorded, we can estimate that the total investment of English merchants in the island that year approached £150,000 Barbados money currency, an extremely large sum of money.[24]

These English merchant investors did not limit their activity in Barbados to investing in plantations. They also played a critical role in supplying the island with slaves by participating in the African trade; they not only supplied Barbados with the workers essential to the development of the sugar industry but also challenged Dutch hegemony in that branch of commerce. There is evidence of their success on both counts. Ernst van den Boogaart and Pieter C. Emmer, leading experts on the Dutch slave trade, concluded that, from the start of the sugar boom, "the English were ahead of the Dutch in bringing slaves to the Caribbean," a conclusion that is supported by the recent work of several other scholars.[25] Between February 1645 and January 1647, the Dutch governors at Elmina on the Gold Coast of West Africa reported the arrival and departure of nineteen English ships, capable of carrying perhaps 2,000 slaves. During the

1640s, deliveries to Barbados averaged roughly 2,300 slaves per year, indicating that, just at the end of the First Civil War, when one might think that Dutch inroads into Barbadian trade should have peaked, English merchants had the capacity to supply a substantial share of the slaves Barbadians imported.[26] Claims that it was the Dutch who were responsible for bringing plantation agriculture and slavery to Barbados are greatly exaggerated.

English merchants also played an important role within Barbados by introducing some institutional changes that were key to the future of the sugar industry both in Barbados and throughout the Americas. Before the Barbadian sugar boom of the mid-seventeenth century, sugar production had always been organized according to what we might call the dispersed system, in which smaller farmers grew sugar that was processed at a large mill owned by a neighboring planter.[27] Events in Barbados changed that arrangement, with profound consequences for the subsequent history of the sugar industry because Barbadians discovered that they could increase efficiency by concentrating growing and milling through an integrated system of production.[28]

How the change came about is a story worth exploring in detail. It is useful to begin with the observation that the integrated plantation was not universal on Barbados throughout its history as a sugar producer. In the middle of the seventeenth century, when the sugar boom was underway on the island, some Barbadian estates were organized in the traditional, dispersed fashion, connected with tenant farming. Before the mid-1640s tenant farming was rare on Barbados. Throughout the 1630s land was cheap and readily available, especially for those willing to move away from the densely settled west coast and into the interior or toward the far northern, southern, and eastern portions of the island (see figure 9.1). Beginning in the mid-1640s, as land prices jumped with the rebounding economy and as rich men began to assemble large holdings to exploit the export boom, recently freed servants who earlier might have acquired a small tract of their own found themselves unable to do so (see table 9.5).[29] Some left the island.[30] Some who stayed became tenants on land owned by the larger planters and grew sugarcane for their landlord's mill. It would be an error to assume an exact equivalency between tenancy and the dispersed system of organizing production. Tenants could and did grow crops other than sugar, while small landowners who could not afford a mill of their own could bring their cane to a neighbor's mill and have it processed for a share of the crop. Just how many small cane farmers there were and what proportion of the crop they raised is impossible to tell. However, the history of Mount Clapham plantation opens a window into the changing organization of the Barbadian sugar industry.[31]

Thomas Noell, the London merchant, who, along with his brothers, Martin,

FIGURE 9.1. Price of Land in Barbados, 1638–1650

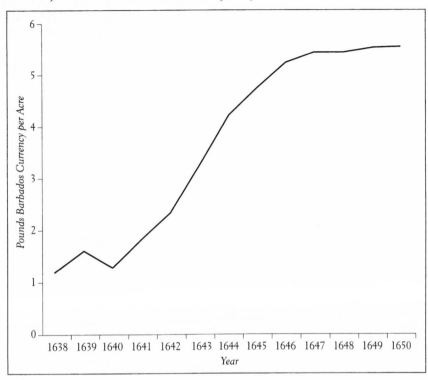

Stephen, and James, known locally as the "four brethren," was among the leading investors in the Barbadian sugar boom. In 1650 Noell acquired Mount Clapham, a 510-acre tract in St. Michael and Christ Church parishes. Four years later, Mount Clapham had a workforce of fifteen servants and slaves. At the optimum ratio of one worker for every two acres of cultivated land, Noell's workforce was barely enough help for a small 100-acre plantation, let alone one of Mount Clapham's size. Noell addressed the labor shortage by recruiting tenants. By June 1654, he had leased out 179 acres to twenty-four tenants for an annual rent of £362 Barbados money currency. The plots ranged in size from three and a half to eighteen acres and were often operated by two or three men in a partnership. The leases ran from six to nine years at average annual rent of £3.00 per acre. Although we cannot assume that all tenants grew sugarcane, it seems likely that those at Mount Clapham did so. Their rents seem too high to be covered by the production of minor crops and we do know that their landlord owned a sugar works.[32]

Mount Clapham was not the only Barbadian sugar estate with insufficient

TABLE 9.5. Price of Barbadian Land, 1638–1650

Years	Average Price of Land on Barbados, per Acre, in Barbadian Money Currency
1638	£1.20
1639	1.60
1640	1.30
1641	1.80
1642	2.30
1643	3.20
1644	4.20
1645	4.70
1646	5.20
1647	5.40
1648	5.40
1649	5.50
1650	5.50

Sources: Recopied Deeds Books, 3/3, pp. 13, 17–21, 30–31, 288, 313, 322–23, 336–38, Barbados Department of Archives, Black Rock, St. James, Barbados. These are the average prices per acre for lands sold in St. James and Christ Church Parishes as compiled and presented in Hilary Beckles, *White Servitude and Black Slavery in Barbados, 1627–1715* (Knoxville: University of Tennessee Press, 1989), p. 156. Given the nature and source of these data, we are somewhat skeptical of their accuracy and comparability, but we feel they convey the general trend in prices reasonably enough. The original prices were in pounds Barbados cotton currency. We have converted them to Barbados money currency with reference to table 9.7.

unfree labor to be operated as an integrated plantation. Richard Dunn reports acreage, servants, and slaves for four Barbadian plantations, between 1640 and 1667, varying in size from 75 to 360 acres. Only two of them had enough servants and slaves to operate as an integrated plantation at the optimum level of two acres per worker. It appears that many mid-century Barbadian plantations were similarly understaffed. Given the high and rising price of land, few planters could afford to leave much of their estate fallow. It made good sense to do as Thomas Noell did and lease out some of their acreage.[33] Just how many did so is, of course, impossible for us to know.

Keeping as much of the plantation as possible in cultivation was not the only concern of large-scale planters that led Noell to seek out tenants. It was also important to the efficient and profitable operation of the sugar mill to make sure that they had enough sugarcane to keep it working steadily. Neither the problem of labor shortages nor tenancy as a solution to it were unique to Barbados in the early stages of the sugar boom. There is evidence of similar developments in other islands during the conversion to sugar monoculture. Thus, when Christopher Jeafferson's effort to rebuild his estate on St. Christopher in the aftermath of the devastating 1681 hurricane stretched his labor force, he turned to share agreements with tenants in order to keep his land fully cultivated and to ensure sufficient cane for his mill.[34]

Integrated plantations were relatively rare in the early stages of the sugar boom. The first seems to have been operated by James Drax, the largest and most important of the early sugar planters. In 1654 visiting French priest, Rev. Antoine Biet, marveled at the sight of Drax's two hundred slaves "working sugar."[35] By the early 1670s, when Mount Clapham estate was reorganized as an integrated plantation, they had become more common as more and more planters had acquired sufficient workers to attain internally the maximum supply of cane for the sugar works. As Richard Dunn's analysis of the 1680 census indicates, by that year the integrated plantation had become a dominant institution in the island's sugar industry. According to that census, just over two hundred planters owned more than sixty slaves each, while another two hundred owned between twenty and sixty slaves. It is likely that the entire first group and at least some of the second were running integrated plantations. In the 1680s, the integrated plantation had become so common that "a plantation of about 200 acres, equipped with two or three sugar mills and a hundred slaves"—doubtless run as an integrated unit—"was considered the optimum size for efficient production."[36]

Nonetheless, the dispersed system did not disappear with the rise of the integrated plantation because hundreds of small farmers remained on the island. The 1680 census reports not only just over 1,000 small planters with more than ten acres of land and fewer than twenty slaves but also about 1,200 "freemen" with less than ten acres. We initially assumed that those farmers grew minor staples or raised provisions exclusively, but David Eltis has recently shown that many of them must have produced some sugar because they exported small amounts of sugar and its by-products.[37]

The case of Mount Clapham suggests that we should focus on labor supply and demography to help clarify the changing organization of sugar production in Barbados. If our reading of the evidence is correct, in the early stages of the sugar boom Barbadians flirted with the dispersed system of organizing their plantations. Not until the 1650s did it become clear that the integrated plantation would dominate the island. A look again at the size of the island's slave population (see table 9.2) suggests why this was the case. Barbados has roughly 166 square miles or 106,000 acres. At a ratio of two acres per worker, in 1640 island planters owned enough slaves to cultivate only a small portion of the island. By 1680 the slave population was big enough to work much of the island in sugar. While planters did not have enough slaves until the 1650s to make much of a dent in the island's available land, potential tenants were available in abundance. Newly freed indentured servants found it increasingly difficult to acquire land of their own as the export boom drove up land prices. Many were willing to sign on as tenants raising cane for someone else's mill, as we noted earlier. By 1680, the number of slaves had become adequate to cultivate most of the island at the same time that the decrease in the number of indentured servants dried up the supply of tenants. Faced with these new conditions, more and more planters decided to pursue the advantages available in the integrated plantation.

One reason for the persistence of the dispersed system is that it took planters some time to work out and implement what would eventually become the integrated plantation's hallmark and the major source of its productivity advantage over the dispersed system: gang labor with its lock-step discipline and liberal use of the whip to force slaves to work as hard as possible. While gangs were ubiquitous on integrated plantations in the late eighteenth century, there is little evidence that work was so organized in the seventeenth century.[38] The interest of London merchants in developing the Barbadian sugar industry was critical to the emergence of the integrated plantation on that island. Because of the London merchants, Barbadians had the key resource, access to the London capital market, which gave them the means to purchase sufficient slaves to run integrated plantations. Barbadian planters were exceptionally "well friended," to use Richard Ligon's apt phrase.[39] Sugar was a capital-intensive crop, no matter how production was organized. The integrated plantation made it even more capital intensive by greatly increasing the investment required to produce sugar. Besides being closely connected to one of the world's largest capital markets, Barbadians also worked within an empirewide legal system favorable to creditors. Once the integrated plantation and the gang system emerged in Barbados, its productivity advantages over other ways of organizing sugar production were so great that it led to a thorough reorganization of the sugar industry, with major consequences for all involved, especially the African slaves who labored in the fields.

As the optimum size of the plantation grew, so did the demand for investment funds to undertake the enterprise. English merchants played a major role in the Barbadian sugar boom of the late 1640s, providing an infusion of funds that helped make that boom possible. Capital provided by English merchants would continue to play an important role for the remainder of the colonial period, but more in the form of short-term commercial credit or longer-term loans to big

FIGURE 9.2. Amsterdam and London Sugar Prices, 1619–1670

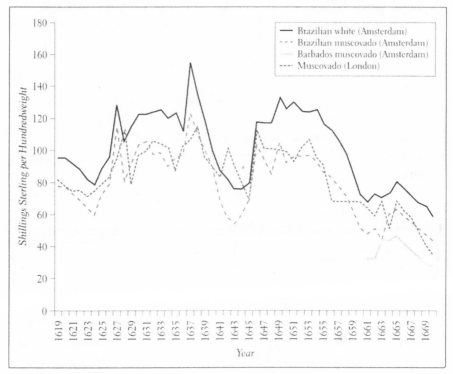

Source: See text and n. 41.

planters than through direct investment.[40] Running a complex sugar-making enterprise at the edge of empire far from its metropolitan center was a difficult and troublesome business. Many of the merchant investors found themselves in tedious squabbles with the relatives and agents they sent out to manage their island affairs. Given the high and rising prices sugar commanded at the beginning of the sugar boom, it was worth the difficulty, and the returns were enough to cover the inefficiencies of absentee management and the headaches it brought, but when prices began to fall after mid-century, merchants began to view their investments in a different light (see figure 9.2).[41] A few held on, still managing their plantations from England into the 1660s. Some moved to the island to run the operations directly, founding in the process some of the great West Indian fortunes of the colonial era. Yet most sold off their holdings, taking the sensible path of limiting their exposure in the sugar industry to financing and marketing.

We could stop here and conclude that the sugar boom was the work of English merchants. Matters were more complex, however. Some of the great

Barbados estates were the result of merchant investment, but many were not. Some of those who eventually acquired "very great and vast estates" "began upon small fortunes" "building according to Barbados custom," with crops that required little capital, land, and labor, reinvesting what they earned in workers, land, and equipment until they were big enough to make sugar.[42] What proportion started small with cotton and tobacco and built great sugar plantations through savings out of current income is impossible to tell, but the diversified export boom of the 1640s brought that dream within reach of many ambitious farmers. It is worth noting that nearly 40 percent of the great planters who dominated the island in 1680 came from families established on the island in the pre-sugar era.[43]

The question of how the sugar boom was financed is illuminated by approaching it through another mystery concerning the sugar boom: why Barbados? Why was that island the site of the first successful effort to produce sugar on a grand scale in the Caribbean? The usual answer, which is partly correct, has turned on security.

Barbadians were fairly safe from attacks by Indians, the Spanish, and pirates, and, despite their occasionally tumultuous politics, fairly safe even from each other. Such security was essential if investors were to risk capital in the amounts necessary to fund the sugar boom. But historians have defined security too narrowly and attended only to its political and military dimensions. Across the two decades after the colony's founding and before sugar took hold, some Barbadians had accumulated estates through farm building and the profitable export of the earlier crops, while at the same time they demonstrated their competence as planters on the Caribbean frontier. Their established properties and established reputations doubtless provided investors in the first West Indian sugar adventure with the confidence, the sense of security, necessary to risk their funds. The importance of this accumulation of property and experience is suggested by the frequency with which the English merchants who invested in the island put their money into established plantations and went partners with resident planters. It is also revealed in a remarkable set of documents in the Barbados Department of Archives, which bring together several of the themes in this essay.

Over two and a half weeks in July and August 1644, Captain George Richardson, master, and Richard Parr, merchant, sold 251 slaves brought to the island in their ship *Marie Bonadventure* of London (see table 9.6). A similar set of documents describing a much smaller sale of twenty-six slaves to eight purchasers from the ship *Mary* of London by John Wadloe in March 1645 describes patterns similar to those discussed here. To begin with the obvious, both ships were

TABLE 9.6. Sale of Slaves from the Ship *Marie Bonadventure* of London, Capt. George Richardson, Master, and Richard Parr, Merchant, at Barbados, 27 July–17 August 1644

A. *Size of Sales*

No. of Slaves per Sale	No. of Purchasers	Total No. of Slaves
1	3 (7.1%)	3 (1.2%)
2	15 (35.7%)	30 (11.9%)
3	2 (4.8%)	6 (2.4%)
4	3 (7.1%)	12 (4.8%)
5	4 (9.5%)	20 (8.0%)
7	4 (9.5%)	28 (11.2%)
8	2 (4.8%)	16 (6.4%)
10	6 (14.3%)	60 (23.9%)
12	1 (2.4%)	12 (4.8%)
30	1 (2.4%)	30 (12.0%)
34	1 (2.4%)	34 (13.5%)
Total	42 (100%)	251

B. *Modes of Dispersal*

Mode	No. of Slaves
Given as gifts	7 (2.8%)
Sold for ready goods	33 (13.1%)
Sold on credit	211 (84.1%)
Total	251 (100%)

C. *Forms of Payment Referenced in Slave Sales*

Form of Payment	No. of Slaves
Tobacco	62 (25.4%)
Cotton	43 (17.6%)
Sugar	42 (17.2%)
Bills of exchange	38 (15.6%)
Indigo	26 (10.7%)
Land	7 (2.9%)
Pork	2 (0.8%)
Unspecified	24 (9.8%)
Total	244

Source: Recopied Deeds Books, 3/1, pp. 691–94, Barbados Department of Archives, Black Rock, St. James, Barbados.

English, not Dutch.[44] Seven slaves from the cargo of the *Marie Bonadventure* were given to Governor Philip Bell, apparently to ensure that the wheels of commerce and law were well greased. Seven more were exchanged for a lot and storehouse in Bridgetown, where Richardson and Parr set up shop. The rest were sold outright. The pace of sales was brisk, testimony to the island's recovering prosperity, and the demand for labor generated by the diversified export boom. Most of the sales were in small lots of one to five slaves, evidence that this was still a society of small planters. Nevertheless, there is also evidence of the transformation just beginning to work its way through island society. The big purchasers were a mixture of established planters and new arrivals; one of the latter, Christopher Thompson, may have been acting for a group of London merchants intent on building sugar plantations.

The way in which the planters paid for the slaves is also instructive.[45] All of the slaves sold were priced in Barbados money currency. What was offered and agreed to in payment mirrored the diversity of the island's exports and the initial use of slaves. Only thirty-three slaves were sold for "ready goods," payment agreed to in various country commodities: indigo, tobacco, cotton, and pork, but not sugar, not in the summer of 1644. A few other purchasers drew bills of exchange on London. The bill of exchange, analogous to the modern check but drawn not against a financial institution but against a commercial firm, provided for the seller the most immediate and secure form of future payment. The buyers of the remaining slaves, 211 in all, 84 percent of the total, arranged for credit from the sellers. Both sides agreed to payment the next April, in 1645, when crops would be in. Some of these promises to pay were secured by mortgages, evidence of how farm building and the export boom helped finance the move toward slavery. Just as those who bought and sold for ready goods agreed to payment in a variety of forms, so too did the planters who arranged for payment in April 1645 stipulate a range of acceptable commodities, but this time they added sugar to that list, indicating that they expected it to be the commodity of the future, although not yet the commodity of the present.

The two largest purchases, by two established planters, William Hilliard and James Drax, epitomize the transformation underway on the island. Hilliard was a long-term resident who had made his fortune in cotton and tobacco and was about to go into sugar in partnership with Samuel Farmer, the son of a Bristol merchant.[46] He bought thirty slaves priced at £660 "sterling" (meaning Barbados money currency), a debt secured by a mortgage against his estate, payable in April 1645 in "Indigoe, Suger, cotton-woole or Tabacco," a list that both encompassed the current state of the Barbadian economy and pointed toward its future. Similarly, the formidable James Drax agreed to the purchase of thirty-four slaves for £726 "sterling," promising for his part to deliver in payment the following spring "Syger or other merchantable commodities." He needed no partner to help secure his credit; there is no mention of a mortgage on his plantation to guarantee his promise to pay. After all, he was James Drax.[47]

It would be useful to conclude this discussion of the sources of funding for the sugar boom by determining which among the sources of support was the most important. Despite the lack of evidence that would permit such an exercise, we do have some thoughts on the issue. First, we suspect that Barbados would have become a major sugar producer with or without the contribution of the London merchants, but, if islanders had to rely on savings out of current income alone, it would have been a much slower process and there would be no talk of "sugar boom," no less a "sugar revolution." While the intervention of the London merchants was not necessary for the rise of the Barbadian sugar industry, it did speed it up, give it that special intensity that set it apart and made it possible for some historians to speak of a "sugar revolution" instead of a more prosaic rise or growth of the sugar industry. The level of Dutch participation remains an enigma.[48]

This essay contends that the idea of a "sugar revolution" is misleading when applied to Barbados. Sugar did not revolutionize Barbados; rather it sped up and intensified a process of experimentation and diversification already underway as resident planters tried first tobacco, then cotton, and then indigo—and, then, ultimately, sugar. The same forces that worked to bring sugar to the fore ultimately altered the sugar industry, too, as the integrated plantation came slowly to replace the dispersed method of organizing sugar production. While the changes connected with the integrated plantation were pervasive—economically, socially, even politically—they did not happen quickly. It took roughly forty years for the integrated plantation to dominate production on the island. Moreover, it took a full century for the emergence of the fully developed integrated plantation on which work was carried out by gang labor. Without minimizing the transforming impact of the institution on the organization of the industry and on the lives of those who made sugar, that transformation was not achieved with the intensity one associates with the concept of a revolution. Perhaps the notion of a "sugar revolution" has outlived its usefulness in Caribbean historiography.

APPENDIX

As table 9.1 indicates, during the first few decades of Barbados's history, the colonists negotiated contracts to buy and sell land and slaves using a variety of currencies. Then, just as now, people could contract for anything in whatever terms all parties agreed. What table 9.1 shows is the shift in the use of the most important of the Barbadian currencies away from cotton and tobacco to sugar, away from pounds Barbados cotton currency and pounds Barbados tobacco currency to pounds Barbados sugar currency. By the late 1640s sugar was the dominant currency, just as it was the dominant commodity grown in the island; it was not to be replaced by Barbados money currency until 1685—at least as the money of account.[49] But what table 9.1 does not indicate necessarily is the form of real money that was used to settle contracts. Then, just as now, both parties could agree, when it came time to pay a debt, to any mode of settlement that was mutually satisfactory. The two parties could contract in Barbados tobacco currency and settle in sugar—or in gold or silver coin—as they pleased. What made this system work was that all of these transactions were monetized, that is, they shared a basic common denominator, Barbados money currency.[50] Each party to all of these contracts was aware of the interrelationship between each of the commodities used as currency and Barbados money currency—so aware, in fact, that they rarely felt the need to spell out that relationship.[51] Therein lies a major problem for historians.

Barbados money currency was denominated in traditional English terms using pounds, shillings, and pence, the same notational system as that used in the mother country. Twelve pence (abbreviated "d.") equaled one shilling ("s."); twenty shillings equaled a pound ("£"). Money sums in both England and its colonies regularly took the form "£12 12s. 6d."; historians of our own era have taken to decimalizing these sums for convenience sake, just as Great Britain itself did for its own currency in February 1971. (Thus historians translate £12 12s. 6d. Barbados currency into £12.62.[52]) Money in England was called "sterling," both the money of account and the real money. Frequently, in order to differentiate among their various currencies, in order to distinguish pounds Barbados money currency from pounds tobacco currency or pounds cotton currency, colonists talked of Barbados pounds sterling—the word "sterling" in these instances meaning simply "money."[53] Colonists elsewhere in the English empire did likewise.[54] This can sometimes breed confusion by causing the reader to think that colonists were talking about English currency when the reference was to money in Barbados—or, say, Ireland or Maryland. Ideally the early colonists would have helped out historians by consistently using the phrase Barbados money currency. Eventually they did, adopting at first "Barbados current money" and, later, more simply "Barbados currency." As a starting point, however, we need to appreciate that transactions between residents of the colony for settlement in the island denominated in pounds, shillings, and pence were expressed in local currency, in Barbados money currency, just as was the case in other colonies—and just as similar transactions are conducted within countries today.[55] Rarely if ever do people carry on business within their own economy using anything other than their own local currency. Barbados "sterling" meant Barbados money currency.

What distinguished Barbados money currency and other colonial currencies from English currency was a difference in value. A modern analogy is, conveniently, the dollar. Many countries call their currency "dollars," but Hong Kong dollars, Canadian dollars, and United States dollars differ considerably in their worth. With Ireland, Virginia, Bermuda, Massachusetts, and Maryland as examples upon which to pattern themselves, Barbadians early established Barbados money currency and set its value both in practice and in law. The most common way for contemporaries and for historians to determine the comparative worth of Barbados money currency and English currency was and is with reference to the values given in both to the same thing, most usually to the Spanish silver coin known to the entire Atlantic World, the *peso do ocho reales*, the piece of eight, the dollar. In English currency, during the seventeenth and eighteenth centuries, the piece of eight was worth four shillings, six pence. In colonial currencies its worth in practice and in law varied from colony to colony and changed over time, but in the colonies it was always valued at more shillings and pence than in England, meaning that, by definition, colonial currencies were worth less than English currency. In other words, it took more colonial currency than it did English currency to buy a piece of eight, or anything else. The ratio between the two currencies set in law and common practice was called the par of exchange.

The earliest Barbadian legislation that set the par of exchange between Barbados money currency and English currency for which we have the text of the law equated £111.11 Barbados money currency with £100 English currency. It did so by establishing the value of the piece of eight at five shillings Barbados money currency. The act "for advancing and raising the value of pieces of eight," dated 14 April 1666, was simply the latest in a series of such laws that extended back in time to before May 1646.[56] While the texts of those earlier acts do not survive and we do not know, therefore, what these earlier acts prescribed, there is some suggestion that the commercial exchange rate was between £105 to £108 Barbados money currency per £100 English currency until the 1666 act was passed.[57] A valuation of the piece of eight at four shillings, ten pence, Barbados currency would have meant a par of about £107.50. It seems reasonable to assume that par was at that level across the entire three decades prior to the mid-1660s. The act of 1666 raised it to a level at which it remained until the middle of the next century. Par was only a benchmark, however. The commercial rate of exchange fluctuated in day-to-day trading depending on a variety of factors, just as it does in foreign exchange transactions in the twenty-first century. While we know that such variations were in play from the start, our first evidence of the commercial rate of exchange dates only from the early 1680s when adequate data are finally available.[58]

Whatever the gaps in historians' knowledge and understanding, Barbadians were well aware of all this and behaved accordingly, buying and selling what they would and settling their accounts when called upon to do so. What they lacked was enough coin. Sterling in England, as a money of account, was neatly paralleled by a real money in that the Royal Mint produced coins called pennies and shillings as well as half-crowns and crowns and others. The English colonists in the New World had no coin they could call their own. What they did about their lack of a real coinage was both inventive and reasonably successful. For one, they used other nations' coins. The gold and silver coins of Spain and Portugal, minted in their metropolitan mints and in their colonial mints, circulated in the English colonies. Prime among those coins was, as mentioned, the silver piece of eight. Everyone knew how much it was worth in Barbadian money currency, so much so that the piece of eight and its bits usually passed by tale (that is, by simple count) rather than by weight.[59]

What the Barbadians also did, taking a lesson from their fellows on the North American continent, was to monetize the major commodities they produced. In practice this meant that they assigned a value in Barbados money currency to units of, initially, tobacco, then cotton, and finally sugar. These values had the power of law and the force of common custom behind them.[60] Whatever the market price for tobacco was in open commerce, as a currency tobacco had a set value for contract purposes.[61] The most important thing to realize about this system is that contracts originally established with reference to commodity money could, and were, settled in whatever terms the two parties could agree on at time of settlement, then just as today. A contract involving 6,000 pounds in Barbados tobacco currency when it was negotiated—which, at six pence currency a pound, equaled £150 Barbados money currency—could be settled by the payment of that £150 in any other legal tender in play at the time of settlement. As a very simple example, one can settle a long-standing obligation expressed in United States dollars (the national money of account) using any form of United States currency, whatever its kind or denomination, however newly minted or newly printed in may be. So well did commodity money work that Marylanders and Virginians continued to employ tobacco in this way until the second decade of the nineteenth century.[62]

So well did this system work for Barbadians in the first three decades of their existence that in no instance known to us did they bother to state it in explicit terms. Thus historians are left to infer what happened from scraps of evidence that are sometimes very difficult to interpret. Critical to their and our understanding of this system is a knowledge of the comparative values of the various commodity monies and Barbados money currency. It is here that the record fails us almost completely.[63]

Sugar is the easiest to track because it was the last to be established as commodity money and the one that was in place for the longest time. The Barbados legislature passed several acts in the early 1650s that accessed fees in sugar as a commodity money. In one act it rated sugar "in money" at three pence Barbados money currency per pound or twenty-five shillings per one hundred pounds, that is, per hundredweight.[64] Very limited evidence from other sources suggests that this was half the rate that applied before those acts, until 1652.[65] It was reduced again by the end of the 1650s, to two pence per pound,[66] and dropped another half-penny early in the 1660s.[67] Thus, as an operating assumption for this essay, we rate the currency value of sugar at six pence Barbados money currency from the 1630s through 1651; at three pence per pound from 1652 through 1659; at two pence per pound through 1661; and at one and one-half pence per pound (a penny, ha'penny) after that (see table 9.7).

We have even less evidence of the currency valuation of the predecessors and alternatives to sugar as a commodity money, cotton and tobacco—and none at all for indigo or pork. There is some indication that the first two were considered of equal value as monies of account through at least the end of the 1630s.[68] Business and legal records concerning commercial practices in Barbados in 1636 and 1637 deposited among the proceedings of the English High Court of Admiralty put the value of both at six pence a pound.[69] A set of valuations from twelve years later rated one pound of tobacco at two pounds of cotton and at three pounds of sugar and their price in Barbados money currency at two pence, four pence, and six pence each, respectively.[70] That suggests that the value of cotton as a commodity money had been reduced sometime between 1636 and 1648, but we have no indication of just when that happened. A 1669 law that set the fees for cases heard in the Court of Common Pleas indicated another drop in the currency value of cotton sometime between 1648 and that year by equating two pounds of sugar with one pound of cotton.[71] Again for the purposes of this essay, we accept a currency valuation for cotton of six pence Barbados money currency per pound for the 1630s, four pence per pound for the 1640s and

TABLE 9.7. Value of Barbados Commodity Currency in Terms of Barbados Money Currency, from the 1630s

Years	Cotton	Tobacco	Sugar
1630s	6d. per pound	6d. per pound	6d. per pound
1640s	4d. per pound	2d. per pound	6d. per pound
1650–1651	4d. per pound		6d. per pound
1652–1659	4d. per pound		3d. per pound
1660–1661	3d. per pound		2d. per pound
After 1662			1.5d. per pound

Source: Appendix.

1650s, and three pence per pound in the 1660s (see table 9.7).[72] However much the valuation of cotton as a commodity money decreased, tobacco fell even more. Valued at half as much as cotton in 1648, it had been at parity with cotton as late as 1640. For the purposes of this essay we accept a currency valuation for tobacco of six pence Barbados money currency for the 1630s and two pence per pound for the 1640s (see table 9.7).[73]

The records of the sale of slaves from the ship *Marie Bonadventure* of London in 1644 allow for a test of these several assumptions.[74] Some of the enslaved Africans had their sale prices recorded in Barbados money currency at an average of £22 Barbados "sterling" each. Others sold for £21.35. And still others had their sale prices set down in different terms. For those people whose sale price was recorded as 2,500 pounds of tobacco, we can now conclude that their sale price was nearly £21 Barbados money currency. Those who were sold for 1,000 pounds of cotton plus 1,000 of tobacco sold in currency terms for £25 each. Given that different lots of slaves sold for different prices normally, we find perhaps undue satisfaction in our procedures from the comfortable proximity of all of those numbers. On the basis of these data, one could suggest that the average price for the sale of slaves off the ship *Marie Bonadventure* was £22.30 Barbados money currency or, at a par of exchange for this period of £107.50 Barbados money currency per £100 sterling, the equivalent of £20.75 sterling.[75] With this one can compare Ligon's statement that "the best man Negroe" slave sold for £30 "sterling" (that is, Barbados money currency) in the late 1640s.[76] The difference in price may have reflected a rise over the five years or, more likely, the usual differential of about one-third between the price of prime field hands and the price of average newly arrived—and as yet unseasoned—enslaved Africans.[77]

Abbreviations

BDA	Barbados Department of Archives (Black Rock, St. James, Barbados)
BL	British Library (London)
CO	Colonial Office Records
GD	Gifts and Deposits
GL	Guildhall Library (London)
HCA	High Court of Admiralty Records
NAS	National Archives of Scotland (Edinburgh)
PRO/TNA	Public Record Office/The National Archives (London)
RB	Recopied Deeds Book

1. Barry W. Higman has summarized the literature on the sugar revolution in "The Sugar Revolution," *Economic History Review*, 2d series, 53 (May 2000): 213–38, and in "The Making of the Sugar Revolution," in *In the Shadow of the Plantation: Caribbean History and Legacy—In Honour of Professor Emeritus Woodville K. Marshall*, ed. Alvin O. Thompson (Kingston, Jamaica: Ian Randle, [2002]), pp. 40–71. For two excellent summary discussions of Barbados and the other English Caribbean colonies in the seventeenth century, see Hilary McD. Beckles, "The 'Hub of Empire': The Caribbean and Britain in the Seventeenth Century," in *The Origins of Empire: British Overseas Enterprise to the Close of the Seventeenth Century*, ed. Nicholas [P.] Canny, vol. 1 of *The Oxford History of the British Empire*, ed. William Roger Louis (Oxford: Oxford University Press, 1998), pp. 218–45; and Nuala [B.] Zahedieh, "Overseas Expansion and Trade in the Seventeenth Century," in ibid., pp. 384–422. We are grateful to Prof. Jerome S. Handler of the Virginia Foundation for the Humanities for a critical reading of this work. Note that, except where indicated, all dates reported herein are as in their original form, meaning they conform to the old-style Julian calendar, which the English used until 1752. The two exceptions to this rule also apply. The new year is taken as beginning with 1 January (not 25 March) and dates from 1 January through 24 March specify the year with reference to both rubrics, e.g., 2 February 1656/57.

2. The most recent version of this standard account is Yda Schreuder, "The Influence of the Dutch Colonial Trade on Barbados in the Seventeenth Century," *Journal of the Barbados Museum and Historical Society* 48 (2002): 43–63.

3. Letter from Vines, at Barbados, to Winthrop, at Boston, 19 July 1647, *Winthrop Papers*, [ed. Worthington Chauncey Ford et al.], in progress (Boston: Massachusetts Historical Society, 1929 to date), 5:172. Vines (fl. 1616–51), a long-time resident of Maine, had only recently emigrated to Barbados. For him, see *Genealogical Dictionary of Maine and New Hampshire*, comp. Sybil Noyes, Charles Thornton Libby, and Walter Goodwin Davis, 5 vols. (Portland, Maine: Southworth-Anthoensen Press, 1928–39), 5:705–6. The New Englanders to whom he wrote recognized and sought to realize the opportunities for themselves in these developments, to sell Barbadians "foode at very deare rates"—and other things, too. John J. McCusker and Russell R. Menard, *The Economy of British America, 1607–1789*, [2d ed.] (Chapel Hill: Published for the Institute of Early American History and Culture, Williams-

burg, Virginia, by the University of North Carolina Press, 1991), passim. See also Bernard Bailyn, *The New England Merchants in the Seventeenth Century* (Cambridge, Mass.: Harvard University Press, [1955]). For an estimate of 40–50 percent as the level of profit for Barbadian sugar planters around 1650, see J[ohn] R. Ward, "The Profitability of Sugar Planting in the British West Indies, 1650–1834," *Economic History Review*, 2d series, 31 (May 1978): 208.

4. We consider this essay an important contribution to the larger discussion among students of the Western Hemisphere sugar industry who are concerned about the interrelationships between such things as the nature of the labor force and technology as sugar production changed over time. One issue of considerable moment in the controversy is the long-standing assumption that planters were a conservative, restraining element, and adverse to change. Formative contributions to the debate have been the paper by Peter Boomgaard and Gert J. Oostindie, "Changing Sugar Technology and the Labour Nexus: The Caribbean, 1750–1900," *Nieuwe West-Indische Gids*, 63 (nos. 1–2, 1989): 3–22, and the response by Michael Craton, "Commentary: The Search for a Unified Field Theory," in ibid., pp. 135–42. More recently, there is a useful summary of the arguments by Karen S. Dhanda, "Labor and Place in Barbados, Jamaica, and Trinidad: A Search for a Comparative Unified Field Theory Revisited," *Nieuwe West-Indische Gids* 75 (nos. 3–4, 2001): 229–56. Compare the more general discussion of these issues in John J. McCusker, "The Economy of the British West Indies, 1763–1790: Growth, Stagnation, or Decline?," in *Essays in the Economic History of the Atlantic World* (London and New York: Routledge, 1997), pp. 324–26.

5. Almost from the beginning, Barbadians grew some sugarcane and used the sugarcane juice as the base material for the distillation of rum. See John J. McCusker, *Rum and the American Revolution: The Rum Trade and the Balance of Payment of the Thirteen Continental Colonies, 1650–1775*, 2 vols. (New York, Garland Publishing, 1989), 1:198–220. Some muscovado sugar was processed and sold locally in the 1630s. See n. 61, below. Minor quantities of manufactured sugar were exported. Imported from Barbados into London in 1634 was roughly 14 hundredweight (cwt.) of sugar valued for customs duty purposes at £41 and from Virginia roughly 178 cwt. of sugar valued at £534, the latter identified as "Muscovadoes and Barbadoes sugar." Port of London, Port Book, Surveyor General of Tunnage and Poundage Overseas, Imports by Denizens, Christmas 1633–Christmas 1634, Records of the Exchequer, King's Remembrancer, E 190/38/5, PRO/TNA, as compiled by A[nnie] M. Millard, "The Import Trade of London, 1600–1640," 3 vols. (Ph.D. diss., University of London, 1956), pp. 308–9. See also table 35 in Millard, "Analyses of Port Books Recording Merchandises Imported into the Port of London by English and Alien and Denizen Merchants for Certain Years between 1588 and 1640" [1960], unpublished manuscript on deposit in the Library, PRO/TNA.

6. Table 9.1 shows result of counting all sales in RB 3/1–3/3, BDA. We found more than 2,500 sales in which some mode of settlement was mentioned. The table reports the percentage frequency for each mode.

Emily Mechner traced the forms of money mentioned in nearly 1,200 Barbadian deeds executed between 1648 and 1674, with results almost identical to those reported here. See Mechner, "Pirates and Planters: Trade and Development in the Caribbean, 1492–1680"

(Ph.D. diss., Harvard University, 1999), p. 57. By the decade 1664–74 over a quarter of the deeds were expressed in Barbados "sterling" (that is, Barbados money currency). For Barbados currency during this period, see Appendix.

7. See Appendix. See also John J. McCusker, *Money and Exchange in Europe and America, 1600–1775: A Handbook*, [2d ed. rev.] (Chapel Hill: Published for the Institute of Early American History and Culture, Williamsburg, Virginia, by the University of North Carolina Press, 1992), passim. The "second edition" of *Money and Exchange in Europe and America* referred to here was issued in 1992. Although the publication information in the book failed to say so, there were many and some rather significant corrections in that reissuance of the original version. We call attention to it only to alert the interested reader to check terms and data before using the earlier version.

8. Writing at the turn of the decade, Thomas Peake made the point: "Their chief Trade is Tobacco, and a kind of course Sugar, which we call Barbados-Sugar, and will not keep long; not that the Countrie is unapt for better, but, as 'tis rather supposed, because the Planters want either skill or stock to improve things to the best." [Thomas Peake], *America, Or An Exact Description of the West-Indies: More Especially of Those Provinces which are under the Dominion of the King of Spain* (London: Edw[ard] Dod, 1655), p. 471. By "stock" Peake meant money capital, just as by "skill" he meant human capital.

9. The best estimates of slave imports to Barbados are those in Philip D. Curtin, *The Atlantic Slave Trade: A Census* (Madison: University of Wisconsin Press, 1969), pp. 54–55, and David Eltis, "The Volume and Structure of the Transatlantic Slave Trade: A Reassessment," *William and Mary Quarterly*, 3rd series, 58 (January 2001): 45. See also McCusker, *Rum and the American Revolution*, 2:648–51. We accept Eltis's estimate that in excess of 24,000 slaves were imported into Barbados beween 1627 and 1650. Philip Morgan, in private communications in May and June 2003, informed us of his estimates for reexported slaves, which reduced the total to a net figure through 1650 of about 22,000 people. We are grateful for his help. See Philip D. Morgan, *The West Indies, ca. 1500–1800* (forthcoming).

10. See Richard S. Dunn, *Sugar and Slaves: The Rise of the Planter Class in the English West Indies, 1624–1713* (Chapel Hill: Published for the Institute of Early American History and Culture, Williamsburg, Virginia, by the University of North Carolina Press, 1972), p. 55.

11. Letter from Downing, on board ship, to Winthrop, *Winthrop Papers*, [ed. Ford et al.], 5:43, 44. Downing (1623–84), a nephew of John Winthrop Sr., was reporting on a visit to Barbados made during an extended voyage to the West Indies on board a ship in which he served as chaplain. For Downing, see John Beresford, *The Godfather of Downing Street, Sir George Downing, 1623–1684: An Essay in Biography* (London: R. Cobden-Sanderson; Boston: Houghton Mifflin, 1925).

12. See nn. 75–76, below.

13. There were some enslaved Africans—and Arawak Indians—on the island in 1630 and 1640 but—at a guess—probably fewer than 100 in the first year and fewer than 500 in the second year. See, for example, the letter from Henry Winthrop, at Barbados, to Emmanuel Downing, at London, 22 August 1627, and Winthrop to John Winthrop, at London, 15 October 1627, *Winthrop Papers*, [ed. Ford et al.], 1:356–57, 361–62. One indication of the extent of the enslavement of native American Indians is the fear expressed to John Win-

throp Jr. by the Pequot Indians of Connecticut, because the English there "thretne [threaten] to send them away to the Sugar Country." The date was 1650. Ibid., 6:63. We are grateful to Joshua Micah Marshall for this reference.

14. These findings directly contradict Eric Eustace Williams, *Capitalism & Slavery* (Chapel Hill: University of North Carolina Press, 1944), p. 11: "No sugar, no negroes." Some who share our conclusion include Carl Bridenbaugh and Roberta Bridenbaugh, *No Peace Beyond the Line: The English in the Caribbean, 1624–1690* (New York: Oxford University Press, 1972), p. 33; and, importantly, William A. Green, "Supply versus Demand in the Barbadian Sugar Revolution," *Journal of Interdisciplinary History* 17 (Winter 1988): 403–18.

15. See the sources summarized in table 3.4 in McCusker and Menard, *Economy of British America*, p. 62, which shows that both the English economy and the North American economies, having slumped in the late 1630s and early 1640s, had begun a recovery by the mid-1640s. The contraction of 1638–44 is referred to in that work as the "first depression in American history" (ibid., p. 65), but see the argument by Menard that the severe economic slump in Virginia from the late 1620s through 1633 is a better candidate for that dubious distinction—and, not incidentally, of considerable relevance to the early tobacco economy of Barbados. Russell R. Menard, "A Note on Chesapeake Tobacco Prices, 1618–1660," *Virginia Magazine of History and Biography* 84 (October 1976): 402.

16. See McCusker, *Rum and the American Revolution*, 1:198–99, and the sources cited there.

17. Memorandum endorsed "toucheing Barbados," by Harley, apparently enclosed in his letter to his brother Sir Edward Harley, dated Barbados, September 1663, fols. 1r–1v, Papers of Sir Robert Harley, vol. 27, packet no. 20, Portland MSS., Loan 29/27, BL. Compare "An Accompt of the English Suger plantations" in the same packet. For the covering letter, see Great Britain, Historical Manuscripts Commission, *The Manuscripts of the Duke of Portland Preserved at Welbeck Abbey*, [ed. Francis H. Blackburne Daniell et al.], 10 vols. (London: H. M. Stationery Office, 1891–1931), 3:277–78. For Harley (1626–73), see Basil Duke Henning, *The History of Parliament: The House of Commons, 1660–1690*, 3 vols. (London: Published for the History of Parliament Trust by Secker & Warburg, 1983), 2:497–98.

18. Compare John Scott, "The Description of Barbados," ca. 1667, Sloane MS 3662, fols. 60r–59v, BL; [Dalby Thomas], *An Historical Account of the Rise and Growth of the West-India Collonies, And of the Great Advantages they are to England, in respect to Trade* (London: Jo[seph] Hindmarsh, 1690), pp. 36–37; Bridenbaugh and Bridenbaugh, *No Peace Beyond the Line*, pp. 63–68, 76–97. John Scott's reliability as an observer and reporter of events, once held in some doubt, has been powerfully reestablished through the efforts of several twentieth-century authors culminating in the biography by Lilian T. Mowrer, *The Indomitable John Scott: Citizen of Long Island, 1632–1704* (New York: Farrar, Straus and Cudahy, 1960), esp. pp. 171–78, 408–9.

19. The case for Dutch mercantile involvement with the English colonies generally was well made over a century ago by George Louis Beer, *The Origins of the British Colonial System, 1578–1660* (New York: The Macmillan Company, 1908), pp. 352–59. Compare John J. McCusker, *Mercantilism and the Economic History of the Early Modern Atlantic World* (Cambridge: Cambridge University Press, forthcoming).

The Dutch presence in Barbadian commerce was already important enough by 1634 that Governor Henry Hawley "instituted an anchorage duty of one pound, sterling [that is, Barbados money currency], on every foreign ship coming to Barbados." It was estimated in the late 1630s to have yielded as much as £60 per year. J. Harry Bennett [Jr.], "Peter Hay, Proprietary Agent in Barbados, 1636–41," *Jamaican Historical Review* 5 (November 1965): 17. As in other English colonies during this period, the pervasiveness of the Dutch commercial presence extended to the use of Dutch currency. See the contract for the hiring of a ship's master in April 1649 discussed in Larry D. Gragg, "Shipmasters in Early Barbados," *Mariner's Mirror: The Journal of the Society for Nautical Research* 77 (May 1991): 108, citing RB 3/1, pp. 535–37, BDA.

Perhaps the most egregious example of Dutch commercial impudence was the venture organized and conducted by none other than Peter Stuyvesant himself, director general of New Netherland, who arrived at Bridgetown in mid-January 1655 with three vessels from New Amsterdam loaded with goods to trade. There were already five Dutch ships in harbor when he got there. English Admiral William Penn, on his way to the conquest of Jamaica, arrived soon thereafter and challenged the Dutch presence as a breech of the Navigation Acts. The subsequent trial ended in a complete victory for the Dutch when the Barbadian jury "found for the strangers against parliament and [the] state." Having finished their business, Stuyvesant and company sailed for Curaçao on 21 March 1655. Letter from Edward Winslow, at Barbados, to John Thurloe, at London, 16 March 1654/55, in *A Collection of the State Papers of John Thurloe, Esq.; Secretary, First, to the Council of State, and Afterwards to the Two Protectors, Oliver and Richard Cromwell . . . Containing Authentic Memorials of the English Affairs from the Year 1638, to the Restoration of King Charles II*, ed. Thomas Birch, 7 vols. (London: Executor of Fletcher Gyles, 1742), 3:249–52. Edward Winslow (1595–1655), one of the founding leaders of Plymouth Colony, returned to England in 1646 and was afterward employed by Oliver Cromwell on a variety of missions, including this appointment as one of the five commissioners to effect Cromwell's "Western Design." Compare Vincent T. Harlow, *History of Barbados, 1625–1685* (Oxford: Clarendon Press, 1926), pp. 85–87. At Curaçao Stuyvesant installed Matthias Beck as his vice director. Within the year Daniel Searle, governor of Barbados, wrote Beck to propose "vrij handelinge" (free trade) between the two islands. Meeting of the Council of Curaçao, 21 February 1656 (new style), *Publikaties en andere wetten alsmede de oudste resoluties betrekking hebbende op Curaçao, Aruba, Bonaire*, ed. J[acobus] Th. de Smidt, T. van der Lee, and J[acob] A. Schiltkamp, West Indisch Plakaatboek, no. 2, Werken der Stichting tot Uitgaaf der Bronnen van het Oud-Vaderlandse Recht, no. 2, 2 vols. (Amsterdam: S. Emmering, 1978), 1:59–61.

20. Even though there are no known instances of Dutch individuals buying Barbados plantations in the 1640s, and only a few instances of people with Dutch names entering the (extant) public records of Barbados, there were certainly some in the colony. We suspect that most of the "small but very influential" number of Dutch (and German?) people whom the Bridenbaughs, *No Peace Beyond the Line*, p. 16, place on Barbados in 1640 were residents of Bridgetown. In 1649 one observer reported "a great company" of Dutchmen dying on Barbados during an epidemic. Letter from Richard Vines, at Barbados, to John Winthrop, at Boston, 29 April 1649, *Winthrop Papers*, [ed. Ford et al.], 5:219–20. Still later two Dutch

owners of plantations in St. George's Parish represented it in the Barbados House of Assembly, Benjamin Heyzar in 1655 and Constant Sylvester in 1661. Ronald G. Hughes, "Barbadian Sugar Plantations, 1640 to 1846" (seminar paper, Department of History, University of the West Indies, Cave Hill, Barbados, 1978), appendix 7, [p. 1]. We used the copy of this paper in the collections of the Library, University of the West Indies, Cave Hill, Barbados. In these instances and in others, it is also possible, even likely, that some of those whom the English residents of Barbados labeled "Dutch" were, in fact, German—or may even have been Dutch-speaking (or German-speaking) Sephardic Jews from Amsterdam, directly or by way of Brazil. See n. 48, below.

The Dutch were similarly active elsewhere in the Caribbean and on the Continent in ways similar to their role on Barbados. At least one Amsterdam merchant owned a plantation on Antigua. Jean le Roux, who died in the late 1630s or early 1640s, transferred ownership of his plantation to his widow, Bertramine Bourse, who owned it until as late as 1649. *Bronnen tot de geschiedenis van het bedrijfsleven en het gildewezen van Amsterdam*, ed. J[ohannes] G. van Dillen, Rijks Geschiedkundige Publicatiën, Grote Serie, vols. 69, 78, 144, 3 vols. (The Hague: Martinus Nijhoff, 1928–74), 3:99, 388, 541. One Jacob Clas, a Dutch refuge from Brazil, founded a plantation on Guadeloupe in 1654. Gérard Lafleur, "La distillerie Bologne [de Basse-Terre]: Du sucre au rhum," *Bulletin de la Société d'Histoire de la Guadeloupe* 103 (no. 1, 1995): 75–81, 107–10. Compare Lafleur, *Les protestants aux Antilles française du Vent sous l'Ancien Régime*, Bulletin de la Société d'Histoire de Guadeloupe, nos. 71–74 (Basse-Terre, Guadeloupe: Société d'Histoire de la Guadeloupe, [1988]), pp. 45–57 and passim. Several Dutch merchants also settled in the Chesapeake during this same period. See April L. Hatfield, "Dutch Merchants in the Seventeenth-Century Chesapeake," paper presented at the conference on "The Emergence of the Atlantic Economy," College of Charleston, Charleston, South Carolina, 14 October 1999. Compare Hatfield, *Atlantic Virginia: Intercolonial Relations in the Seventeenth Century* (Philadelphia: University of Pennsylvania Press, [2004]).

For an impressively insightful perspective on the involvement in the English colonies by Europeans generally, see Claudia Schnurmann, *Atlantische Welten: Engländer und Niederländer in amerikanisch-atlantischen Raum, 1648–1713*, Wirtschafts- und Sozialhistorisch Studien, vol. 9 (Cologne and Vienna: Böhlau Verlag, 1998). See especially her investigation of "Barbados im niederländischen Handelsnetz, 1640–1655," pp. 179–91. Coordination between Amsterdam and Hamburg merchants was significant at this time; many Dutch firms operated in both cities. See n. 48, below. In the seven years 1644–50, an average of three ships per years sailed between Hamburg and Barbados. Martin Reißmann, *Die hamburgische Kaufmannschaft des 17. Jahrhunderts in sozialgeschichtlicher Sicht*, Beiträge zur Geschichte Hamburgs, Band 4 (Hamburg: Hans Christians Verlag, 1975), pp. 75, 371. The peak year was 1647 when six vessels cleared inward and outward. Compare G[eorge] D. Ramsay, "Hamburg and the English Revolution," in *Wirtschaftskräfte und Wirtschaftswege: Festschrift für Hermann Kellenbenz*, ed. Jürgen Schneider, Beiträge zur Wirtschaftsgeschichte, 5 vols. ([Stuttgart]: Franz Steiner, 1978), 2:433. At least one German merchant's will was probated at Barbados during the 1640s: Conrad Stryhold/Conrade Stridehall, of Bridgetown. Will dated 31 January 1648–49, RB 3/1, p. 124, BDA, as summarized in *Barbados Records*, comp. and ed.

Joanne Mcree Sanders, 5 vols. (Houston: Sanders Historical Publications, 1979–82), I: Wills and Administrations, 1639–1680, p. 342.

There is much of relevance to all of this in the account by Heinrich von Uchteritz, a minor German nobleman and soldier of fortune who, captured by Cromwell's armies at the Battle of Worcester in 1651, was subsequently transported to Barbados and sold into servitude. See Uchteritz, *Kurtze Reise Beschreibung . . . Worinnen vermeldet, was er auf derselben für Unglück und Glück gehabt, sonderlich wie er gefangen nach West Indien geführet, zur Sclaverey verkaufft, und auff der Insel Barbados durch den Namen seines Herrn Vettern Johan Christoff von Uchteritz, vff Medewitz und Spansdorff erbgesessen Cammer Juncker auff Gottorff, wunderlich errettet und erlöset worden* (Schleswig: Johan Holwein, 1666); "A German Indentured Servant in Barbados in 1652: The Account of Heinrich von Uchteritz," ed. and trans. Alexander Gunkel and Jerome S. Handler, *Journal of the Barbados Museum and Historical Society* 33 (May 1970): 91–100. For the origins of Uchteritz's dilemma, see Malcolm Atkin, *Cromwell's Crowning Mercy: The Battle of Worcester, 1651* ([Thrupp, Stroud, Gloucestershire, England]: Sutton Publishing, [1998]), esp. pp. 123–28. Uchteritz described the arrival at Barbados in 1652 of "several ships . . . from Germany with merchandise to be traded, according to custom." He subsequently tried to arrange passage home in one of the German ships but the German merchants had to remain "in Barbados because of their business." Instead they contracted passage for him in one of "several [Dutch?] ships [that] had come from Brazil" and were on their way to Amsterdam. Uchteritz, *Kurtze Reise*, pp. 9, 15.

21. For Noell (ca. 1620–65), see Gerald E. Aylmer, *The State's Servants: The Civil Service of the English Republic, 1649–1660* (London: Routledge & Kegan Paul, [1973]), pp. 250–51. See also Richard Pares, *Merchants and Planters*, Economic History Review, supplement no. 4 (Cambridge: Published for the Economic History Review at the University Press, 1960), pp. 59–60. For biographical details about other merchant investors in Barbados, see Maurice Ashley, *Financial and Commercial Policy under the Cromwellian Protectorate* (Oxford: Oxford University Press, 1934); Robert Brenner, *Merchants and Revolution: Commercial Change, Political Conflict, and London's Overseas Traders, 1550–1653* (Princeton: Princeton University Press, [1993]); and Richard B. Sheridan, *Sugar and Slavery: An Economic History of the British West Indies, 1623–1775* (St. Lawrence, Barbados: Caribbean University Press; Baltimore: Johns Hopkins University Press, [1974]). For Thompson, see also John R. Pagan, "Growth of the Tobacco Trade between London and Virginia, 1614–40," *Guildhall Studies in London History* 3 (April 1979): 260–63.

22. The years 1646 and 1647 witnessed the culmination of several related developments. (Compare tables 9.1, 9.3, 9.4, and 9.5. See also figures 9.1 and 9.2.) The Portuguese revolt against the Dutch in Brazil began in 1645; the quantity of sugar exported from Pernambuco, which had been slowly declining from a high in 1641, dropped in 1646 to one-third of what it had been the year before, one-sixth of what it had been five years earlier. Hermann [J. E.] Wätjen, *Das holländische Kolonialreich in Brasilien: Ein Kapitel aus der Kolonialgeschichte des 17. Jahrhunderts* (The Hague: Martinus Nijhoff; Gotha, Germany: Friedrich Andreas Perthes, 1921), pp. 316–23. Compare McCusker, *Rum and the American Revolution*, 1:106–11. The year 1646 saw European sugar prices rise to levels they had not attained since the late

1630s (see figure 9.2). The year 1647 is memorable, also, for increased foreign shipping activity at Barbados. See n. 20, above.

23. On the subject of land sales, compare F[rank] C. Innes, "The Pre-Sugar Era of European Settlement in Barbados," *Journal of Caribbean History* 1 (November 1970): 1–22, and P[eter] F. Campbell, "Aspects of Barbados Land Tenure, 1627–1663," *Journal of the Barbados Museum and Historical Society* 37 (no. 2, 1984): 112–58.

24. We assumed that the mean value of all thirty transfers was the same as the mean value as those transfers that recorded a price. As one gauge of the size of this investment, £150,000 Barbados money currency, the equivalent to about £140,000 sterling, was equal to 10 percent of the average annual total revenue of the English government in the late 1640s and the 1650s (1643–59). James Scott Wheeler, "English Army Finance and Logistics, 1642–1660" (Ph.D. diss., University of California, Berkeley, 1980), p. 174. For the rate of exchange between Barbados money currency and sterling, see Appendix.

25. Boogaart and Emmer, "The Dutch Participation in the Atlantic Slave Trade, 1596–1650," in *The Uncommon Market: Essays in the Economic History of the Atlantic Slave Trade*, ed. Henry A. Gemery and Jan S. Hogendorn (New York: Academic Press, 1979), pp. 353–75. On this subject, see also Johannes Menne Postma, *The Dutch in the Atlantic Slave Trade, 1600–1815* (Cambridge: Cambridge University Press, 1990).

26. See Larry D. Gragg, " 'To Procure Negroes': The English Slave Trade to Barbados, 1627–60," *Slavery and Abolition* 16 (April 1995): 65–84, who develops these ideas in detail, offers evidence about who was involved in the trade, and demonstrates how it worked out in practice.

27. On the organization of sugar production before the Barbadian sugar boom, see the other essays in this volume, and the text and references in Philip D. Curtin, *The Rise and Fall of the Plantation Complex: Essays in Atlantic History* (Cambridge: Cambridge University Press, 1990), pp. 3–69. As Stuart B. Schwartz, *Sugar Plantations in the Formation of Brazilian Society: Bahia, 1550–1835*, Cambridge Latin American Studies, vol. 52 (Cambridge: Cambridge University Press, 1985), p. 10, has told us: this "form of organization . . . was a reasonable social and economic response to a situation in which land was relatively cheap but capital was scarce." He was speaking of sugar production on both the Atlantic Islands and in northeastern Brazil.

28. For a preliminary version of the argument in this and the next paragraph, which provides more detailed documentation, see Russell R. Menard, "Law, Credit, the Supply of Labour, and the Organization of Sugar Production in the Colonial Greater Caribbean: A Comparison of Brazil and Barbados in the Seventeenth Century," in *The Early Modern Atlantic Economy*, ed. John J. McCusker and Kenneth Morgan (Cambridge: Cambridge University Press, [2000]), pp. 154–62. See also Menard, "Toward African Slavery in Barbados: The Origins of a West Indian Plantation Regime," in *Lois Green Carr—The Chesapeake and Beyond—A Celebration: A Collection of Discussion Papers Presented at a Conference, May 22–23, 1992, University of Maryland, University College Conference Center, College Park, Maryland*, ed. by John J. McCusker et al. (Crownsville: Maryland Historical and Cultural Publications, 1992), pp. 19–27.

29. The best series of Barbadian land prices that we have for this period appears in Hilary McD. Beckles, *White Servitude and Black Slavery in Barbados, 1627–1715* (Knoxville: University of Tennessee Press, [1989]), p. 156. See table 9.5 and figure 9.1. Compare his comments on p. 155.

Scattered references from other sources suggest that these prices applied broadly. For example, " . . . there is not an Acre of land in the whole Island to bee purchased under five pound sterling" (that is, £5 Barbados money currency). Letter from William Hay (otherwise William Powrey), at Barbados, to Archibald Hay, at London, 8 April 1646, Papers Relating to the Island of Barbados, 1636–48, Hay of Haystoun Muniments, 1507–1911, GD 34, NAS, and as quoted in part in J. H[arry] Bennett [Jr.], "The English Caribbees in the Period of the Civil War, 1642–1646," *William and Mary Quarterly*, 3rd series, 24 (July 1967): 359–77. Hay/Powrey was a member of the Barbados Council.

Prices continued to rise for a few years and then flattened for the next couple of decades. In 1657 a major plantation sold for £8.00 per acre while ten years later another sold for £8.33 per acre. See n. 32, below. On 9 October 1667, John Peers sold Valentine Hawtaine a thirty-acre estate in Christ Church Parish for 40,000 pounds Barbados sugar currency. George H. Hawtayne, "Records of Old Barbados," *Timehri: The Journal of the Royal Agricultural and Commercial Society of British Guiana*, new series, 10 (June 1896): 101. With reference to table 9.7, we can determine that, at 1.5d. Barbados money currency per pound of sugar, it sold for £8.33 Barbados money currency per acre. In 1675 the same plantation brought £250 Barbados money currency, precisely the same amount per acre (ibid.).

30. Alfred D. Chandler [Jr.], "The Expansion of Barbados," *Journal of the Barbados Museum and Historical Society* 13 (May–August 1946): 106–30.

31. Details on Mount Clapham are provided in Menard, "Law, Credit, the Supply of Labour, and the Organization of Sugar Production." See also Beckles, *White Servitude and Black Slavery in Barbados*, p. 157. Their source is the record of the sale of the plantation by Noell to Governor Daniel Searle, 15 January 1653/54, RB 3/3, 109–13, BDA.

32. Three years later, in 1657, Mount Clapham sold for 328,000 pounds Barbados sugar currency. Hughes, "Barbadian Sugar Plantations," appendix 11, [p. 1]. With reference to table 9.7, we can determine that, at 3d. Barbados money currency per pound of sugar, it sold for £8.00 Barbados money currency per acre.

33. Dunn, *Sugar and Slaves*, pp. 68–69. We are aware, of course, that not every acre of land on the island could be turned to sugar growing; that some of the "minor crops" continued to be grown, sold, and even exported; that some food crops were always produced on sugar plantations, large and small; and that slaves were allowed, even encouraged to grow some of their own foods on plots allocated to them. It was simply that, once the sugar boom had set in, sugar growers turned every effort to maximizing the production of the staple crop. For the importance of the "plantation yard," see Jerome S. Handler, "Plantation Slave Settlements in Barbados, 1650s to 1834," in *In the Shadow of the Plantation*, ed. Thompson, pp. 121–61.

34. *A Young Squire of the Seventeenth Century: From the Papers* (A.D. 1676–1686) *of Christopher Jeaffreson*, ed. John Cordy Jeaffreson, 2 vols. (London: Hurst and Blackett, 1878).

35. Antoine Biet, *Voyage de la France équinoxiale en l'isle de Cayenne, entrepris par les*

François en l'année M. DC. LII (Paris: François Clovzier, 1664), 295: "C'estoit une merveille de voir deux cens Esclaves travailler au sucre." Although we had recourse to the book itself, we acknowledge with thanks Jerome S. Handler alerting us to it with his article "Father Antoine Biet's Visit to Barbados in 1654," *Journal of the Barbados Museum and Historical Society* 32 (May 1967): 56–76.

Richard Ligon, *A True & Exact History of the Island of Barbados. Illustrated with a Mapp of the Island, as also the Principall Trees and Plants there, set forth in their due Proportions and Shapes, drawne out by their severall and respective scales. Together with the Ingenio that makes the Sugar, with the Plots of the severall Houses, Roomes, and other places, that are used in the whole processe of Sugar-making*, [1st ed.] (London: Humphrey Moseley, 1657), p. 85 et seq., spoke of Drax as one of those who started large-scale sugar manufacturing on Barbados. Ligon was a first-hand observer of much of the late 1640s sugar boom, arriving there in September 1647 and departing three years later. Peter F. Campbell, "Richard Ligon," *Journal of the Barbados Museum and Historical Society* 37 (no. 3, 1985): 215–38.

36. We are, of course, relying here on Dunn's analysis of the 1680 Barbadian census in *Sugar and Slaves*, pp. 90–99; the quotation is on p. 96. The original census is in CO 1/44, fols. 142r–379v, PRO/TNA. There is a modern, typewritten copy with an index, "Census of the Island of Barbados, 1679," MS 2202, GL.

37. David Eltis, *The Rise of African Slavery in the Americas* (Cambridge: Cambridge University Press, 2000), 203–7.

38. The origins of the gang system is a subject much in need of research. In *Rise of African Slavery in the Americas*, 221–23, Eltis notes that it did not appear immediately with the introduction of sugar. Writing in the 1650s, Ligon described gang labor as employed on a large prototypical plantation (perhaps based on his friend James Drax's plantation); twenty years later Henry Drax gave his overseer careful instructions for its use on the equally large but real plantations he owned. For the former, see Ligon, *History of the Island of Barbados*, p. 114. There are two versions of the second source, one in MS Rawl. A. 348, Manuscripts from the Collection of Richard Rawlinson, 1690–1755, Bodleian Library, University of Oxford, the other printed in William Belgrove, *A Treatise upon Husbandry or Planting* (Boston: D. Fowle, 1755), pp. 51–86 (see pp. 64–66). Both versions are undated but Judge Nathaniel Lucas (1761–1828) dated the latter—and, by implication, the former—to 1670. See "The Lucas Manuscript Volumes in the Barbados Public Library," [ed. C. A. L. Gale], *Journal of the Barbados Museum and Historical Society* 21 (November 1953): 21, 25. Compare Jerome S. Handler, *Supplement to "A Guide to Source Materials for the Study of Barbados History, 1627–1834"* (Providence: John Carter Brown Library and The Barbados Museum and Historical Society, 1991), pp. 56–58, who dated them to 1670–79. Professor Handler tells us in a private communication that these are the only two seventeenth-century references to gang labor in Barbados known to him. For the broader context of this discussion, see Philip D. Morgan, "Task and Gang Systems: The Organization of Labor on New World Plantations," in *Work and Labor in Early America*, ed. Stephen Innes (Chapel Hill: Published for the Institute of Early American History and Culture, Williamsburg, Virginia, by the University of North Carolina Press, [1988]), pp. 189–220.

39. Ligon, *History of the Island of Barbados*, p. 117.

40. Compare S[imon] D. Smith, "Merchants and Planters Revisited," *Economic History Review*, 2d series, 55 (August 2002): 434–65.

41. The series displayed in figure 9.2 are a preliminary compilation of data that will be presented in final form in John J. McCusker, *The Price of Sugar in the Early Modern Atlantic World* (in progress). The Amsterdam series record the price for the two benchmark sugars of the era, Brazilian raw muscovado sugar and Brazilian clayed white sugar, as well as the new, Barbados muscovado sugar. The London series is for muscovado sugar, presumably, after the mid-1640s, Barbados muscovado sugar also. Gaps in the series have been filled using straight-line interpolations. The Amsterdam prices have been converted from current money to bank money and then from Dutch currency to English sterling following procedures described in and data drawn from McCusker, *Money and Exchange in Europe and America*, pp. 42–60; and McCusker and Simon Hart, "The Rate of Exchange on Amsterdam in London, 1590–1660," *Journal of European Economic History* 71 (1979), as revised and updated in McCusker, *Essays in the Economic History of the Atlantic World*, pp. 102–16. Finally, the original Amsterdam prices, which were expressed in Dutch *ponden*, have been converted into English cwt. by reference to the discussions in McCusker, "Weights and Measures in the Colonial Sugar Trade: The Gallon and the Pound and Their International Equivalents," *William and Mary Quarterly*, 3rd series, 30 (October 1973), as revised and updated in McCusker, *Essays in the Economic History of the Atlantic World*, pp. 76–101. Compare the analysis of the effect on European sugar prices of the Dutch invasion of Brazil and the subsequent reconquest, in Robert Carlyle Batie, "A Comparative Economic History of the Spanish, French, and English on the Caribbean Islands during the Seventeenth Century" (Ph.D. diss., University of Washington, 1972), pp. 84–89.

As figure 9.2 demonstrates for the period 1619–70, the price of sugar at London tracked the price at Amsterdam closely throughout the seventeenth and eighteenth centuries with some obvious exceptions (for example, during times of warfare). Between 1619 and 1670, the trend lines for muscovado sugar at the two cities ran precisely parallel to each other with the price at London averaging roughly 4s. 6d. per cwt. more there than in Amsterdam. Compare the London series with the prices quoted by Ligon, *History of the Island of Barbados*, pp. 92, 95–96, 112, who, probably speaking of the late 1640s, said that muscovado sugar sold for 3d. per pound and clayed white at 6d. per pound on Barbados. At 100 pounds per hundredweight (cwt.), this was the equivalent of 25s. and 50s. per cwt. At London, Ligon continued, the worst of the Barbados muscovado sugars sold for 70s. and the best for 124s.; the worst clayed white sold for 12d. per pound and the best for 20d. per pound, that is, 112s. per cwt. and 186.7s. per cwt. (In England the cwt. of sugar measured 112 pounds.) In 1661, according to a petition authored by "the Planters, Merchants, Mariners and Traders in the Island of Barbadoes," the price of muscovado in London averaged 21s. per cwt., down from 70s. some time before, perhaps a reference to Ligon. Petition, undated but presented to the Council for Foreign Plantations, 12 July 1661, pp. 12–14, CO 1/15, fol. 61v, PRO/TNA. Compare "the Humble address and petticion of the president Council and Assembly of the Island of Barbados" to the Council for Foreign Plantations, 11 May 1661, CO 31/1, p. 46, PRO/TNA. Barbados prices were expressed in Barbados money currency; prices in London were expressed in sterling. For the rate of exchange between Barbados money currency and sterling, see Appendix.

By 1655 the Portuguese recognized Barbados as a serious competitor to their Brazilian sugar industry. And so it was. See the "carta de Sua Majestade [King João IV] que acusa a de cima," 30 December 1655, as printed in [Rio de Janeiro, Bibliotheca Nacional], *Documentos Historicos*, in progress (Rio de Janeiro: Braggio & Reis et al., 1928 to date), 66:127.

42. Ligon, *History of the Island of Barbados*, 43; J. Harry Bennett [Jr.], "Cary Heylyar, Merchant and Planter of Seventeenth-Century Jamaica," *William and Mary Quarterly*, 3rd series, 21 (January 1964): 59–60.

43. Dunn, *Sugar and Slaves*, pp. 57–58.

44. Compare the record of the sale of slaves by Dutch merchant John Severne (Jon Severijn?) in 1640. RB 3/1, pp. 27–28, BDA.

45. See the discussion in K[enneth] G. Davies, *The Royal African Company* (London: Longmans, Green, 1957), pp. 316–25, which begins: slaves sold "in the West Indies might be paid for in coin, in kind, or in paper." The "paper" of which he wrote was the bill of exchange. Compare Appendix.

46. For more about Hilliard (d. 1659 or 1660), see P[eter] F. Campbell, "Two Generations of Walronds," *Journal of the Barbados Museum and Historical Society* 38 (no. 3, 1989): 278–81; and Campbell, *Some Early Barbadian History* ([St. Michael, Barbados: n.p., 1993]), pp. 48, 91–94. For his partnership with Farmer, see RB 3/2, pp. 219–23, BDA; and Campbell, "Richard Ligon," pp. 221–27. For John Farmer, merchant of Bristol, admitted to freedom April 1639, see [Bristol (England), Council], *Bristol Merchants, Shipwrights, Ships-Carpenters, Sailmakers, Anchorsmiths, Seamen, Tobacco-Cutters, Tobacco-Rollers, Tobacconists, Tobacco Pipe-Makers—A Transcript Chronologically Arranged, from the Bristol Burgess Books: 1607–1700*, compiled by N[orman] C. P. Tyack ([Bristol, England: n.p., 1930]), pp. 5, 11.

Ligon lived much of his time on Hilliard's plantation and he drew upon his experiences there for his description of how sugar was produced on Barbados. Ligon, *History of the Island of Barbados*, pp. 22–23. Ligon described the plantation in some detail, dwelling especially on its condition in 1647 when Hilliard sold half of it to Thomas Modyford. Early in the 1640s, Ligon says, it could have brought 16s. an acre; in 1647 Modyford paid £28 per acre. Ligon, *History of the Island of Barbados*, pp. 22, 86; N[icholas] Darnell Davis, *The Cavaliers & Roundheads of Barbados, 1650–1652, with Some Account of the Early History of Barbados* (Georgetown, British Guiana: "Argosy" Press, 1887), pp. 80–82. Modyford had just arrived in 1647, flush with capital, intent on making his fortune in sugar. What he got for £7,000 was half of a fully articulated sugar plantation including land with standing crops (some of them provisions), all necessary buildings and equipment, nearly one hundred slaves (at least some of them people bought off the *Marie Bonadventure*), twenty-eight indentured servants, and several dozen work animals. Modyford lived and worked on "Buckland Plantation" for the next decade, growing in wealth and influence to become the colony's governor for a short time in 1660. For Modyford, see Carlton Rowe Williams, "Sir Thomas Modyford, 1620–1679: 'That Grand Propagator of English Honour and Power in the West Indies'" (Ph.D. diss., University of Kentucky, 1978).

47. For Colonel (later Sir) James Drax (fl. 1627–61) of Drax Hall, see Bridenbaugh and Bridenbaugh, *No Peace Beyond the Line*, pp. 137–39 and passim. Hilliard and Drax were linked not only by sugar planting and the purchase of slaves as well as by numerous shared connections in England but also seem to have had family ties, having married sisters, Barbara and Meliora Horton. Campbell, "Two Generations of Walronds," pp. 278–79. For the origins and character of the early Barbadian planter elite, see Ronald G. Hughes, "The Barbadian Sugar Magnates, 1643–1783: Some Jottings," *Journal of the Barbados Museum and Historical Society* 35 (no. 3, 1977): 211–22.

48. Although we have found no evidence of direct Dutch investment in Barbados sugar planting, we are fully prepared to believe that some of the funds lent Barbados planters by London merchants were Dutch in origin. The anecdotal evidence of Dutch participation is too strong to dismiss out of hand. Dutch willingness to lend to London-based enterprises is well known. That London partners of Barbadian planters served as the intermediaries, lending Dutch money and their own credit, seems a reasonable way to reconcile what contemporaries recounted and what the archives attest. Violet Barbour, *Capitalism in Amsterdam in the Seventeenth Century*, Johns Hopkins University Studies in Historical and Political Science, series 67, no. 1 (Baltimore: Johns Hopkins University Press, 1950), pp. 122–27 and passim, demonstrates the availability of and the receptiveness to Dutch capital in London, some of it coming either directly from Amsterdam, some by way of Dutch houses in places such as Hamburg. The Dutch lent not only to government but also to businesses, both in direct investments and in portfolio investments. Higher savings rates in the Netherlands made for lower interest rates there; lower savings rates in other places, such as England, made money more expensive. The Dutch capitalized on the difference, borrowing at home at 3 and 4 percent, lending abroad at 6 to 8 percent and more. Ibid., pp. 85–88, 109, 122–25. That Dutch capital shied away from English markets during the Civil Wars and the Interregnum may explain, in part, the absence of evidence of Dutch participation in the early years of the sugar boom. Ibid., pp. 124–25. But it was there in abundance before 1640, never fully retreated, and returned with a vengeance after the Restoration.

By the late-1640s, some Jewish merchants had migrated from Brazil to Barbados and more were to arrive in the next decade. Their connections were in Amsterdam and they, too, could have served as channels of Amsterdam investment funds to island planters. The role of Jews in Barbados has been the subject of continuing discussion, though most of what has been written is anecdotal and filiopietistic. There is a major opportunity for someone to pursue systematically the role of Jewish merchants in the emerging English colonies of the seventeenth century. See P. A. Farrar, "The Jews in Barbados," *Journal of the Barbados Museum and Historical Society* 9 (May 1942): 130–33; Bridenbaugh and Bridenbaugh, *No Peace Beyond the Line*, pp. 326–27. The most useful work is by Eustace M. Shilstone, *Monumental Inscriptions in the Burial Ground of the Jewish Synagogue at Bridgetown, Barbados* ([New York: American Jewish Historical Society, 1956]), especially the author's introduction and the preface by Wilfred S. Samuel (pp. iii–xiii). The most recent work is Mordehay Arbell, *The Jewish Nation in the Caribbean: The Spanish-Portuguese Jewish Settlements in the Caribbean and the Guianas* (Jerusalem and New York: Gefen Publishing House, [2001]), pp. 191–217. Gedalia Yogev, *Diamonds and Coral: Anglo-Dutch Jews and Eighteenth-Century Trade*

([Leicester, England]: Leicester University Press; [New York]: Holmes & Meier, 1978) is a distinct cut above the rest but concentrates on the eighteenth century, as the title says—nevertheless, see pp. 60–66. We find it powerfully suggestive that there was a Sephardic Jewish community already established in Port Royal, Jamaica, when Cromwell's forces captured the island in 1655. Meyer Kayserling, "The Jews in Jamaica and Daniel Israel Lopez Laguna," *Jewish Quarterly Review*, 1st series, 12 (1900): 712.

49. Sugar was still a valid legal tender in payment of debts in the last third of the eighteenth century. See nn. 60, 67, below.

50. See, for instance, the contract between George Martin and Humphrey Walrond, 8 August 1647, in which they agree that, on 1 June 1648, Martin was to pay "two hundred pounds in [the money currency of] the said island of Barbados." He could make payment "by and in merchantable goods and commodities of the said island at such rates as the same shall at the tyme usually pass in buying and selling between man and man in the island." RB 3/2, p. 176, BDA. Also important in making commodity money work were laws that guaranteed the merchantable quality of "the several commodities of this Island given and received in payment, that is to say, Cotton, Wool, Sugar, Indico, and Tobacco." See "An Act for the better incouragement of Trade," 18 June 1652, *Acts and Statutes of the Island of Barbados. Made and Enacted since the Reducement of the Same, unto the Authority of the Commonwealth of England. And Set Forth the seventh day of September, in the Year of our Lord God, 1652*, [ed. John Jennings] (London: Will. Bentley, [1654]), pp. 83–88.

51. Thus the 1636 lease that required a payment "of twenty pounds of cleane Cotton, or the value thereof" leaves historians in the dark as just how much was the value of a pound of cotton in that particular year. RB 3/1, p. 534, BDA. Table 9.7 indicates that, during the 1630s, one pound of Barbados cotton currency was valued at six pence Barbados money currency.

On 10 October 1662 the government of Jamaica, explicitly patterning its action on "the laudable practice of Barbadoes and other Plantations," established the legal tender value in Jamaican money currency of sugar currency (3d. per pound), cocoa currency (4d. per pound), and tobacco (3d. per pound). Act of 10 October 1662, "Journal and Laws," 1661–79, fols. 35v–36r, CO 139/1, PRO/TNA. See also "The Council Book of Jamaica," 1661–72, pp. 58–60, CO 140/1, PRO/TNA (quotation on p. 58). Compare the decision of the Council, 18 June 1661, ibid., fols. 11v–12v.

52. See McCusker, *Money and Exchange in Europe and America*, 323.

53. While this usage can be inferred from a variety of instances, unequivocal examples are less frequent. See, for example, the references to prices on the island of Barbados in Ligon, *History of the Island of Barbados*, passim. Our contention is simply that, when Ligon wrote of prices on the island and expressed them in the traditional English mode and called such prices "sterling," he meant Barbados money currency, in order to distinguish them from prices expressed in Barbados sugar currency. In the same way, when he talked of prices in England, he meant English currency. He certainly knew the difference having come from England and having lived on the island for three years. Similarly the vestry of Christ Church parish were clearly speaking colloquially when they complained about Governor Henry Hawley's scurrilous accusation that, through their "excessive drinking," the colony's "inhabitants had pist out 15000 . . . [pounds] sterl[ing] . . . against the wall" just this last year alone.

Vestry of Christ Church, Instructions to Representatives, [undated but ca. February 1641], Papers Relating to the Island of Barbados, GD 34, NAS, and as quoted in Bennett, "Peter Hay," p. 27. In a similar fashion, the Minutes of the Council from the 1650s regularly and frequently recorded sums of money in pounds "sterling" that the context shows are references to the local money of account, pounds Barbados money currency. See the Minutes of the Council of Barbados, 1654–58, CO 31/0, passim, PRO/TNA. The unusual numbering of these records is because of their anomalous provenance. This is a two-volume typescript made in the late 1920s or early 1930s for Frederick G. Spurdle from original documents in Barbados and that he then presented to the PRO/TNA in 1934. Between that date and the removal of the PRO/TNA to Kew, the volumes were on the open reference shelves in the Round Room, Chancery Lane, at pressmark 20/83–84. Most recently they can be found among the finding aids at Kew. Given their fragile state and the fact that many of the original documents from which they were copied no longer exist, they deserve to be incorporated into the collections of the PRO/TNA. For these records, see Spurdle, *Early West Indian Government: Showing the Progress of Government in Barbados, Jamaica and the Leeward Islands, 1660–1783* (Palmerston North, New Zealand: The Author, [1961]), p. 268; and Herbert C. [F.] Bell and David W. Parker, *Guide to British West Indian Archive Materials, in London and in the Island, for the History of the United States*, Carnegie Institution of Washington, Publication, no. 372 (Washington, D.C.: Carnegie Institution of Washington, 1926), pp. 335–38.

For unequivocal examples of the use of the word "sterling" to mean Barbados money currency, see "An Act for appointing Muscovado-Sugar to pass at the rate of three pence sterling per pound," 22 November 1655, in *Acts, Passed in the Island of Barbados. From 1643, to 1762, inclusive; Carefully revised, innumerable Errors corrected, and the Whole compared and examined, with the original Acts, In the Secretary's Office*, ed. Richard Hall Sr. and Richard Hall Jr. (London: Richard Hall, 1764), p. 467. The volume shelved as CO 30/1, PRO/TNA, is a copy of this work. See also John Poyntz, *The Present Prospect of the Famous and Fertile Island of Tabago* (London: Printed by George Larking for the Author, 1683), p. 43, where he noted that "each piece of eight is valued at five shillings sterlin in Barbadoes; but in the Lee-ward Islands it goes for six." Compare the entries in the Barbados Custom House Journal, 1665–67, pp. 85, 89, Department of Manuscripts and Rare Books, Hispanic Society of America, New York; the payment of "the Moyety of 3£ Starl," to Stephen Starr, carpenter, in the account of the 4-½ percent Duty, 1669–70, signed by Sir Tobias Bridge, 16 April 1670, BL 375, William Blathwayt Collection, Box 1, Huntington Library, San Marino, California.

54. Most likely the last and most assuredly the most curious instance of this convention was the use of "South Carolina sterling" for a few years during the 1780s, after the end of the American Revolution. See John J. McCusker, *How Much Is That in Real Money? A Historical Commodity Price Index for Use as a Deflator of Money Values in the Economy of the United States*, 2d ed., rev. and enlarged (Worcester, Mass.: American Antiquarian Society, 2001), pp. 85–86. In addition to the examples of this practice cited there, see the reference to the state's "Duties on Importation" being assessed in "sterling," meaning, again, South Carolina state currency, as discussed in *The Columbian Herald* (Charleston), 5 December 1785. Compare the 1662 reference to "sterling money of France," meaning in this instance

French money currency. *Records and Files of the Quarterly Courts of Essex County, Massachusetts*, [ed. George Francis Dow], 8 vols. (Salem, Mass.: Essex Institute, 1911–21), 2:385.

55. For more on the subject of this paragraph and the next one, see McCusker, *Money and Exchange in Europe and America*, passim.

56. "An Act for advancing and raising the Value of Pieces of Eight," [14 April 1666], *The Laws of Barbados*, ed. William Rawlin (London: William Rawlin, 1699), p. 55; dated with reference to the subsequent act of 14 November 1668 (ibid., p. 111), which repealed the earlier act. Compare "The Laws of Barbados," a contemporary compilation of the laws in force as of March 1666/67, CO 30/3, p. 114, PRO/TNA.

The volume entitled *Acts, Passed in the Island of Barbados*, ed. Hall and Hall, 459–76, listed several earlier acts. While the Halls quoted the titles of these acts "which have now become obsolete, repealed [or] expired" and indicated the dates of their passage for most of them, they did not print the text of the laws. See acts no. 3 (undated but prior to May 1646), no. 29 (13 May 1646), no. 104 (12 September 1651), no. 201 (27 September 1661), no. 206 (13 March 1661/62), no. 223 (14 April 1666), and no. 276 (14 November 1668). The text of the act listed by the Halls as no. 290 (22 December 1669) can be found in CO 30/3, pp. 122–23, PRO/TNA. We have not located the texts of the other acts.

Robert Chalmers, *A History of Currency in the British Colonies* (London: H. M. Stationery Office, 1893), 48, quoted several passages from an act dated 12 September 1651 for which he gave no source and which cannot be found or verified even after several decades of searching. It is of small matter if only because, like all legislation passed during the twenty-one months from May 1650 to January 1652, it was repealed in December of 1652. *Acts and Statutes of the Island of Barbados*, [ed. Jennings], pp. 131–32. Thus, despite what was said in McCusker, *Money and Exchange in Europe and America*, p. 239, the par of exchange before 1666 is most likely to have been at £107.50 Barbados money currency per £100 English currency.

57. Compare John Ashley, *Memoirs and Considerations Covering the Trade and Revenues of the British Colonies in America*, 2 vols. (London: C. Corbett, E. Comyns, and J. Jolliffe, 1740–43), 1:50–51.

58. McCusker, *Money and Exchange in Europe and America*, pp. 239–45.

59. Ibid., pp. 3–18.

60. The legal tender provision enforcing the use of Barbados sugar currency in the payment of fees due to the island government, originally enacted prior to 1650 and repeated in subsequent acts, was still in force as late as 1764. See "An Act for the regulating and appointing the Fee of the several Officers and Courts of the Island," undated but passed between 1642 and 1650, in *Acts, Passed in the Island of Barbados*, ed. Hall and Hall, pp. 8–11. See also ibid., p. 54. Compare CO 30/3, pp. 76–82, PRO/TNA. The remarks of Otis P. Starkey, *The Economic Geography of Barbados: A Study of the Relationships between Environmental Variations and Economic Development* (New York: Columbia University Press, 1939), p. 64, are generally very useful.

61. Thus, in January 1637, when Andrew Hardie, the master of the *Abraham* of London, bought provisions in Bridgetown, Barbados, he recorded the purchases in the island's currency, Barbados tobacco currency—including some thirty pounds of sugar. On 8 and 11 Feb-

ruary, he paid five pounds of tobacco per pound of sugar; by 25 May the price had dropped to four pounds Barbados tobacco currency per pound of sugar. "A Journall . . . for Shipp Abraham," pp. 9, 11, 25, Accounts and Papers of the Ship *Abraham*, Admiralty Miscellanea, High Court of Admiralty, HCA 30/636, PRO/TNA. With reference to table 9.7, we can determine that, at 6d. Barbados money currency per pound of tobacco, sugar at retail cost 2s. 6d. per pound Barbados money currency in the first two instances and 2s. per pound on the third occasion, four and five times its value as a commodity money.

62. For a valuable discussion of the use of commodity money in the various colonies, see Curtis P. Nettels, *The Money Supply of the American Colonies before 1720*, University of Wisconsin Studies in the Social Sciences and History, no. 20 (Madison: University of Wisconsin Press, 1934), pp. 202–28. Compare, for Barbados, Frank Wesley Pitman, *The Development of the British West Indies, 1700–1763*, Yale Historical Publications, Studies, no. 4 (New Haven: Yale University Press, 1917), pp. 139–40; McCusker, *How Much Is That in Real Money?*, p. 88.

The French priest, Rev. Antoine Biet, who lived on Barbados from February through April 1654, described how commodity money worked in daily life. When Barbadians who lived in the country came to town, they stayed in one of Bridgetown's many inns. The innkeeper kept a running account for each of his or her customers; he or she ran a tab for them. The cost per day including meals ranged from ten to fifteen pounds of sugar (2s. 6d. to 3s. 9d. Barbadian money currency, that is, 3d. per pound) and, when the outstanding sum totaled three or four hundred pounds of sugar ("trois ou quatre cens livres de succre"), the planter settled his account by sending from the plantation the amount owed to the innkeeper. Biet, *Voyage de la France équinoxiale*, p. 289. Compare n. 66, below.

63. Richard Pares was certainly on the right track in his very preliminary discussion of the subject. Pares, *Merchants and Planters*, 32, 77–78, n. 49. He also suggested that pork had served as a commodity money in the first few years of the colony. Compare table 9.6, panel C.

64. "For every Subpœna ten pounds of Suger, or two shill. and six pence in money . . . ," "An Act for the appointing, or regulating of the Fees of the several Officers, and Courts of this Island," 17 September 1652, *Acts and Statutes of the Island of Barbados* [ed. Jennings], pp. 92–101 (quotation on p. 99). For the consistent use of sugar as a commodity money, see the acts set out in ibid., pp. 13–14, 14, 15, 16–17, 20–21, and passim. Compare also the act of 22 November 1655 in *Acts, Passed in the Island of Barbados*, ed. Hall and Hall, p. 467. See as well *Suffolk Deeds*, [ed. William B. Trask et al.], 14 vols. (Boston: Rockwell and Churchill Press, 1880-1906), Liber III, 81, 169; and the records of the Council meetings of 7 February 1653/54, 12 May 1657, and 5 October 1658, Minutes of the Council of Barbados, 1654–1658, pp. 2, 226, 318, CO 31/0, PRO/TNA.

65. *A Volume Relating to the Early History of Boston Containing the Aspinwall Notarial Records from 1644 to 1651*, [ed. William H. Whitmore and Walter K. Watkins], Records Relating to the Early History of Boston, vol. 32 (Boston: [City of Boston, Registry Department], 1903), pp. 140–41 (July 1648), 338–39 (April 1650). The years 1650–52 seem to have been a period of transition to the lower rate, again from very limited evidence. Ibid., pp. 340–41; *Suffolk Deeds*, ed. Trask et al., Liber I, 259. The three-penny rate was still in effect as late

as 29 November 1659 according to the account of charges and disbursements for the ship *Barbados Merchant* printed in Hawtayne, "Records of Old Barbados," p. 108.

66. "An Act for the encouragement of all the Faithful Minister, in the Pastoral-charge," 27 September 1661, in *Acts, Passed in the Island of Barbados*, ed. Hall and Hall, pp. 33–34. Compare the testimony dated 9 December 1661 of James Whitcombe, who lived in Barbados between January 1659/60 and September 1661 concerning the price of muscovado sugar there: "The common price it passed for in taverns and common victualing houses was two pence [Barbados money currency] per pound." *Records and Files of the Quarterly Courts of Essex County*, [ed. Dow], 2:322, n. Compare n. 62, above.

67. At the rate of a penny, ha'penny per pound (1.5d.) or 12s.6d. per cwt., it persisted for several decades more. See, for instance, the account of James Book "in the Barbadas," 26 December 1662, Company of Royal Adventurers of England Trading into Africa, Entry Book of Invoices, 1660–63, Treasury Office Records, T 70/1221, PRO/TNA; Barbados Custom House Journal, 1665–67, pp. 73, 85, Hispanic Society of America; "An Act for regulating . . . the Fees of the several Offices in this Island," 5 November 1668, *Laws of Barbados*, ed. Rawlin, pp. 76–80; *Acts, Passed in the Island of Barbados*, ed. Hall and Hall, pp. 65–69; CO 30/3, pp. 107–13, 137–41, PRO/TNA. In the latter compilation the act is dated 9 November 1668. Compare Ashley, *Memoirs and Considerations Covering the Trade and Revenues of the British Colonies*, 1:50–51. See also Harlow, *History of Barbados*, pp. 312, 316–17.

68. See, for instance, the discussion of the Governor in Council, 21 July 1636, concerning cases to be heard in the Court of Common Pleas in *Acts, Passed in the Island of Barbados*, ed. Hall and Hall, p. 33. See also Harlow, *History of Barbados*, pp. 16, 21, n. 5. The two commodity moneys were rated as equal to each other throughout most of the tenure of Peter Hay, the proprietors' Receiver General. Papers Relating to the Island of Barbados, GD 34, NAS. Compare Bennett, "Peter Hay," pp. 9–29. See n. 73, below.

69. "Journall . . . for Shipp Abraham," p. 1 (28–29 January 1636/37), HCA 30/636, PRO/TNA; deposition of Thomas Irish, master of the ship *Faulcon* of London, 6 March 1636/37, High Court of Admiralty, Instance and Prize Courts, Examinations and Answers, 1637–38, fols. 65v–66r, HCA 13/53, PRO/TNA.

70. *Volume . . . Containing the Aspinwall Notarial Records*, [ed. Whitmore and Watkins], pp. 140–41. The same ratio between sugar currency and tobacco currency is cited in a reference dated 29 March 1649, RB 3/3, p. 489, BDA, and in a list of accounts receivable, dated 28 March 1651, *Suffolk Deeds*, ed. Trask et al., Liber I, 293.

71. "An Act appointing Bench Actions, and the manner of proceeding therein," 22 December 1669, *Laws of Barbados*, ed. Rawlin, 87–88. Land prices in particular seem to have continued for a long time to be quoted in Barbados cotton currency. Compare table 9.5.

72. The statement by Innes, "Pre-Sugar Era," p. 19, that Barbados cotton currency was in use as late as 1693 is incorrect, based on a misreading of the date of sale for Windsor plantation in St. George Parish as reported in the source he cited. The correct date of the sale is 1643 (2/6 February 1643/44). RB 3/1, p. 299, BDA; Hughes, "Barbadian Sugar Plantations," "Prefatory Note," [pp. 1–2].

73. The colony's proprietary trustees reacted to the rapidly changing values of the different commodity currencies during the early 1640s, thereby reflecting what was happening. By

1644 they had come to realize that rents assessed in tobacco or cotton currency were yielding considerably diminished returns. "Cotton or tubacco . . . is a very uncertaine Commodite . . . and cannot produce any certaine rent." They sought to reassess rents in pounds money currency. Letter from Sir James Hay and Archibald Hay, at Oxford, to the Governor, Council, and Assembly of Barbados, 17 April 1644, Papers Relating to the Island of Barbados, GD 34, NAS, and as quoted, in part, in Bennett, "English Caribbees in the Period of the Civil War," p. 371.

74. RB 3/1, pp. 691–94, BDA.

75. In comparison, the average price for fifty-six indentured servants sold at Bridgetown on 28–29 January 1636/37 was 493.75 pounds tobacco currency or £12.34 Barbados money currency. "Journall . . . for Shipp Abraham," p. 1, HCA 30/636, PRO/TNA. In 1652 the price was 800 pounds of sugar, or £10 Barbados money currency. Uchteritz, "A German Indentured Servant in Barbados in 1652," ed. and trans. Gunkel and Handler, pp. 92, 95. Thus the average price paid for seventy exiled English captives in May 1656, 1,550 pounds of sugar or £18.75 Barbados money currency, supports their contention that they had been sold into lifetime servitude. Marcellus Rivers and Oxenbridge Foyle, *Englands Slavery, or Barbados Merchandize; Represented In a Petition to the High and Honourable court of Parliament by Marcellus Rivers and Oxenbridge Foyle Gentlemen, on the behalf of themselves and three-score and ten more Free-born Englishmen sold (uncondemned) into slavery: Together with letters written to some Honourable Members of Parliament* (London: n.p., 1659), p. 5; *Proceedings and Debates of the British Parliaments Respecting North America*, ed. Leo Francis Stock, Carnegie Institute of Washington, Publication no. 338, 5 vols. (Washington, D.C.: Carnegie Institute of Washington, [1924–41]), 1:249.

76. Ligon, *History of the Island of Barbados*, p. 46. The inventory and valuation of the estate of William Powrey (otherwise William Hay) included nineteen able Negroes at 1,100 pounds of sugar each. Dated 8 June 1649, in Papers Relating to the Island of Barbados, GD 34, NAS. With reference to table 9.7, we can determine that, at 6d. Barbados money currency per pound of sugar, this was the equivalent of £27.50 per person (£25.58 sterling). Five years later, in February or March 1654, the party of Frenchmen of which Rev. Antoine Biet was a member sold six of the slaves they had brought with them from Cayenne for 2,000 pounds of sugar each ("le prix de deux milliers de sucre par teste"). Biet, *Voyage de la France équinoxiale*, pp. 274–75. With reference to table 9.7, we can determine that, at 3d. Barbados money currency per pound of sugar, this was the equivalent of £25.00 per person (£23.26 sterling).

77. This was almost precisely the same as the ratio between the cost at Grenada in 1770 of a newly arrived slave, £37 sterling, and the valuation given a prime field hand, £53. See the "Inventory & Valuation" of the estate on the island of Grenada owned by Alexander Johnstone, dated 1 December 1770, item 41/32, Records of the Westerhall Estate, West Indian Documents, University Library, University of Bristol.

Contributors

ALEJANDRO DE LA FUENTE is Associate Professor in the Department of History and the Center for Latin American Studies at the University of Pittsburgh. He taught at the University of South Florida from 1996 to 2000. He is the author of *A Nation for All: Race, Inequality, and Politics in Twentieth-Century Cuba* (2001). His work on slavery and race in Cuba has been published by journals in Cuba, Brazil, Spain, the Netherlands, England, Germany, and the United States.

HERBERT KLEIN is Gouverneur Morris Professor of History at Columbia University. He is the author of two books on the Atlantic slave trade: *The Middle Passage: Comparative Studies in the Atlantic Slave Trade* (1978) and *The Atlantic Slave Trade* (1999), as well as two works on African slavery in the Americas. He has published four books on Bolivian history, the most recent of which are *Haciendas and Ayllus: Rural Society in the Bolivian Andes in the 18th and 19th Centuries* (1993) and *A Concise History of Bolivia* (2002). He has also studied immigration and slavery in Brazil and recently coauthored *Slavery and the Economy of São Paulo, 1750–1850* (2002). He has also written *The American Finances of the Spanish Empire, 1680–1809* (1998) on colonial fiscal history and is currently working on *A Population History of the United States*.

JOHN J. MCCUSKER is Ewing Halsell Distinguished Professor of American History and Professor of Economics at Trinity University in San Antonio, Texas. He has lectured and taught in Belgium, Canada, China, England, Finland, France, Ireland, Italy, and the United States. In his research and writing, he focuses on the economy of the Atlantic world in the seventeenth and eighteenth centuries. He has published many works on the subject, including the prize-winning *Economy of British America, 1607–1789* (1985), which he coauthored with Russell R. Menard.

RUSSELL R. MENARD has been a member of the Department of History at the University of Minnesota since 1975. He is coauthor with John J. McCusker of *The Economy of British America, 1607–1789* (1985) and coauthor with Lois Carr and Lorena Walsh of *Robert Cole's World: Agriculture and Society in Early Maryland* (1991). His most recent book is *Migrants, Servants, and Slaves: Unfree Labor in Colonial British America* (2001). He is currently working on *Sweet Negotiations: A New Look at the Barbadian Sugar Industry, 1640–1775*.

WILLIAM D. PHILLIPS JR. is Professor of History and Director of the Center for Early Modern History at the University of Minnesota. Among his publications are *Enrique IV and the Crisis of Fifteenth-Century Castile, 1425–1480* (1978); *Slavery from Roman Times to the Early Transatlantic Trade* (1985); and the edited volume *Testimonies from the Columbian Lawsuits* (2000). He coauthored with Carla Rahn Phillips *The Worlds of Christopher Columbus* (1992), a *New York Times* notable book of the year and second-place winner of the Spain in America prize, and *Spain's Golden Fleece* (1997).

GENARO RODRÍGUEZ MOREL, a native of the Dominican Republic, resides in Seville, where he conducts research for the Dominican Academy of History. Among his publications are *Cartas del cabildo de la cuidad de Santo Domingo en el siglo XVI* (1999); *Cartas de los cabildos eclesiásticos de Santo Domingo y Concepci de la Vega* (2000); and "Controles comerciales y alternativas de mercado en La Española," in *Actas: Congreso Internacional 500 años de la fundación de la Casa de Contratación* (2003). He is currently completing a study of the sugar economy of Hispaniola.

STUART B. SCHWARTZ is George Burton Adams Professor of History at Yale University. His research has concentrated on colonial Brazil and Spanish America. He is the author of *Bureaucracy and Society in Colonial Brazil* (1973); *Sugar Plantations in the Formation of Brazilian Society* (1985); and *Slaves, Peasants, and Rebels* (1988) and coauthor of *Early Latin America* (1983). He was editor of *Implicit Understandings* (1994) and coeditor of *The Cambridge History of Native Peoples of the Americas*, vol. 3 (1999). He is currently completing a study of popular attitudes of religious tolerance in the Luso-Hispanic world.

EDDY STOLS is Professor at Catholic University Leuven in Belgium. He has also taught at Leiden University and the University of the State of São Paulo Marília. He is the author of *De Spaanse Brabanders de handelsbetrekkingen van Zuidelijke Nederlanden met de Iberische Wereld, 1598–1638* (1971), and *Brazilië: 500 jahr geschiedenis Din dribbelpas* (1996), and coeditor of *Flandre et Portugal* (1991) and *Flandre et Amérique latine* (1993).

ALBERTO VIEIRA is Coordinator of the Center for Atlantic Historical Studies in Madeira. His research has concentrated on Madeira and the Atlantic islands. Among his publications are *O Comércio Inter-Insular (Madeira, Açores e Canárias), séculos XV–XVII* (1987); *Os escravos no arquipélago da Madeira, séculos XV–XVII* (1991); *Guia para a história e investigação das ilhas Atlânticas* (1995); *História do vinho de Madeira: Textos e documentos* (1993); and *Portugal y las islas del Atlántico* (1992).

Index

and Peru, 205–6, 208; and Mexico, 207; and West Africa, 210; government's role in, 213–16; economics of free trade era, 216–23; and African slave traders, 219; and introduction of American foods, 223–24; origins of slaves, 226–27; and slave mortality, 228, 229, 230–32

Augusta, João de, 65

Augusti, Quirico degli, 247

Aveiro, Duke of, 159

Ayala de Obidos, Josefa de, 254

Ayllón, Lucas Vázquez de, 90, 92

Azores, 204

Babylon, sugar mill described as, 3–4, 21, 26 (n. 46)

Baéz, Andrés, 53

Baez, Manuel, 139

Baeza, Miguel de, 244

Baghdad, 29

Bahia, Brazil: and sugar production, 3, 161, 162, 181, 208; and sugar prices, 12; production levels in, 18, 161, 163, 165, 170; and Madeiran sugar specialists, 75; local conditions in, 159; sugar mills in, 161–62; and Dutch occupation, 162, 164, 166; and technology, 163; and sugar trade, 174; sugar production techniques in, 176–80, 197 (n. 40); and sugar mill owners, 182; and cane farmers, 184, 186

Balbani, Jan, 269

Ballester, Miguel de, 87, 96, 98

Ballesteros, Francisco, 12

Barbados: *ingenio* in, 2; productivity per slave in, 5; and sugar revolution, 7, 8, 10, 18, 289, 290, 306; Brazil compared to, 12, 18; technology transfer and, 18; managerial strategies in, 18–19; as intermediary between Old and New World, 67; and Atlantic slave trade, 124, 209, 290, 296–97; Brazilian competition with, 166, 170, 194; and sugar production, 209, 289, 291–92, 297, 303, 313 (n. 5); Dutch involvement in, 289, 290, 291, 294–95, 297, 306, 315–16 (n. 19), 324 (n. 48); restructuring of sugar industry, 290, 294, 297–98, 300, 301; English involvement in, 291, 295, 296–97; currency in, 306–11, 325–26 (nn. 51, 53), 327–28 (nn. 56, 61, 62)

Barbary coast, 71, 265, 267

Bardeci, Lope de, 102

Barred, Garcia de, 92

Barrett, Ward, 4–5, 11, 12, 19–20

Basiliers, Jasper, 266

Bateson, Richard, 295

Battus, Carolus, 249, 250

Baugin, Lubin, 254

Beatrice of Brussels, 248

Beckles, Hilary, 293

Beert, Osias, 254

Béguin, Jean, 245

Béguines, 248–49, 250

Bell, Philip, 305

Belon, Pierre, 250–51

Benedictines, 180, 181, 192

Benson, Jorge, 266

Berbers, 210

Berrio, Hernando de, 90

Berthe, Jean-Pierre, 11, 147

Berudo, Juanoto, 69

Bethencourt, Gaspar, 54

Beuckelaer, Joachim de, 253

Bevers, Daniel, 274

Bies, Johann, 261

Biet, Antoine, 300, 328 (n. 62)

Bissan, Juan, 69

Blackburn, Robin, 7

Bocollo, Cristóvão, 65

Bonnefons, Nicolas de, 246

Borja, Cardinal, 252

Bosch, Jerome, 253

Bosse, Abraham, 254

Brandão, João, 242–43

Braudel, Fernand, 239, 258

Bray, Joseph de, 258

Brazil: and *engenho*, 2; planters' accounting practices in, 4; and plantation complex, 7; sugar prices in, 12, 13; production levels in, 12, 17, 18, 49, 50, 151 (n. 25), 161, 163–64, 171, 178; and slaves, 12, 17, 18, 59, 148, 158, 169–70, 176, 177, 180, 183, 186, 187, 188–89, 208; sugar production techniques in, 13–14, 176–80; and sugar production, 15, 135, 138, 208, 214, 215; productivity levels of slaves in, 19, 20; and feudalism, 22 (n. 4); planters' market creation in, 23 (n. 13); Madeira as model for settlement of, 50, 75; Madeiran sugar replaced by

Brazilian sugar, 67; and sugar trade, 71, 72–73, 172–76, 243; and monopoly, 73; and land distribution, 74–75; and European markets, 108, 161, 162, 164, 166, 170, 172, 208; Cuba compared to, 116; and sugar mills, 135–36, 161, 167, 176; settlement of, 158; indigenous population of, 160–61, 188; Dutch occupation of northeastern Brazil, 162, 166–72, 193, 195 (n. 20), 208–9; Dutch attacks on, 164, 166; and sugar production structure, 180–81; key elements of sugar economy in, 180–94; and credit, 191–92, 193; and Antwerp sugar market, 260, 262. *See also* Bahia, Brazil; Pernambuco, Brazil

Brazil Company, 73

Bredero, Gerbrand Adriaensz, 256

Breton merchants: and Madeira, 64

Breughel, Jan, the Elder, 251, 254, 257

Breughel, Pieter, the Elder, 251, 253

Breughel, Pieter, the Younger, 251, 253

British West Indies, 116, 124, 138, 209, 234 (nn. 17, 19), 290

Bruges merchants, 260–61

Bruges sugar market, 260, 264

Brulez, Wilfrid, 259

Caballero, Alvaro, 104, 106

Caballero, Diego, 96, 98, 99

Cabezas Altamirano, Juan de las, 142

Cabrera, Manuel Lobo, 49, 58, 60

Cáceres y Ovando, Alonso de, 134

Cadamosto, Luigi da, 48

Calendar of Córdoba, 29

Calvo, Francisco, 65

Calvo de la Puerta, Martín, 131, 132, 139, 143

Camacho y Pérez Galdós, Guillermo, 53, 68

Campi, Vincenzo, 254

Campos Moreno, Diogo de, 161

Canary Islands: indigenous population of, 6, 43, 51, 205; spread of sugar production to, 7, 8; and plantation complex, 7, 10; sugar industry in, 16; climate of, 27; and Iberian sugar market, 35; and spread of sugar production to Caribbean, 42; as sugar islands, 42; property system of, 42–43, 44, 47, 50–51, 52, 57; and sugar production, 48–49, 70, 121; as consumer of

European production, 62; as intermediary between Old and New Worlds, 62–63, 74; and Italy, 63; and Spain, 63; and sugar trade, 64, 69, 70, 71, 72, 244, 265; and sugar as means of exchange, 70; specialists from, 89, 139; skilled workers from, 100, 159; and cane farmers, 183; Spanish settlement of, 204; and Antwerp sugar market, 260, 261

Candele, Louis le, 262

Cane farmers: in Madeira, 55, 57, 70, 182, 183; in Brazil, 182–87, 192–93, 198 (n. 54), 199 (n. 59); in Barbados, 297

Cape Bojador, 209

Cape Verde Islands, 204, 209, 210

Capital: European capital, 1, 15, 159–60, 213; merchant capital, 1, 63; and staple theory, 6; sources of, 9; in New Spain, 12. *See also* Foreign capital; Local capital

Capital accumulation: and sugar plantations, 5

Capitalism: and sugar production, 1–2, 3, 5; agrarian origins of, 5; development of, 5, 22 (n. 12); and sugar trade, 239, 275

Cardona, Hugo de, 34

Carême, Antonin, 252

Caribbean islands: and sugar production, 15, 49; spread of sugar production to, 42; and Brazil, 170, 172, 177. *See also specific islands*

Caribs, 11, 88, 107

Carmelites, 180, 181, 248

Carminatís, Jácome de, 69

Carrasco, Constantino, 54

Cartagena de Indias, 123, 124, 130

Carvajal, Hernando de, 90

Carvajal, Pedro de, 133

Casas, Melchor, 132–33, 134, 142, 143

Casteau, Lancelot du, 246, 250

Castelhano, João Rodrigues, 66

Catarina (queen), 243

Catherine of Bragança, 172

Catholic Counter-Reformation, 248, 255

Cats, Jacob, 253

Cembalos, Francisco, 92

Centurión, Melchior, 92

Cesare, João Antonio, 65

Charles II (king), 172

Charles V (emperor), 72, 74, 92, 238, 263

and sugar production, 138, 154 (n. 78);
and Brazil, 159, 177, 179, 180, 184

Fixed costs, 4

Flanders, 70, 71, 72, 105, 248, 250, 251, 261

Flegel, Georg, 254

Flemish merchants: and Canary Islands, 63,
64, 68, 69, 261, 266; and Madeira, 64, 65,
66, 260–61, 266; and Española, 92

Florentine merchants, 65, 69

Florida, 119

Fonte, Miguel, 54

Foreign capital: role of, 9, 16; in Brazil, 12,
191; in Madeira, 47, 63; in Canary Islands,
63; and sugar trade, 63; in Española, 92–
93; in Barbados, 295, 301–2, 303

Fraginals, Moreno, 147

Fragoso, João, 191

France: as sugar consumer, 15, 170; and
Atlantic slave trade, 205, 214–15, 217, 220,
221, 230; and fruit preserves, 245; and
cookbooks, 245, 246, 247, 248; and sugar
in diet, 250; and Antwerp, 265; and sugar
refining, 270–71, 274

Franciscans, 181

Francken, Jerome, 253

Franco, Diego, 90

François I (king), 238, 247

Frederick II (emperor), 33

Free blacks, 206–7

French Caribbean, 166

French colonies, 7, 18, 166, 170, 205, 209, 290

French merchants, 64, 65, 66, 68, 103, 261

Frutuoso, Gaspar, 47, 48, 54, 56

Fuerteventura, 43, 51

Fuggers, 33

Fulcher of Chartres, 31

Galen, 30, 245, 258

Galloway, J. H., 8

Gamboa, Inés de, 129–30

Gandía, 35

Garcia, Cristóbal de, 53

Garcia de Moguer, Cristóbal, 59

Garzoni, Giovanna, 254

Gautier, Paris, 274

Genoese merchants, 64, 65, 68–69, 74, 82
(n. 71)

German merchants, 67, 260, 261, 317–18
(n. 20)

Germany, 247–48, 260, 264–65, 271–72

Ginger, 11, 107

Giraldes, Lucas, 159

Giraldi, Lucas, 65

Goa, Dinis de, 52

Godin, Jacques, 266

Godinho, Cristóvão, 256

Godinho, Vitorino Magalhães, 45, 63

Góis, Damião de, 243

Góis, Luis de, 270

Gold exports, 201, 202, 203, 205, 210, 212–13,
214, 223, 225

Gold mining, 11, 85, 88, 117, 118, 125, 172, 214

Gomera, 43, 45, 48, 49, 51, 54, 57, 59, 63

Gómez, Francisco, 102

Gómez de Rojas Manrique, 130, 132, 153
(n. 52)

Gómez Reinel, Pedro, 121

Gonçalves, João, 65

Gonçalves, Maria, 58

Gonçalves, Vicente, 263

Gonçalves da Câmara, Pedro, 54

Gonçalves Salvador, José, 72–73

Gonçalves Zarco, João, 42, 43, 51

González, Juan, 53

González, Pedro, 140, 143

González Cordero, Pedro, 143

Gorjón, Hernando, 87, 90, 92, 98, 106

Governmental role: in early Atlantic sugar
economy, 9, 11; in Española, 89–92, 93,
111 (n. 33); 121; in Cuba, 121–22, 126–27,
134–35, 141; in Brazil, 158, 160, 161, 163,
166, 167, 170, 171–72, 175, 180–81, 183, 188,
191; in Atlantic slave trade, 213–16

Gramaye, Gerard, 259

Granada, 29, 31, 34, 108, 244

Granado, Diego, 244

Gran Canaria: as royal island, 43, 51; and
sugar production, 48; production levels
in, 49, 56, 63; settlement of, 50; sugar
mills in, 54; slaves in, 57, 58, 59, 61; and
sugar prices, 62; and sugar merchants, 68,
69, 72; and sugar trade, 72

Greenfield, Sidney, 74

Groenenberg, Jakob, 261

Grosse Ravensburger Gesellschaft, 260

Guadeloupe, 7, 209

Guicciardini, Ludovico, 259

Guinisy, Baltasar, 269

Noell, James, 298
Noell, Martin, 295, 297–98
Noell, Stephen, 298
Noell, Thomas, 297–98, 299
Nola, Ruperto de, 244
Normans, 32–33
North Africa, 30, 202, 205, 225, 226, 260
Núñez, Juana, 142

Ochoa de la Vega, Diego, 132
Odiarne, Beatrice, 296
Oñate, Pedro de, 122, 133
Order of Christ, 42, 51
Orejón, Francisco, 90
Orihuela, Juan de, 90
Ornelas e Vasconcelos, Ayres, 59
Orta Yusta, Ginés de, 133
Ortelius, Abraham, 240
Ortiz, Fernando, 134
Ottoman Turks, 33, 35
Oviedo, Gonzalo Fernández de, 97, 101, 102, 106, 134

Pacheco, María, 131
Pacific islands, 28
Paiva, João de, 74
Palacios, Francisco de, 254
Palencia, Juan de, 102
Palestine, 31
Palomar, Francisco, 54
Parr, Richard, 303, 304, 305
Partnerships, 93, 96–97, 142–43, 181, 216, 305
Pasamonte, Estéban de, 90, 92
Pasamonte, Juan de, 92
Pasamonte, Miguel de, 87, 89, 90, 92
Pascal, Jacques, 246
Paternalism, 4
Peeters, Clara, 254
Peixoto Viegas, João, 172, 179
Penoyer, William, 295
Pepper, 66, 201, 202, 239, 259
Pereda, Antonio de, 254
Pereira, Abraham de, 273
Pereira, Estevão, 192–93, 200 (n. 72)
Pereira, Fernando Jasmins, 48
Pereira, Isaac de, 273
Pereira da Silva, Wenceslão, 187
Perestrelo, Bartolomeu, 42–43

Pérez, Juan, 143
Pérez, Manuel, 138
Pérez de Borroto, Beatriz, 132
Pérez de Borroto, Francisco, 132
Pérez Vidal, J., 53
Pernambuco, Brazil: production levels in, 18, 162, 163, 167, 178; productivity level of slaves in, 19; and Madeiran sugar specialists, 75; local conditions in, 159; and sugar production, 161, 180, 318 (n. 22); sugar mills in, 161–62; Dutch occupation of, 162, 166–70, 197 (n. 36), 208–9, 289, 318 (n. 22); and sugar trade, 174; and cane farmers, 184, 185, 187; and slaves, 208
Peru, 12, 86, 121, 135, 163, 204, 205–6, 208
Philip II (king), 49, 108, 238, 263
Philippines, 8
Philip the Good, 261
Phillip III (king), 243–44
Phillips, William, 15–16
Piemontese, Alessio, 247, 249
Pizarro, Francisco, 205
Plantation: meaning of, 2–3
Plantation system: and sugar revolution, 1–2, 6–7; and capitalism, 5; development of, 13, 208; in Española, 86, 88, 105, 107, 109; in Cuba, 116, 126, 148, 149 (n. 6); in Barbados, 297, 300, 301, 306
Planters, 4, 11, 16
Pliny, 52
Pohl, Hans, 264, 269
Politics: influence of sugar production on, 11; and Atlantic sugar market, 18; and Canary Islands sugar merchants, 69; and Española's sugar economy, 85, 87; and Brazil's sugar mill owners, 158; and Brazil's sugar economy, 164, 166–67, 169, 171, 182, 193; and African slave trade, 202; sugar confectionery associated with, 256–57; in Barbados, 290, 303, 306
Pons, Jacques, 246
Ponte, Cristóbal de, 59, 68
Ponteverde, Juan de, 56
Population: growth of, 6; transformation of, 7, 11, 125–26
Portugal: and capitalism, 2; and spread of sugar production, 35; and trade with Madeira, 63, 70–71, 74; and Atlantic slave trade, 124, 202–4, 209–11, 214, 220, 226;

and Cuba, 125; and sugar production supplies, 140; and Brazil, 158, 160, 161, 166, 170, 171–72, 174–75, 199; and Spanish Hapsburgs, 162, 164, 166, 197–98 (n. 43); and sugar refining, 179; and African trade, 202–3, 205, 210, 214, 218; and Asian trade, 205, 208; and African autonomy, 212; and sugar in diet, 240–44, 277–78 (n. 12); and Flanders, 261; currency of, 309
Portuguese colonies, 208
Portuguese exploration, 63, 202
Portuguese merchants, 64, 66, 68, 121, 259–60, 262
Post, Frans, 253
Poyo Valenzuela, Juan del, 142, 144
Precursor sugar economies, 7
Príncipe, 210
Production levels: comparison of, 9; in Española, 11, 16, 99–103, 107–8, 112–13 (n. 66), 151 (n. 25); in Cuba, 11, 18, 122, 123, 151 (n. 25); in Brazil, 12, 17, 18, 49, 50, 151 (n. 25), 161, 163–64, 171, 178; in Madeira, 54–55; in Canary Islands, 56
Production unit size, 9
Productivity per unit of land, 4–5
Productivity per worker, 4–5, 9, 19–20
Productivity per mill, 165
Protectionist policies, 10, 75
Pudsey, Cuthbert, 13–14, 158, 176
Puerto Rico, 8, 10–11, 15, 16, 117, 121, 151 (n. 25), 183

Queensland, 8

Rabelais, François, 256
Raes, Antoinette, 262
Ramirão, João, 177
Rau, Virginia, 45
Ravassa, Gaspar Esteva, 37
Ravensburger Handelsgesellschaft, 33–34
Recio, Antón, 131–32
Recio Márquez, Lucia, 131
Reiniers, Theodoor, 274
Rem, Lucas, 67
Ribera, Antonio de, 131, 132, 134, 139, 142
Riberol, Bautista, 68
Riberol, Francisco, 54, 68
Richardson, George, 303, 304, 305

Rodrigues, António, 60
Rodrigues, Domingos, 245
Rodrigues, Jorge, 60
Rodrigues, Lionel, 74
Rodrigues, Manuel, 60
Rodríguez, Juana, 135
Rodríguez, Jusepe, 139
Rodríguez Morel, Genaro, 10, 16, 19
Rodríguez Quintero, Juan, 140
Rodríguez Tabares, Hernán, 133, 138
Rogell, Mafei, 65
Roiz, Cristóvão, 75
Rojas, Alonso de, 128–29, 131, 132, 153 (n. 52)
Rojas, Baltasar de, 128, 129, 132, 138, 142
Rojas, Catalina de, 129, 132
Rojas, Lucas de, 128, 129, 132, 142
Rojas, Manuel de, 133
Rouzée d'Arras, Pedro, 262, 263
Rovelasca, João Batista de, 264
Rumpolt, Marx, 248
Ryff, Walter, 247–48

Sá, Mem de, 75, 160–61, 181
Sacchia, Bartolomeo, 247
Safra, 176–77, 178
Saillot, Nicolas, 268
St. Christopher, 209, 300
St. Domingue (Haiti), 15
Salazar, Gaspar de, 142
Salvador, Frei Vicente do, 164
Sánches de Moya, Francisco, 119, 121
Sánchez Cotán, Juan, 254
Santafé, Nicolau, 33, 34
Santo Domingo, 11, 52, 86, 183, 209, 262. *See also* Española
São Tomé: and plantation complex, 7, 10; spread of sugar production to, 8, 74; sugar production capacity in, 10, 18, 24 (n. 25), 204; slaves in, 59, 188, 204; and sugar trade, 71, 243, 265; local conditions in, 159; Portuguese settlement of, 204, 209, 210; and Atlantic slave trade, 211, 213; and Antwerp sugar market, 260, 262
Schetz, Baltasar, 262
Schetz, Erasmo, 159–60, 173, 262
Schetz, Gaspar, 262, 263
Schetz, Melchior, 262, 263
Schmidl, Ulrich, 262
Schwartz, Stuart B., 12, 151 (n. 25)

Sugar market: expansion of, 71–72; of Antwerp, 239, 258–67; and tea, 240

Sugar merchants: political power of, 16; in Madeira, 64–67; in Canary Islands, 68–69, 72; in Española, 105; in Brazil, 175–76, 187. *See also specific nationalities*

Sugar mill owners: social and political role of, 9; in Madeira, 16, 70; in Cuba, 121–22, 125, 126–33; in Brazil, 158, 159, 167–68, 170, 177, 180–82, 184–87

Sugar mills: quasi-industrial nature of, 3–4; output of, 9; as symbolic of Atlantic economy, 21; in Iberia, 36–38; in Madeira, 52–54, 57, 58, 135; in Canary Islands, 53, 54; in Española, 86–97, 135; in Cuba, 116, 117, 119, 122, 123, 124, 125, 133–48; in Brazil, 135–36, 161, 167, 176–80, 197 (n. 38); in Barbados, 299

Sugar prices: comparison of, 12, 13; and sugar as commodity, 28, 275; in Madeira, 61–62, 71; in Canary Islands, 61–62, 261; in Española, 105; in Cuba, 121, 125, 152 (n. 43); in Brazil, 162, 164, 165–66, 170, 172, 175, 193; and Antwerp sugar market, 259; in Barbados, 302; in Amsterdam, 302, 322–23 (n. 41); in England, 302, 322–23 (n. 41)

Sugar-producing regions: serial development of, 13, 18

Sugar production: and slavery, 1, 2, 3; and capitalism, 1–2; and European markets, 6; spread of from Mediterranean region, 7, 8, 13; in Iberia, 8, 15–16, 27–38; in Cuba, 10, 11, 115, 116–26, 138–39; in São Tomé, 10, 18, 24 (n. 25), 204; and local markets, 12; basic techniques of, 13–14, 30, 52, 101, 176–80, 197 (n. 40); in Brazil, 15, 135, 138, 208, 214, 215; and Muslims, 28–31; in Madeira, 47–48, 49, 50, 55, 58, 73, 74, 75, 121, 204; in Canary Islands, 48–49, 70, 121; supplies for, 89–90, 93, 96–99, 140–41, 155–56 (nn. 92, 93), 158, 169, 178, 191, 197–98 (n. 43), 296; in Barbados, 209, 289, 291–92, 297, 303, 313 (n. 5)

Sugar quotas: in Madeira, 66

Sugar refining: and Muslims, 29; in England, 179, 271; in Antwerp, 267–72, 274; art of, 267–75; and technology, 269; in France, 270–71, 274; and women, 275

Sugar revolution: and plantation system, 1–2, 6–7; and plantation accounting, 5; and precursor sugar economies, 7; in Barbados, 7, 8, 10, 18, 289, 290, 306; in Cuba, 7, 8, 15; and labor force composition, 7, 9, 23 (n. 16); and early Atlantic sugar economy, 10

Sugar route, 42

Sugar trade: in Madeira, 63–64, 66–67, 69–75, 243, 265; in Canary Islands, 64, 69, 70, 71, 72, 244, 265; in Brazil, 71, 72–73, 172–76, 243; in Cuba, 117, 122, 124, 125, 199; and capitalism, 239, 275

Suriname, 7, 194

Susio, Ludovico di, 254

Swaen, Michiel de, 256

Sweden, 178

Syria, 31

Tapia, Cristóbal de, 89, 96, 97

Tapia, Francisco de, 98

Tarifa, marqués of, 36

Taxation of imports, 89–90, 105, 122

Taxation of sugar: and early Atlantic sugar economy, 9, 18; and Muslims of Khuzistan, 28–29; in Madeira, 50, 54, 70, 78 (n. 24); in Canary Islands, 78 (n. 24); in Española, 112 (n. 66); in Cuba, 122, 124–25; in Brazil, 166, 170, 171–72, 175

Technology: European technology, 1; and sugar plantations, 3; innovation of, 4, 19, 163; and staple theory, 6; transference of, 9, 16–17, 20; in New Spain, 11–12; in Madeira, 52; in Española, 97–99; in Cuba, 135–37, 139, 145, 156–57 (n. 107); in Brazil, 162–63, 178, 193, 197 (n. 40); and sugar refining, 269; and sugar production, 313 (n. 4). See also *Ingenios* (water mills); Three-roller vertical mills; *Trapiches* (animal-powered mills)

Teive, Diogo de, 52, 182

Teixeira, António, 53, 54

Tenant farmers. *See* Sharecroppers

Tenerife: as royal island, 43, 51; land rights in, 45, 47; and sugar production, 48; production levels in, 49, 56, 63; settlement of, 50–51; sugar mills in, 54; slaves in, 57, 58–59, 61; and sugar prices, 61, 62; and

Welsers, 33, 67, 69, 261, 262
West Africa, 58, 63, 203, 205, 210, 212
Westerbaen, Jacob, 258
West Indies: and sugar production, 1, 209, 290; planters' accounting practices in, 4; and Iberian sugar market, 35; British West Indies, 116, 124, 138, 209, 234 (nn. 17, 19), 290; and Cuba, 124; and sugar mills, 135; and slaves, 209
West Indies Company, 272, 274
William of Tyre, 32
Williams, Eric, 1, 5

Winthrop, John, 290
Winthrop, John, Jr., 292
Women: and internal African slave trade, 202; and fruit preserves, 240–43; and sugar refining, 275

Yanes, Gonzalo, 59

Zacchia, Paulo, 258
Zuazo, Alonso, 98, 99, 106
Zurara, Gomes Eanes de, 48